Advanced Automotive Fault Diagnosis

Advanced Automotive Fault Diagnosis

Second edition

Tom Denton *BA, MSAE, MIMI, MIRTE, Cert. Ed.*
Associate Lecturer, Open University, UK

AMSTERDAM • BOSTON • HEIDELBERG • LONDON • OXFORD • NEW YORK
PARIS • SAN DIEGO • SAN FRANCISCO • SINGAPORE • SYDNEY • TOKYO
Butterworth-Heinemann is an imprint of Elsevier

Elsevier Butterworth-Heinemann
Linacre House, Jordan Hill, Oxford OX2 8DP, UK
30 Corporate Drive, Suite 400, Burlington MA 01803, USA

First published 2000
Reprinted 2002, 2003, 2004
Second edition 2006

Copyright © 2006, Tom Denton. Published by Elsevier Ltd. All rights reserved.

The right of Tom Denton to be identified as the author of this work has been asserted in accordance with the Copyright, Designs and Patents Act 1988

No part of this publication may be reproduced, stored in a retrieval system or transmitted in any form or by any means electronic, mechanical, photocopying, recording or otherwise without the prior written permission of the publisher

Permission may be sought directly from Elsevier's Science & Technology Rights Department in Oxford, UK: phone (+44) (0) 1865 843830; fax (+44) (0) 1865 853333; email: permissions@elsevier.com. Alternatively you can submit your request online by visiting the Elsevier web site at http://elsevier.com/locate/permissions, and selecting *Obtaining permission to use Elsevier material*

Notice
No responsibility is assumed by the publisher for any injury and/or damage to persons or property as a matter of products liability, negligence or otherwise, or from any use or operation of any methods, products, instructions or ideas contained in the material herein. Because of rapid advances in the medical sciences, in particular, independent verification of diagnoses and drug dosages should be made

British Library Cataloguing in Publication Data
A catalogue record for this book is available from the British Library

Library of Congress Cataloging-in-Publication Data
A catalog record for this book is available from the Library of Congress

ISBN-13: 978-0-75-066991-7
ISBN-10: 0-75-066991-8

For information on all Butterworth-Heinemann publications visit our website at www.books.elsevier.com

Typeset by Charon Tec Ltd, Chennai, India
www.charontec.com
Printed and bound in the United Kingdom
06 07 08 09 10   9 8 7 6 5 4 3 2 1

Working together to grow
libraries in developing countries

www.elsevier.com | www.bookaid.org | www.sabre.org

ELSEVIER   BOOK AID International   Sabre Foundation

Contents

Preface	*ix*
Introduction to the second edition	*xi*
Acknowledgments	*xiii*
Glossary	*xv*

1		**Introduction**	**1**
	1.1	'If it ain't broke, don't fix it!'	1
	1.2	Safe working practices	1
	1.3	Terminology	2
	1.4	Report writing	3
2		**Diagnostic techniques**	**6**
	2.1	Introduction	6
	2.2	Diagnostic process	6
	2.3	Diagnostics on paper	8
	2.4	Mechanical diagnostic techniques	9
	2.5	Electrical diagnostic techniques	12
	2.6	Fault codes	17
	2.7	Systems	18
	2.8	On- and off-board diagnostics	20
	2.9	Data sources	21
3		**Tools and equipment**	**25**
	3.1	Basic equipment	25
	3.2	Oscilloscopes	28
	3.3	Scanners/fault code readers	30
	3.4	Engine analysers	32
4		**Oscilloscope diagnostics**	**40**
	4.1	Introduction	40
	4.2	Sensors	40
	4.3	Actuators	49
	4.4	Ignition system	54
	4.5	Other components	58
	4.6	Summary	60
5		**On-board diagnostics**	**61**
	5.1	On-board diagnostics – a first perspective	61
	5.2	What is on-board diagnostics?	63
	5.3	Petrol/gasoline on-board diagnostic monitors	66
	5.4	On-board diagnostics – a second perspective	74
	5.5	Summary	84
6		**Sensors and actuators**	**86**
	6.1	Introduction	86
	6.2	Sensors	86
	6.3	Actuators	94

7	**Engine systems**	**98**
7.1	Introduction	98
7.2	Engine operation	98
7.3	Diagnostics – engines	101
7.4	Fuel system	103
7.5	Diagnostics – fuel system	106
7.6	Introduction to engine management	108
7.7	Ignition	108
7.8	Diagnostic – ignition systems	116
7.9	Emissions	120
7.10	Diagnostics – emissions	122
7.11	Fuel injection	122
7.12	Diagnostics – fuel injection systems	126
7.13	Diesel injection	127
7.14	Diagnostics – diesel injection systems	130
7.15	Engine management	132
7.16	Diagnostics – combined injection and fuel control systems	138
7.17	Engine management and faultfinding information	143
7.18	Air supply and exhaust systems	144
7.19	Diagnostics – exhaust and air supply	147
7.20	Cooling	148
7.21	Diagnostics – cooling	150
7.22	Lubrication	151
7.23	Diagnostics – lubrication	153
7.24	Batteries	155
7.25	Diagnosing battery faults	156
7.26	Starting	159
7.27	Diagnosing starting system faults	163
7.28	Charging	163
7.29	Diagnosing charging system faults	167
8	**Chassis systems**	**169**
8.1	Brakes	169
8.2	Diagnostics – brakes	172
8.3	Anti-lock brakes	174
8.4	Diagnostics – anti-lock brakes	177
8.5	Traction control	178
8.6	Diagnostics – traction control	180
8.7	Steering and tyres	181
8.8	Diagnostics – steering and tyres	186
8.9	Suspension	190
8.10	Diagnostics – suspension	195
9	**Electrical systems**	**197**
9.1	Electronic components and circuits	197
9.2	Multiplexing	200
9.3	Lighting	203
9.4	Diagnosing lighting system faults	207
9.5	Auxiliaries	209
9.6	Diagnosing auxiliary system faults	214
9.7	In car entertainment (ICE) security and communications	216
9.8	Diagnosing ICE, security and communication system faults	219
9.9	Body electrical systems	221
9.10	Diagnosing body electrical system faults	225
9.11	Instrumentation	226
9.12	Diagnosing instruments system faults	230

9.13	Heating, ventilation and air conditioning (HVAC)	231
9.14	Diagnostics – HVAC	237
9.15	Cruise control	239
9.16	Diagnostics – cruise control	240
9.17	Air bags and belt tensioners	241
9.18	Diagnostics – air bags and belt tensioners	244
10	**Transmission systems**	**247**
10.1	Manual transmission	247
10.2	Diagnostics – manual transmission	251
10.3	Automatic transmission	253
10.4	Diagnostics – automatic transmission	256
11	**Conclusion, web resources and developments**	**260**
11.1	Introduction	260
11.2	Web contacts	260
11.3	Future developments in diagnostic systems	262
11.4	Software	263
11.5	Summary	265

Index   *267*

Preface

The aspect I still enjoy most about working on vehicles is being able to diagnose a fault that has beaten others! This skill takes a few years to develop but it is worth the effort. Diagnostic work is much like that of a detective solving a difficult crime, all the clues are usually there – if you know where to look. I think it was Sherlock Holmes (a fictional detective if you have never heard of him!) who said:

> When you have eliminated all which is impossible, then whatever remains, however improbable, must be the truth.

This is a great thought for a diagnostic technician to keep in mind.

To help you learn 'where to look' for the clues and to eliminate the impossible, this book combines some aspects of automotive technology covered in my other books. However, it goes much further with a new approach to the art of diagnostics as a science.

The skills needed to be a good diagnostic technician are many and varied. For one job you may need to listen to a rumbling noise as the car corners, for another you may need to interpret an oscilloscope waveform or a diagnostic trouble code.

Vehicles continue to become more complicated, particularly in the area of electronics. The need for technicians with good diagnostic skills therefore remains. This could be you and you should be paid well!

Look on the bright side of having complicated technology on vehicles – fewer 'home mechanics' and more work for you and me!

Tom Denton
2006

PS. Comments and contributions are welcome at my web site: www.automotive-technology.co.uk. You will also find lots of useful information, updates, news and details about my other books as well as automotive training software and web links.

Introduction to the second edition

The book has grown! But then it was always going to, because the complexity of automotive systems has grown and the associated diagnostic skills must follow.

The main change for this edition is that I have included two completely new chapters. The first is all about on-board diagnostics (OBD) and the second covers oscilloscope diagnostics in some detail. Both of these subjects are very relevant to all aspects of the automotive repair trade, light or heavy vehicle.

I have tried wherever possible to make the content relevant to all types of vehicle whether used in the UK, USA or anywhere else in the world. After all, most vehicles have an engine that makes the wheels go round – even if the steering wheel changes sides…

There has been a significant rationalisation of motor vehicle qualifications in the UK since the first edition. The result is that this book has become even more appropriate because of the higher technical content. The order of the material has been changed a bit so that it lines up more with current qualifications. For example, engine management and all engine electrics (batteries, etc.) are now part of the Engines chapter.

The book is ideal for all MV qualifications, in particular:

- City & Guilds 4101 Technical Certificates and NVQs
- IMI Technical Certificates and NVQs
- Level 4 diagnostic units
- BTEC/Edexcel National and Higher National qualifications
- International MV qualifications such as C&G 3905
- ASE certification in the US
- Supplementary reading for MV degree level course.

Of course, you may already be qualified and just need a few pointers!

You may also simply want to learn more about how your car works – and how to fix it when it doesn't!

I hope you enjoy this book, but most of all I hope it helps you to become a better diagnostic technician – something you should be very proud to be.

Acknowledgments

I am very grateful to the following companies who have supplied information and/or permission to reproduce photographs and/or diagrams:

AA Photo Library
Autodata
Autologic Data Systems
BMW UK
Bosch Press Photos
Eberspaecher GmbH
Ford Motor Company
GenRad
Hella UK
Institute of Road Transport Engineers
Jaguar Cars
Kavlico Corp
Lucas Service
LucasVarity
Mazda Cars UK
NGK Spark Plugs UK
Pioneer Radio
Renault UK
Ricardo
Robert Bosch GmbH
Robert Bosch UK
Rover
Saab UK
Scandmec UK
Snap-on Tools
Sun Electric UK
Sykes-Pickavant
Valeo UK
ZF Servomatic

Many if not all the companies here have good web pages. You will find a link to them from my site. Thanks again to the listed companies. If I have used any information or mentioned a company name that is not noted here, please accept my apologies and acknowledgments.

An extra thanks to Dave Rogers (AVL) and Alan Malby (Ford Motor Company) for their kind assistance with the OBD chapter.

Also, if I forget to mention my family: Vanda, Malcolm and Beth, I will be in trouble...

Glossary

Authors Note: To keep the glossary to a reasonable size, I decided to limit the entries to useful acronyms that are specified by the society of automotive engineers (SAE) and on-board diagnostic version two (OBD2) recommendations. I have provided free access to online glossaries (UK, US and Spanish) that include several thousand words.
www.automotive-technology.co.uk

Two key words never to forget:

Symptom(s): What the user/operator/repairer of the system (vehicle or whatever) notices
Fault: An error in the system that causes symptom(s)

OBD2/SAE acronyms

ABS:	antilock brake system
A/C:	air conditioning
AC:	air cleaner
AIR:	secondary air injection
A/T:	automatic transmission or transaxle
AP:	accelerator pedal
B+:	battery positive voltage
BARO:	barometric pressure
CAC:	charge air cooler
CFI:	continuous fuel injection
CL:	closed loop
CKP:	crankshaft position sensor
CKP REF:	crankshaft reference
CMP:	camshaft position sensor
CMP REF:	camshaft reference
CO:	carbon monoxide
CO_2:	carbon dioxide
CPP:	clutch pedal position
CTOX:	continuous trap oxidizer
CTP:	closed throttle position
DEPS:	digital engine position sensor
DFCO:	deceleration fuel cut-off mode
DFI:	direct fuel injection
DLC:	data link connector
DTC:	diagnostic trouble code
DTM:	diagnostic test mode
EBCM:	electronic brake control module
EBTCM:	electronic brake traction control module
EC:	engine control
ECL:	engine coolant level
ECM:	engine control module
ECT:	engine coolant temperature
EEPROM:	electrically erasable programmable read only memory
EFE:	early fuel evaporation
EGR:	exhaust gas recirculation
EGRT:	EGR temperature
EI:	electronic ignition
EM:	engine modification
EPROM:	erasable programmable read only memory
EVAP:	evaporative emission system
FC:	fan control
FEEPROM:	flash electrically erasable programmable read only memory
FF:	flexible fuel
FP:	fuel pump
FPROM:	flash erasable programmable read only memory
FT:	fuel trim
FTP:	federal test procedure
GCM:	governor control module
GEN:	generator
GND:	ground
H_2O:	water
HC:	hydrocarbon
HO_2S:	heated oxygen sensor
HO_2S_1:	upstream heated oxygen sensor
HO_2S_2:	up or downstream heated oxygen sensor
HO_2S_3:	downstream heated oxygen sensor
HVS:	high voltage switch
HVAC:	heating ventilation and air conditioning system
IA:	intake air
IAC:	idle air control
IAT:	intake air temperature
IC:	ignition control circuit
ICM:	ignition control module
IFI:	indirect fuel injection
IFS:	inertia fuel shutoff
I/M:	inspection/maintenance
IPC:	instrument panel cluster
ISC:	idle speed control

KOEC:	key on, engine cranking	PSP:	power steering pressure
KOEO:	key on, engine off	PTOX:	periodic trap oxidizer
KOER:	key on, engine running	RAM:	random access memory
KS:	knock sensor	RM:	relay module
KSM:	knock sensor module	ROM:	read only memory
LT:	long term fuel trim	RPM:	revolutions per minute
MAF:	mass airflow sensor	SC:	supercharger
MAP:	manifold absolute pressure sensor	SCB:	supercharger bypass
		SDM:	sensing diagnostic mode
MC:	mixture control	SFI:	sequential fuel injection
MDP:	manifold differential pressure	SRI:	service reminder indicator
MFI:	multi-port fuel injection	SRT:	system readiness test
MIL:	malfunction indicator lamp	ST:	short term fuel trim
MPH:	miles per hour	TB:	throttle body
MST:	manifold surface temperature	TBI:	throttle body injection
MVZ:	manifold vacuum zone	TC:	turbocharger
NOX:	oxides of nitrogen	TCC:	torque converter clutch
NVRAM:	non-volatile random access memory	TCM:	transmission or transaxle control module
O2S:	oxygen sensor	TFP:	throttle fluid pressure
OBD:	on-board diagnostics	TP:	throttle position
OBD I:	on-board diagnostics generation one	TPS:	throttle position sensor
		TVV:	thermal vacuum valve
OBD II:	on-board diagnostics, second generation	TWC:	three way catalyst
		TWC + OC:	three way + oxidation catalytic converter
OC:	oxidation catalyst		
ODM:	output device monitor	VAF:	volume airflow
OL:	open loop	VCM:	vehicle control module
OSC:	oxygen sensor storage	VR:	voltage regulator
PAIR:	pulsed secondary air injection	VS:	vehicle sensor
PCM:	powertrain control module	VSS:	vehicle speed sensor
PCV:	positive crankcase ventilation	WOT:	wide open throttle
PNP:	park/neutral switch	WU-TWC:	warm up three way catalytic converter
PROM:	program read only memory		
PSA:	pressure switch assembly		

1
Introduction

1.1 'If it ain't broke, don't fix it!'

1.1.1 What is needed to find faults?

Finding the problem when complex automotive systems go wrong is easy, if you have the necessary knowledge. This knowledge is in two parts:

- understanding of the system in which the problem exists;
- the ability to apply a logical diagnostic routine.

It is also important to be clear about two definitions:

- **symptom(s)** what the user/operator/repairer of the system (vehicle or whatever) notices;
- **fault** the error in the system that causes the symptom(s).

'If it is not broken then do not go to the trouble of repairing it,' is the translation of this main section heading! It's a fair comment but if a system is not operating to its optimum then it should be repaired. This is where the skills come in to play. It is necessary to recognise that something is not operating correctly by applying your knowledge of the system, and then by applying this knowledge further and combining it with the skills of diagnostics, to be able to find out why.

Each main chapter of this book includes a basic explanation of the vehicle system followed by diagnostic techniques that are particularly appropriate for that area. Examples of fault-finding charts are also included. In the main, references will be to generic systems rather than to specific vehicles or marques. For specific details about a particular vehicle or system the manufacturer's information is the main source. Alternatively 'Autodata' produce a fine range of books; visit www.autodata.com for more details.

The knowledge requirement and the necessity for diagnostic skills are further illustrated in the next chapter.

Figure 1.1 Diagnostics in action

Figure 1.1 shows a diagnostic procedure in action!

1.1.2 Heavy or light vehicles?

An important note about diagnostics is that the general principles or techniques can be applied to *any* system, physical or otherwise. As far as heavy or light vehicles are concerned then this is definitely the case. As discussed earlier, there is a need for knowledge of the particular system, but diagnostic skills are transferable.

1.2 Safe working practices

1.2.1 Introduction

Safe working practices in relation to diagnostic procedures and indeed any work on a vehicle are essential – for your safety as well as that of others. You only have to follow two rules to be safe:

- use your common sense – don't fool about;
- if in doubt – seek help.

Further, always wear appropriate personal protective equipment (PPE) when working on vehicles.

The following section lists some particular risks when working with electricity or electrical systems, together with suggestions for reducing them. This is known as risk assessment.

1.2.2 Risk assessment and reduction

The following table notes some identified risks involved with working on vehicles. It is by no means exhaustive but serves as a good guide.

Identified risk	Reducing the risk
Battery acid	Sulphuric acid is corrosive so always use good PPE – in this case overalls and if necessary rubber gloves. A rubber apron is ideal, as are goggles if working with batteries a lot
Electric shock	Ignition HT (high tension, which simply means high voltage) is the most likely place to suffer a shock, up to 25 000 V is quite normal. Use insulated tools if it is necessary to work on HT circuits with the engine running. Note that high voltages are also present on circuits containing windings due to back emf (electromotive force) as they are switched off; a few hundred volts is common. Mains supplied power tools and their leads should be in good condition and using an earth leakage trip is highly recommended
Exhaust gases	Suitable extraction must be used if the engine is running indoors. Remember it is not just the carbon monoxide (CO) that might make you ill or even kill you, other exhaust components could cause asthma or even cancer
Fire	Do not smoke when working on a vehicle. Fuel leaks must be attended to immediately. Remember the triangle of fire – (heat/fuel/oxygen) – don't let the three sides come together
Moving loads	Only lift what is comfortable for you; ask for help if necessary and/or use lifting equipment. As a general guide, do not lift on your own if it feels too heavy!
Raising or lifting vehicles	Apply brakes and/or chock the wheels and when raising a vehicle on a jack or drive on lift. Only jack under substantial chassis and suspension structures. Use axle stands in case the jack fails
Running engines	Do not wear loose clothing; good overalls are ideal. Keep the keys in your possession when working on an engine to prevent others starting it. Take extra care if working near running drive belts
Short circuits	Use a jump lead with an in-line fuse to prevent damage due to a short when testing. Disconnect the battery (earth lead off first and back on last) if any danger of a short exists. A very high current can flow from a vehicle battery, it will burn you as well as the vehicle
Skin problems	Use a good barrier cream and/or latex gloves. Wash skin and clothes regularly

1.3 Terminology

1.3.1 Introduction

The terminology included in the following tables is provided to ensure that we are talking the same language. These tables are provided just as a simple reference source.

1.3.2 Diagnostic terminology

Symptom	The effect of a fault noticed by the driver, user or technician
Fault	The root cause of a symptom/problem
Diagnostics	The process of tracing a fault by means of its symptoms, applying knowledge and analysing test results
Knowledge	The understanding of a system that is required to diagnose faults
Logical procedure	A step by step method used to ensure nothing is missed
Report	A standard format for the presentation of results

1.3.3 General terminology

System	A collection of components that carry out a function
Efficiency	This is a simple measure of any system. It can be scientific for example if the power out of a system is less then the power put in, its percentage efficiency can be determined ($P_{out}/P_{in} \times 100\%$). This could, for example, be given as 80%. In a less scientific example, a vehicle using more fuel than normal is said to be inefficient
Noise	Emanation of sound from a system that is either simply unwanted or is not the normal sound that should be produced
Active	Any system that is in operation all the time (steering for example)
Passive	A system that waits for an event before it is activated (an air bag is a good example)
Short circuit	An electrical conductor is touching something that it should not be (usually another conductor or the chassis)
Open circuit	A circuit that is broken (a switched off switch is an open circuit)
High resistance	In relation to electricity, this is part of a circuit that has become more difficult for the electricity to get through. In a mechanical system a partially blocked pipe would have a resistance to the flow of fluid
Worn	This word works better with further additions such as: worn to excess, worn out of tolerance, or even, worn, but still within tolerance!

Quote	To make an estimate of or give exact information on the price of a part or service. A quotation may often be considered to be legally binding
Estimate	A statement of the expected cost of a certain job (e.g. a service or repairs) An estimate is normally a best guess and is not legally binding
Dodgy, knackered or @#%&*!	Words often used to describe a system or component, but they mean nothing! Get used to describing things so that misunderstandings are eliminated

1.4 Report writing

1.4.1 Introduction

As technicians you may be called on to produce a report for a customer. Also, if you are involved in research of some kind it is important to be able to present results in a professional way. The following sections describe the main headings that a report will often need to contain together with an example report based on the performance testing of a vehicle alternator.

Laying out results in a standard format is the best way to ensure that all the important and required aspects of the test have been covered. Keep in mind that the report should convey clearly to another person what has been done. Further, a 'qualified' person should be able to extract enough information to be able to repeat the test – and check your findings! Use clear simple language remembering that in some cases the intended audience may not be as technically competent as you are.

1.4.2 Main headings of a report

The following suggestions for the headings of a professional report will cover most requirements but can of course be added to or subtracted from if necessary. After each heading I have included brief notes on what should be included.

Contents

If the report is more than about five pages, a list of contents with page numbers will help the reader find his/her way through it.

Introduction

Explain the purpose of what has been done and set the general scene.

Test criteria

Define the limits within which the test was carried out. For example, temperature range or speed settings.

Facilities/resources

State or describe what equipment was used. For example: 'A *"Revitup"* engine dynamometer, model number *C3PO* was used for the consumption test'.

Test procedures

Explain here exactly what was done to gain the results. In this part of the report it is very important not to leave out any details.

Measured results

Present the results in a way that is easy to interpret. A simple table of figures may be appropriate. If the trend of the results or a comparison is important, a graph may be better. Pictures of results or oscilloscope waveforms may be needed. If necessary a very complex table of results from which you draw out a few key figures, could be presented as an appendix. You should also note the accuracy of any figures presented ($\pm 0.5\%$ for example).

Analysis of results

This is the part where you should comment on the results obtained. For example, if say a fuel consumption test was carried out on two vehicles, a graph comparing one result to the other may be appropriate. Comments should be added if necessary, such as any anomaly that could have affected the results (change of wind direction for example).

Conclusions/comments/observations

Note here any further tests that may be necessary. Conclude that device X does perform better than device Y – if it did! If appropriate, add observations such as how device Y performed better under the set conditions, but under other circumstances the results could have been different. Comment on the method used if necessary.

Forecast

If necessary comment on how the 'item' tested will continue to perform based on the existing data.

Advanced automotive fault diagnosis

Appendices
Detailed pages of results that would 'clog up' the main report or background material such as leaflets relating to the test equipment.

1.4.3 Example report
An example report is presented here relating to a simple alternator test where its actual output is to be compared to the rated output. Minimal details are included so as to illustrate the main points.

Introduction
A *'Rotato'* 12 V alternator was tested under normal operating conditions to check its maximum output. The manufacturer's specifications stated that the alternator, when hot, should produce 95 A at 6000 rev/min.

Test criteria
- Start at room temperature.
- Run alternator at 3000 rev/min, 30 A output for 10 minutes.
- Run alternator at 6000 rev/min, maximum output. Check reading every 30 seconds for 10 minutes.
- Run alternator at 6000 rev/min, maximum output for a further 20 minutes to ensure that output reading is stable.

Facilities/resources
A *'Krypton'* test bench model *R2D2* was used to drive the alternator. The test bench revcounter was used and a *'Flake'* digital meter fitted with a 200 A shunt was used to measure the output. A variable resistance load was employed.

Test procedures
The alternator was run for 10 minutes at 3000 rev/min and the load adjusted to cause an output of 30 A. This was to ensure that it was at a nominal operating temperature. The normal fan was kept in place during the test.

Speed was then increased to 6000 rev/min and the load adjusted to achieve the maximum possible output. The load was further adjusted as required to keep the maximum possible output in case the load resistance changed due to temperature. Measurements were taken every 30 seconds for a period of 10 minutes.

Measured results
Speed held constant at 6000 (± 200) rev/min.
Room temperature (18°C).

Time (± 1 s)	0	30	60	90	120	150	180	210	240	270
Output (± 0.2 A)	101	100	99	99	98	98	98	98	98	98

Time (± 1 s)	300	330	360	390	420	450	480	510	540	570	600
Output (± 0.2 A)	97	97	96	96	96	96	96	96	96	96	96

To ensure that the alternator output had stabilised it was kept running for a further 20 minutes at full output. It continued to hold at 96 A.

Analysis of results
Figure 1.2 shows the results in graphical format.

Conclusions
The manufacturer's claims were validated. The device exceeded the rated output by 6% at the start

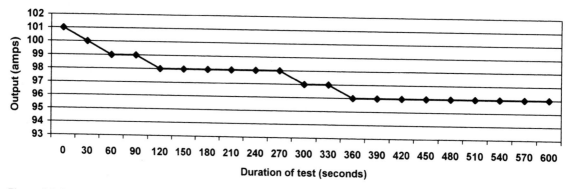

Figure 1.2 Results presented in graphical format

of the test and under continuous operation at full load, continued to exceed the rated output by 1%.

The maximum duration of this test was 20 minutes. It is possible, however, that the device would increase in temperature and the output may fall further after prolonged operation. Further tests are necessary to check this. Overall the device performed in excess of its rated output.

(Sign and date the report)
Tom Denton, Wednesday, 25th Jan. 2006

Knowledge check questions

To use these questions, you should first try to answer them without help but if necessary, refer back to the content of the chapter. Use notes, lists and sketches as appropriate to answer them. It is not necessary to write pages and pages of text!

1. State the meaning of the terms 'fault' and 'symptom'.
2. Explain how to reduce the risk of a short circuit when testing electrical systems.
3. List the main headings that could be used for a standard report.
4. State the two main pieces of knowledge necessary to diagnose faults.
5. Describe the potential dangers of running an engine in an enclosed space without exhaust extraction.

2
Diagnostic techniques

2.1 Introduction

2.1.1 Logic

Diagnostics or faultfinding is a fundamental part of an automotive technician's work. The subject of diagnostics does not relate to individual areas of the vehicle. If your knowledge of a vehicle system is at a suitable level, then you will use the same logical process for diagnosing the fault, whatever the system.

2.1.2 Information

Information and data are available for carrying out many forms of diagnostic work. The data may come as a book or on CD. This information is vital and will ensure that you find the fault – particularly if you have developed the diagnostic skills to go with it. Faultfinding charts and specific examples are presented in later chapters.

The general type of information available is as follows:

- engine diagnostics, testing and tuning;
- servicing, repairs and times;
- fuel and ignition systems and carburettor;
- auto electrics;
- ABS diagnostics;
- component location;
- body repairs, tracking and tyres.

2.1.3 Where to stop?

This is one of the most difficult skills to learn. It is also one of the most important. The secret is twofold:

- know your own limitations – it is not possible to be good at everything;
- leave systems alone where you could cause more damage or even injury – for example air bag circuits.

Often with the best of intentions, a person new to diagnostics will not only fail to find the fault but introduce more faults into the system in the process.

I would suggest you learn your own strengths and weaknesses; you may be confident and good at dealing with mechanical system problems but less so when electronics is involved. Of course you may be just the opposite of this.

Remember that diagnostic skill is in two parts – the knowledge of the system and the ability to apply diagnostics. If you do not yet fully understand a system – leave it alone!

2.2 Diagnostic process

2.2.1 Six-stage process

A key checklist – the six stages of fault diagnosis – is given in Table 2.1.

Here is a very simple example to illustrate the diagnostic process. The reported fault is excessive use of engine oil.

1. Question the customer to find out how much oil is being used (is it excessive?).
2. Examine the vehicle for oil leaks and blue smoke from the exhaust.
3. If leaks are found the engine could still be burning oil but leaks would be a likely cause.
4. A compression test, if the results were acceptable, would indicate the leak to be the most likely fault. Clean down the engine and run for a while. The leak will show up better.
5. Change a gasket or seal, etc.
6. Run through an inspection of the vehicle systems particularly associated with the engine.

Table 2.1

1. Verify the fault
2. Collect further information
3. Evaluate the evidence
4. Carry out further tests in a logical sequence
5. Rectify the problem
6. Check all systems

Double check the fault has been rectified and that you have not caused any further problems.

The stages of faultfinding will be used extensively to illustrate how a logical process can be applied to any situation.

2.2.2 The art of diagnostics

The knowledge needed for accurate diagnostics is in two parts:

1. understanding of the system in which the problem exists;
2. having the ability to apply a logical diagnostic routine.

The knowledge requirement and use of diagnostic skills can now be illustrated with a very simple example. After connecting a hose pipe and turning on the tap, no water comes out of the end! Your knowledge of this system tells you that water should come out providing the tap is on, because the pressure from a tap pushes water through the pipe, and so on. This is where your diagnostic skills become essential. The following stages are now required.

1. Confirm that no water is coming out by looking down the end of the pipe!
2. Does water come out of the other taps, or did it come out of this tap before you connected the hose?
3. Consider what this information tells you; for example, if the answer is '*Yes*' the hose must be blocked or kinked.
4. Walk the length of the pipe looking for a kink.
5. Straighten out the hose.
6. Check that water now comes out and that no other problems have been created.

Much simplified I accept, but the procedure you have just followed made the hose work and it is also guaranteed to find a fault in *any* system. It is easy to see how it works in connection with a hose pipe and I'm sure anybody could have found that fault (well most people anyway)! The higher skill is to be able to apply the same logical routine to more complex situations. The routine (Table 2.1) is represented by Figure 2.1. The loop will continue until the fault is located. I will now explain each of these steps further in relation to a more realistic automotive workshop situation – not that getting the hose to work is not important!

Often electrical faults are considered to be the most difficult to diagnose – but this is not true. I will use a vehicle cooling system fault as an example here, but electrical systems will be covered in detail in later chapters. Remember that the diagnostic procedure can be applied to any problem, mechanical, electrical or even personal!

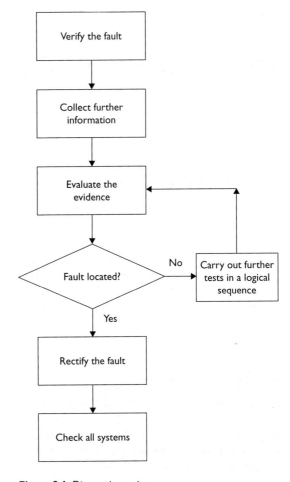

Figure 2.1 Diagnostic routine

However, let's assume that the reported fault with the vehicle is overheating. As is quite common in many workshop situations that's all the information we have to start with. Now work through the six stages.

- **Stage 1** Take a quick look to check for obvious problems such as leaks, broken drive belts or lack of coolant. Run the vehicle and confirm that the fault exists. It could be the temperature gauge for example.
- **Stage 2** Is the driver available to give more information? For example, does the engine overheat all the time or just when working hard? Check records, if available, of previous work done to the vehicle.
- **Stage 3** Consider what you now know. Does this allow you to narrow down what the cause of the fault could be? For example, if the vehicle

overheats all the time and it had recently had a new cylinder head gasket fitted, would you be suspicious about this? Don't let two and two make five, but do let it act as a pointer. Remember that in the science of *logical* diagnostics, two and two always makes four! However, until you know this for certain then play the best odds to narrow down the fault.

- **Stage 4** The further tests carried out would now be directed by your thinking at stage three. You don't yet know if the fault is a leaking head gasket, the thermostat stuck closed or some other problem. Playing the odds, a cooling system pressure test would probably be the next test. If the pressure increases when the engine is running then it is likely to be a head gasket or similar problem. If no pressure increase then move on to the next test and so on. After each test go back to stage 3 and evaluate what you know, *not* what you don't know!
- **Stage 5** Let's assume the problem was a thermostat stuck closed – replace it and top up the coolant, etc.
- **Stage 6** Check that the system is now working. Also check that you have not caused any further problems such as leaks or loose wires. This example is simplified a little, but like the hose pipe problem it is the sequence that matters, particularly the '*stop and think*' at stage 3. It is often possible to go directly to the cause of the fault at this stage, providing that you have an adequate knowledge of how the system works.

2.2.3 Summary

I have introduced the six-stage process of diagnostics, not so it should necessarily be used as a checklist but to illustrate the process that must be followed. Much more detail is required still, particular in relation to stages 3 and 4. The purpose of this set process is to ensure that 'we' work in a particular, logical way.

I would like to stress the need for a logical process again – with a quotation! '*Logic is the beginning of wisdom not the end.*' (Spock to Valeris, Star Trek II)

2.3 Diagnostics on paper

2.3.1 Introduction

This section is again a way of changing the way you approach problems on a vehicle. The key message is that if you stop and think before 'pulling the car to pieces', it will often save a great deal of time. In other words, some of the diagnostic work can be done 'on paper' before we start on the vehicle. To illustrate this, the next section lists symptoms for three separate faults on a car and for each of these symptoms, three possible faults. All the faults are possible but in each case choose the 'most likely' option.

2.3.2 Examples

Symptoms	Possible faults
A The brake/stop lights are reported not operating. On checking it is confirmed that neither of the two bulbs or the row of high-mounted LEDs is operating as the pedal is pressed. All other systems work correctly	1. Two bulbs and 12 LEDs blown 2. Auxiliary systems relay open circuit 3. Brake light switch not closing
B An engine fitted with full management system tends to stall when running slowly. It runs well under all other conditions and the reported symptom is found to be intermittent	1. Fuel pump output pressure low 2. Idle control valve sticking 3. Engine speed sensor wire loose
C The off side dip beam headlight not operating. This is confirmed on examination and also noted is that the off side tail light does not work	1. Two bulbs blown 2. Main lighting fusible link blown 3. Short circuit between off side tail and dip beam lights

The most likely fault for example A, is number 3. It is possible that all the lights have blown but unlikely. It could not be the auxiliary relay because this would affect other systems.

For example B, the best answer would be number 2. It is possible that the pump pressure is low but this would be more likely to affect operation under other conditions. A loose wire on the engine speed sensor could cause the engine to stall but it would almost certainly cause misfire under other conditions.

The symptoms in C would suggest answer 1. The short circuit suggested as answer 3 would be more likely to cause lights and others to stay on rather that not work, equally the chance of a short between these two circuits is remote if not impossible. If the lighting fusible link were blown then none of the lights would operate.

The technique suggested here relates to stages 1 to 3 of the 'the six stages of fault diagnosis' process. By applying a little thought before even taking a screwdriver to the car, a lot of time can be saved. If the problems suggested in the previous table were real we would at least now be able to start looking in the right area for the fault.

2.3.3 How long is a piece of string?

Yes I know, twice the distance from the middle to one end! What I am really getting at here though is the issue about what is a valid reading/measurement and what is not – when compared to data. For example if the 'data book' says the resistance of the component should be between 60 and 90 Ω, what do you do when the measured value is 55 Ω? If the measured value was 0 Ω or 1000 Ω then the answer is easy – the component is faulty! However, when the value is very close you have to make a decision. In this case (55 Ω) it is very likely that the component is serviceable.

The decision over this type of issue is difficult and must in many cases be based on experience. As a general guide however, I would suggest that if the reading is in the right 'order of magnitude', then the component has a good chance of being OK. By this I mean that if the value falls within the correct range of 1s, 10s, 100s or 1000s etc. then it is probably good.

Do notice that I have ensured that words or phrases such as 'probably', 'good chance' and 'very likely' have been used here! This is not just to make sure I have a get out clause; it is also to illustrate that diagnostic work can involve 'playing the best odds' – as long as this is within a logical process.

2.4 Mechanical diagnostic techniques

2.4.1 Check the obvious first!

Start *all* hands on diagnostic routines with 'hand and eye checks'. In other words look over the vehicle for obvious faults. For example, if automatic transmission fluid is leaking on to the floor then put this right before carrying out complicated stall tests. Here are some further suggestions that will at some point save you a lot of time.

- If the engine is blowing blue smoke out of the exhaust – consider the worth of tracing the cause of a tapping noise in the engine!
- When an engine will not start – check that there is fuel in the tank!

2.4.2 Noise, vibration and harshness

Noise, vibration and harshness (NVH) concerns have become more important as drivers have become more sensitive to these issues. Drivers have higher expectations of comfort levels. Noise, vibration and harshness issues are more noticeable due to reduced engine noise and better insulation in general. The main areas of the vehicle that produce NVH are:

- tyres;
- engine accessories;
- suspension;
- driveline.

It is necessary to isolate the NVH into its specific area(s) to allow more detailed diagnosis. A road test as outlined later is often the best method.

The five most common sources of non-axle noise are exhaust, tyres, roof racks, trim and mouldings, and transmission. Ensure that none of the following conditions is the cause of the noise before proceeding with a driveline strip down and diagnosis.

1. In certain conditions, the pitch of the exhaust may sound like gear noise or under other conditions like a wheel bearing rumble.
2. Tyres can produce a high pitched tread whine or roar, similar to gear noise. This is particularly the case for non-standard tyres.
3. Trim and mouldings can cause whistling or whining noises.
4. Clunk may occur when the throttle is applied or released due to backlash somewhere in the driveline.
5. Bearing rumble sounds like marbles being tumbled.

2.4.3 Noise conditions

Noise is very difficult to describe. However, the following are useful terms and are accompanied by suggestions as to when they are most likely to occur.

- Gear noise is typically a howling or whining due to gear damage or incorrect bearing preload. It can occur at various speeds and driving conditions, or it can be continuous.
- 'Chuckle' is a rattling noise that sounds like a stick held against the spokes of a spinning bicycle wheel. It usually occurs while decelerating.

- Knock is very similar to chuckle though it may be louder and occurs on acceleration or deceleration.

Check and rule out tyres, exhaust and trim items before any disassembly to diagnose and correct gear noise.

2.4.4 Vibration conditions

Clicking, popping or grinding noises may be noticeable at low speeds and be caused by the following:

- inner or outer CV joints worn (often due to lack of lubrication so check for split gaiters);
- loose drive shaft;
- another component contacting a drive shaft;
- damaged or incorrectly installed wheel bearing, brake or suspension component.

The following may cause vibration at normal road speeds:

- out-of-balance wheels;
- out-of-round tyres.

The following may cause shudder or vibration during acceleration:

- damaged power train/drive train mounts;
- excessively worn or damaged out-board or in-board CV joints.

The cause of noise can often be traced by first looking for leaks. A dry bearing or joint will produce significant noise.

1. Inspect the CV joint gaiters (boots) for cracks, tears or splits.
2. Inspect the underbody for any indication of grease splatter near the front wheel half shaft joint boots.
3. Inspect the in-board CV joint stub shaft bearing housing seal for leakage at the bearing housing.
4. Check the torque on the front axle wheel hub retainer.

2.4.5 Road test

A vehicle will produce a certain amount of noise! Some noise is acceptable and may be audible at certain speeds or under various driving conditions such as on a new road.

Carry out a thorough visual inspection of the vehicle before carrying out the road test. Keep in mind anything that is unusual. A key point is to *not* repair or adjust anything until the road test is carried out. Of course this does not apply if the condition could be dangerous or the vehicle will not start!

Establish a route that will be used for all diagnosis road tests. This allows you to get to know what is normal and what is not! The roads selected should have sections that are reasonably smooth, level and free of undulations as well as lesser quality sections needed to diagnose faults that only occur under particular conditions. A road that allows driving over a range of speeds is best. Gravel, dirt or bumpy roads are unsuitable because of the additional noise they produce.

If a customer complains of a noise or vibration on a particular road and only on a particular road, the source of the concern may be the road surface. Test the vehicle on the same type of road.

Make a visual inspection as part of the preliminary diagnosis routine prior to the road test; note anything that does not look right.

1. Tyre pressures, but do not adjust them yet.
2. Leaking fluids.
3. Loose nuts and bolts.
4. Bright spots where components may be rubbing against each other.
5. Check the luggage compartment for unusual loads.

Road test the vehicle and define the condition by reproducing it several times during the road test. During the road test recreate the following conditions.

1. **Normal driving speeds of 20 to 80 km/h (15 to 50 mph)** with light acceleration, a moaning noise may be heard and possibly a vibration is felt in the front floor pan. It may get worse at a certain engine speed or load.
2. **Acceleration/deceleration** with slow acceleration and deceleration, a shake is sometimes noticed through the steering wheel seats, front floor pan, front door trim panels, etc.
3. **High speed** a vibration may be felt in the front floor pan or seats with no visible shake, but with an accompanying sound or rumble, buzz, hum, drone or booming noise. Coast with the clutch pedal down or gear lever in neutral and engine idling. If vibration is still evident, it may be related to wheels, tyres, front brake discs, wheel hubs or wheel bearings.
4. **Engine rev/min sensitive** a vibration may be felt whenever the engine reaches a particular speed. It may disappear in neutral coasts. Operating the engine at the problem speed while the vehicle is stationary can duplicate

the vibration. It can be caused by any component, from the accessory drive belt to the clutch or torque converter, which turns at engine speed when the vehicle is stopped.

5. **Noise and vibration while turning** clicking, popping or grinding noises may be due to the following: damaged CV joint; loose front wheel half shaft joint boot clamps; another component contacting the half shaft; worn, damaged or incorrectly installed wheel bearing; damaged power train/drive train mounts.

After a road test, it is often useful to do a similar test on a lift. When carrying out the shake and vibration diagnosis or engine accessory vibration diagnosis on a lift, observe the following precautions.

- If only one drive wheel is allowed to rotate, speed must be limited to 55 km/h (35 mph) indicated on the speedometer. This is because the actual wheel speed will be twice that indicated on the speedometer.
- The suspension should not be allowed to hang free. If a CV joint were run at a high angle, extra vibration as well as damage to the seals and joints could occur.

Support the front suspension lower arm as far out-board as possible. This will ensure that the vehicle is at its correct ride height. The procedure is outlined by the following steps.

1. Raise and support the vehicle.
2. Explore the speed range of interest using the road test checks as previously discussed.
3. Carry out a coast down (overrun) in neutral. If the vehicle is free of vibration when operating at a steady indicated speed and behaves very differently in drive and coast, a transmission concern is likely.

Note, however, that a test on the lift may produce different vibrations and noises than a road test because of the effect of the lift. It is not unusual to find a vibration on the lift that was not noticed during the road test. If the condition found on the road can be duplicated on the lift, carrying out experiments on the lift may save a great deal of time.

2.4.6 Engine noises

How do you tell a constant tapping from a rattle? Worse still, how do you describe a noise in a book? I'll do my best! Try the following table as a non-definitive guide to the source or cause of engine or engine ancillary noises.

Noise description	Possible source
Tap	Valve clearances out of adjustment, cam followers or cam lobes worn
Rattle	A loose component, broken piston ring or component
Light knock	Small end bearings worn, cam or cam follower
Deep knock or thud	Big end bearings worn
Rumble	Main bearings worn
Slap	Worn pistons or bores
Vibration	Loose or out of balance components
Clatter	Broken rocker shaft or broken piston rings
Hiss	Leak from inlet or exhaust manifolds or connections
Roar	Air intake noise, air filter missing, exhaust blowing or a seized viscous fan drive
Clunk	Loose flywheel, worm thrust bearings or a loose front pulley/damper
Whine	Power steering pump or alternator bearing
Shriek	Dry bearing in an ancillary component
Squeal	Slipping drive belt

2.4.7 Sources of engine noise

The following table is a further guide to engine noise. Possible causes are listed together with the necessary repair or further diagnosis action as appropriate.

Sources of engine noise	Possible cause	Required action
Misfiring/backfiring	Fuel in tank has wrong octane/cetane number, or is wrong type of fuel	Determine which type of fuel was last put in the tank
	Ignition system faulty	Check the ignition system
	Engine temperature too high	Check the engine cooling system
	Carbon deposits in the combustion chamber start to glow and cause misfiring	Remove the carbon deposits by using fuel additives and driving the vehicle carefully
	Timing incorrect, which causes misfiring in the intake/exhaust system	Check the timing

Valve train faulty	Valve clearance too large due to faulty bucket tappets or incorrect adjustment of valve clearance	Adjust valve clearance if possible and renew faulty bucket tappets – check cam condition
	Valve timing incorrectly adjusted, valves and pistons are touching	Check the valve timing and adjust if necessary
	Timing belt broken or damaged	Check timing belt and check pistons and valves for damage – renew any faulty parts
Engine components faulty	Pistons Piston rings Cylinder head gasket Big end and/or main bearing journals	Disassemble the engine and check components
Ancillary components	Engine components or ancillary components loose or broken	Check that all components are secure, tighten/ adjust as required Renew if broken

2.5 Electrical diagnostic techniques

2.5.1 Check the obvious first!

Start *all* hands on diagnostic routines with 'hand and eye checks'. In other words look over the vehicle for obvious faults. For example, if the battery terminals are loose or corroded then put this right before carrying out complicated voltage readings. Here are some further suggestions that will at some point save you a lot of time.

- A misfire may be caused by a loose plug lead – it is easier to look for this than interpret the ignition waveforms on a scope.
- If the ABS warning light stays on – look to see if the wheel speed sensor(s) are covered in mud or oil.

2.5.2 Test lights and analogue meters – warning!

A test lamp is ideal for tracing faults in say a lighting circuit because it will cause a current to flow which tests out high resistance connections. However, it is this same property that will damage delicate electronic circuits – so don't use it for any circuit that contains an electronic control unit (ECU). Even an analogue voltmeter can cause enough current to flow to at best give you a false reading and at worst damage an ECU – so don't use it!

A digital multimeter is ideal for all forms of testing. Most have an internal resistance in excess of $10\,M\Omega$. This means that the current they draw is almost insignificant. An LED test lamp or a logic probe is also acceptable.

2.5.3 Generic electrical testing procedure

The following procedure is very generic but with a little adaptation can be applied to any electrical system. Refer to manufacturer's recommendations if in any doubt. The process of checking any system circuit is broadly as follows.

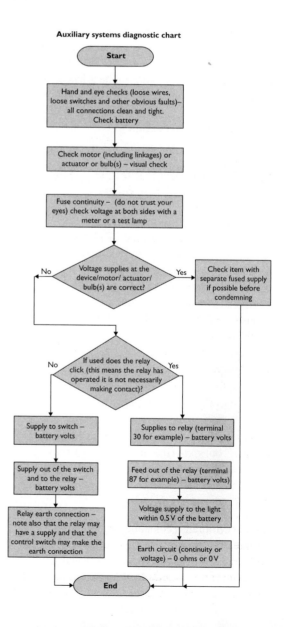

Diagnostic techniques **13**

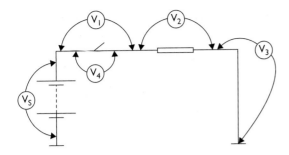

Figure 2.2 Volt drop testing

2.5.4 Volt drop testing

Volt drop is a term used to describe the difference between two points in a circuit. In this way we can talk about a voltage drop across a battery (normally about 12.6 V) or the voltage drop across a closed switch (ideally 0 V but may be 0.1 or 0.2 V).

The first secret to volt drop testing is to remember a basic rule about a series electrical circuit: *The sum of all volt drops around a circuit always adds up to the supply.*

The second secret is to ensure that the circuit is switched on and operating – or at least the circuit should be 'trying to operate'!

In Figure 2.2 this means that $V_1 + V_2 + V_3 = V_s$. When electrical testing, therefore, and if the battery voltage measured as say 12 V, a reading of less than 12 V at V_2 would indicate a volt drop between the terminals of V_1 and/or V_3. Likewise the correct operation of the switch, that is it closes and makes a good connection, would be confirmed by a very low reading on V_1.

What is often described as a 'bad earth' (when what is meant is a high resistance to earth), could equally be determined by the reading on V_3.

To further narrow the cause of a volt drop down a bit, simply measure across a smaller area. The voltmeter V_4, for example, would only assess the condition of the switch contacts.

2.5.5 Testing for short circuits to earth

This fault will normally blow a fuse – or burn out the wiring completely! To trace a short circuit is very different from looking for a high resistance connection or an open circuit. The volt drop testing above will trace an open circuit or a high resistance connection.

My preferred method of tracing a short, after looking for the obvious signs of trapped wires, is to connect a bulb or test lamp across the blown fuse and switch on the circuit. The bulb will light because on one side it is connected to the supply for the fuse and on the other side it is connected to earth via the fault. Now disconnect small sections of the circuit one at a time until the test lamp goes out. This will indicate the particular circuit section that has shorted out.

2.5.6 On and off load tests

On load means that a circuit is drawing a current. Off load means it is not! One example where this may be an issue is when testing a starter circuit. Battery voltage may be 12 V off load but only 9 V when on load.

A second example is the supply voltage to the positive terminal of an ignition coil via a high resistance connection (corroded switch terminal for example). With the ignition on and the vehicle not running, the reading will almost certainly be battery voltage because the ignition ECU switches off the primary circuit and no volt drop will show up. However, if the circuit were switched on (with a fused jumper lead if necessary) a lower reading would result showing up the fault.

2.5.7 Black box technique

The technique that will be covered here is known as 'black box faultfinding'. This is an excellent technique and can be applied to many vehicle systems from engine management and ABS to cruise control and instrumentation.

As most systems now revolve around an ECU, the ECU is considered to be a 'black box', in other words we know what it should do but how it does it is irrelevant! Any colour, so long as it's black [Henry Ford (1920s)]. I doubt that he was referring to ECUs though …

Figure 2.3 shows a block diagram that could be used to represent any number of automobile electrical or electronic systems. In reality the arrows from the 'inputs' to the ECU and from the ECU to the 'outputs' are wires. Treating the ECU as a 'black box' allows us to ignore its complexity. The theory is that if all the sensors and associated wiring to the 'black box' are OK, all the output actuators and their wiring are OK and the supply/earth connections are OK, then the fault must be the 'black box'. Most ECUs are very reliable, however, and it is far more likely that the fault will be found in the inputs or outputs.

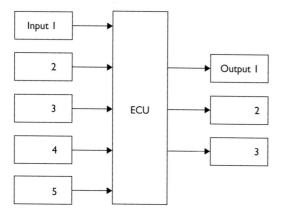

Figure 2.3 System block diagram

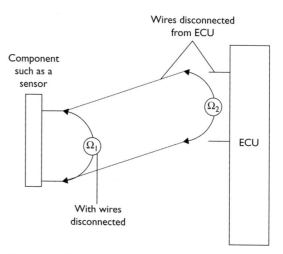

Figure 2.4 Ohmmeter tests

Normal faultfinding or testing techniques can be applied to the sensors and actuators. For example, if an ABS system uses four inductive type wheel speed sensors, then an easy test is to measure their resistance. Even if the correct value were not known, it would be very unlikely for all four to be wrong at the same time so a comparison can be made. If the same resistance reading is obtained on the end of the sensor wires at the ECU then almost all of the 'inputs' have been tested with just a few ohmmeter readings.

The same technique will often work with 'outputs'. If the resistance of all the operating windings in say a hydraulic modulator were the same, then it would be reasonable to assume the figure was correct.

Sometimes, however, it is almost an advantage *not* to know the manufacturer's recommended readings. If the 'book' says the value should be between 800 and 900 Ω, what do you do when your ohmmeter reads 905 Ω? Answers on a postcard please ... (or see Section 2.3.3).

Finally, don't forget that no matter how complex the electronics in an ECU, they will not work without a good power supply and an earth!

2.5.8 Sensor to ECU method

This technique is simple but very useful. Figure 2.4 shows a resistance test being carried out on a component. Ω_1 is a direct measure of its resistance whereas Ω_2 includes the condition of the circuit. If the second reading is the same as the first then the circuit must be in good order.

> **Warning**
> The circuit supply must always be off when carrying out ohmmeter tests.

2.5.9 Flight recorder tests

It is said that the best place to sit in an aeroplane is on the black box flight recorder! Well, apart from the black box usually being painted bright orange so it can be found after a crash, my reason for mentioning it is to illustrate how the flight recorder principle can be applied to automotive diagnostics.

Most hand-held scopes now have flight record facilities. This means that they will save the signal from any probe connection in memory for later play back. The time duration will vary depending on the available memory and the sample speed but this is a very useful feature.

As an example, consider an engine with an intermittent misfire that occurs only under load. If a connection is made to the suspected component (coil HT output for example), and the vehicle is road tested, the waveforms produced can be examined afterwards.

Many engine (and other system) ECUs have built in flight recorders in the form of self-diagnostic circuits. If a wire breaks loose causing a misfire but then reconnects the faulty circuit will be 'remembered' by the ECU.

2.5.10 Faultfinding by luck!

Or is it logic? If four electric windows stopped working at the same time, it would be very unlikely that all four motors had burned out. On the other hand if just one electric window stopped working,

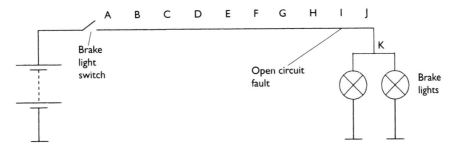

Figure 2.5 Faultfinding by luck!

then it may be reasonable to suspect the motor. It is this type of reasoning that is necessary when faultfinding. However, be warned it is theoretically possible for four motors to apparently burn out all at the same time!

Using this 'playing the odds' technique can save time when tracing a fault in a vehicle system. For example, if both stop lights do not work and everything else on the vehicle is OK, I would suspect the switch (stages 1 to 3 of the normal process). At this stage though, the fault could be anywhere – even two or three blown bulbs. None-the-less a quick test at the switch with a voltmeter would prove the point. Now, let's assume the switch is OK and it produces an output when the brake pedal is pushed down. Testing the length of wire from the front to the back of the vehicle further illustrates how 'luck' comes into play.

Figure 2.5 represents the main supply wire from the brake switch to the point where the wire 'divides' to each individual stop light (the odds say the fault must be in this wire). For the purpose of this illustration we will assume the open circuit is just before point 'I'. The procedure continues in one of the two following ways. Either:

- guess that the fault is in the first half and test at point F;
- we were wrong! Guess that the fault is in the first half of the second half and test at point I;
- we were right! Check at H and we have the fault ... on test number THREE;

or:

- test from A to K in a logical sequence of tests;
- we would find the fault ... On test number NINE!

You may choose which method you prefer!

2.5.11 Colour codes and terminal numbers

This section is really more to be used as a reference source. It is useful, however, to become familiar with a few key wire colours and terminal numbers when diagnosing electrical faults. As seems to be the case for any standardisation a number of colour code systems are in operation! For reference purposes I will just mention two.

Firstly, the British Standard system (BS AU 7a: 1983): this system uses 12 colours to determine the main purpose of the cable and tracer colours to further refine its use. The main colour uses and some further examples are given in the following table.

Colour	Symbol	Destination/use
Brown	N	Main battery feed
Blue	U	Headlight switch to dip switch
Blue/White	UW	Headlight main beam
Blue/Red	UR	Headlight dip beam
Red	R	Side light main feed
Red/Black	RB	Left hand side lights and no. plate
Red/Orange	RO	Right hand side lights
Purple	P	Constant fused supply
Green	G	Ignition controlled fused supply
Green/Red	GR	Left side indicators
Green/White	GW	Right side indicators
Light Green	LG	Instruments
White	W	Ignition to ballast resistor
White/Black	WB	Coil negative
Yellow	Y	Overdrive and fuel injection
Black	B	All earth connections
Slate	S	Electric windows
Orange	O	Wiper circuits (fused)
Pink/White	KW	Ballast resistor wire
Green/Brown	GN	Reverse
Green/Purple	GP	Stop lights
Blue/Yellow	UY	Rear fog light

Secondly there is a 'European' system used by a number of manufacturers and based broadly on the following table. Please note there is no correlation between the 'Euro' system and the British Standard colour codes. In particular note the use of the colour brown in each system! After some practice with the use of colour code systems the job of the technician is made a lot easier when faultfinding an electrical circuit.

Colour	Symbol	Destination/use
Red	Rt	Main battery feed
White/Black	Ws/Sw	Headlight switch to dip switch
White	Ws	Headlight main beam
Yellow	Ge	Headlight dip beam
Grey	Gr	Side light main feed
Grey/Black	Gr/Sw	Left hand side lights
Grey/Red	Gr/Rt	Right hand side lights
Black/Yellow	Sw/Ge	Fuel injection
Black/Green	Sw/Gn	Ignition controlled supply
Black/White/Green	Sw/Ws/Gn	Indicator switch
Black/White	Sw/Ws	Left side indicators
Black/Green	Sw/Gn	Right side indicators
Light Green	LGn	Coil negative
Brown	Br	Earth
Brown/White	Br/Ws	Earth connections
Black	Sw	Reverse
Black/Red	Sw/Rt	Stop lights
Green/Black	Gn/Sw	Rear fog light

A system now in use almost universally is the terminal designation system in accordance with DIN 72 552. This system is to enable easy and correct connections to be made on the vehicle, particularly in after sales repairs. Note that the designations are not to identify individual wires but to define the terminals of a device. Listed below are some of the most popular numbers.

1	Ignition coil negative
4	Ignition coil high tension
15	Switched positive (ignition switch output)
30	Input from battery positive
31	Earth connection
49	Input to flasher unit
49a	Output from flasher unit
50	Starter control (solenoid terminal)
53	Wiper motor input
54	Stop lamps
55	Fog lamps
56	Headlamps
56a	Main beam
56b	Dip beam
58L	Left side lights
58R	Right side lights
61	Charge warning light
85	Relay winding out
86	Relay winding input
87	Relay contact input (change over relay)
87a	Relay contact output (break)
87b	Relay contact output (make)
L	Left side indicators
R	Right side indicators
C	Indicator warning light (vehicle)

The Ford Motor Company now uses a circuit numbering and wire identification system. This is in use worldwide and is known as Function, System-Connection (FSC). The system was developed to assist in vehicle development and production processes. However, it is also very useful in helping the technician with fault finding. Many of the function codes are based on the DIN system. Note that earth wires are now black! The system works as follows.

$$31S\text{-}AC3A \parallel 1.5\ BK/RD$$

Function:

$31 =$ ground/earth

$S =$ additionally switched circuit

System:

$AC =$ headlamp levelling

Connection:

$3 =$ switch connection

$A =$ branch

Size:

$1.5 = 1.5\,\text{mm}^2$

Colour:

$BK =$ black (determined by function 31)

$RD =$ red stripe

The new Ford colour codes table is as follows:

Code	Colour
BK	Black
BN	Brown
BU	Blue
GN	Green
GY	Grey
LG	Light-green
OG	Orange
PK	Pink
RD	Red
SR	Silver
VT	Violet
WH	White
YE	Yellow

Ford system codes are as follows.

Letter	Main system	Examples
D	Distribution systems	DE = earth
A	Actuated systems	AK = wiper/washer
B	Basic systems	BA = charging
		BB = starting
C	Control systems	CE = power steering
G	Gauge systems	GA = level/pressure/temperature
H	Heated systems	HC = heated seats
L	Lighting systems	LE = headlights

M	Miscellaneous systems	MA = air bags
P	Power train control systems	PA = engine control
W	Indicator systems ('indications' not turn signals)	WC = bulb failure
X	Temporary for future features	XS = too much!

As a final point to this section it must be noted that the colour codes and terminal designations given are for illustration only. Further reference should be made for specific details to the manufacturer's information.

2.5.12 Back probing connectors

Just a quick warning! If you are testing for a supply (for example) at an ECU, then use the probes of your digital meter with care. Connect to the back of the terminals; this will not damage the connecting surfaces as long as you do not apply excessive force. Sometimes a pin clamped in the test lead's crocodile/alligator clip is ideal for connecting 'through' the insulation of a wire without having to disconnect it. Figure 2.6 shows the 'back probing' technique.

2.6 Fault codes

2.6.1 Fast and slow

Most modern vehicle management systems carry out self-diagnostic checks on the sensors and actuators that connect to the vehicle ECU(s). A fault in one of the components or its associated circuit causes a code to be stored in the ECU memory.

The codes may be described as 'fast' or 'slow'. Some ECUs produce both types. An LED, dash warning light, scope or even an analogue voltmeter can be used to read slow codes. Normally, slow codes are output as a series of flashes that must then be interpreted by looking up the code in a fault code table. The slow codes are normally initiated by shorting two connections on the diagnostic plug and then switching the ignition on. Refer to detailed data before shorting any pins out!

Fast codes can only be read by using a fault code reader or scanner. Future ECUs will use fast codes. In the same way as we accept that a good digital multimeter is an essential piece of test equipment, it is now necessary to consider a fault code reader in the same way.

If a code reader is attached to the serial port on the vehicle harness, fast and slow codes can be read out from the vehicle computer. These are either displayed in the form of a two, three or four digit output code or in text format if software is used.

2.6.2 Fault code examples

A number of codes and descriptions are reproduced below as an example of the detailed information that is available from a self-diagnosis system. The data relates to the Bosch Motronic 1.7

Figure 2.6 Test the voltage at a connection with care

and 3.1. Fault code lists are available in publications such as those by 'Autodata' and 'Autologic'.

FCR	Description code
000	No faults found in the ECU
001	Fuel pump relay or fuel pump relay circuit
001	Crank angle sensor (CAS) or circuit (alternative code)
002	Idle speed control valve circuit
003	Injector number 1 or group one circuit
004	Injector number 3 or circuit
005	Injector number 2 or circuit
006	Injectors or injector circuit.
012	Throttle position switch or circuit
016	CAS or circuit
018	Amplifier to ECU amplifier circuit
019	ECU
023	Ignition amplifier number 2 cylinder or circuit
024	Ignition amplifier number 3 cylinder or circuit
025	Ignition amplifier number 1 cylinder or circuit
026	ECU supply
029	Idle speed control valve (ISCV) or circuit
031	Injector number 5 or circuit
032	Injector number 6 or injector group two circuit
033	Injector number 4 or circuit
036	Carbon filter solenoid valve (CFSV) or circuit
037	Oxygen sensor (OS) or circuit
041	Mass airflow (MAF) sensor or circuit
046	ECU
048	Air conditioning (AC) compressor or circuit
050	Ignition amplifier cylinder number 4 or circuit
051	Ignition amplifier cylinder number 6 or circuit
054	ECU
055	Ignition amplifier or circuit
062	Electronic throttle control or circuit
064	Ignition timing (electronic)
067	Vehicle speed sensor (VSS) or circuit
067	CAS or circuit
070	OS or circuit
073	Vehicle speed sensor (VSS) or circuit
076	CO potentiometer (non-cat)
077	Intake air temperature sensor (ATS) or circuit
078	Engine coolant temperature sensor (CTS) or circuit
081	Alarm system or circuit
082	Traction control or circuit
083	Suspension control or circuit
085	AC compressor or circuit
100	ECU
200	ECU
201	OS control or circuit
202	ECU
203	Ignition primary or circuit
204	Electronic throttle control signal or circuit
300	Engine

2.6.3 Clearing

Fault codes can be cleared from the ECU memory in two ways:

- using the facilities of a fault code reader (scanner) to clear the memory;
- disconnecting the battery earth lead for about two minutes (does not always work however).

The first method is clearly recommended because disconnecting the battery will also 'reset' many other functions such as the radio code, the clock and even the learnt or adaptive functions in the ECUs.

2.7 Systems

2.7.1 What is a system

System is a word used to describe a collection of related components which interact as a whole. A motorway system, the education system or computer systems are three varied examples. A large system is often made up of many smaller systems which in turn can each be made up of smaller systems and so on. Figure 2.7 shows how this can be represented in a visual form.

One further definition: *A group of devices serving a common purpose.*

Using the systems approach helps to split extremely complex technical entities into more manageable parts. It is important to note, however, that the links between the smaller parts and the boundaries around them are also very important. System boundaries will overlap in many cases.

The modern motor vehicle is a complex system and in itself forms just a small part of a larger transport system. It is the capacity for the motor vehicle to be split into systems on many levels which aids in both its design and construction. The systems approach helps in particular with understanding how something works and, further, how to go about repairing it when it doesn't!

2.7.2 Vehicle systems

Splitting the vehicle into systems is not an easy task because it can be done in many different ways. A split between mechanical systems and electrical systems would seem a good start. However, this

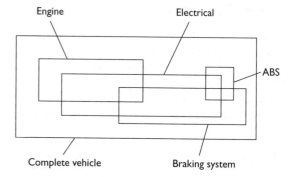

Figure 2.7 Vehicle systems representation

division can cause as many problems as it solves. For example, in which half do we put anti-lock brakes, mechanical or electrical. The answer is of course both! None-the-less, it is still easier if we just consider one area of the vehicle and do not try to comprehend the whole.

Once a complex set of interacting parts such as a motor vehicle has been 'systemised', the function or performance of each part can be examined in more detail. In other words, knowing what each part of the system should do in turn helps in determining how each part actually works. It is again important to stress that the links and interactions between various sub-systems are a very important consideration. Examples of this would be how the power demands of the vehicle lighting system will have an effect on the charging system operation, or in the case of a fault, how an air leak from a brake servo could cause a weak air/fuel ratio.

To further analyse a system, whatever way it has been sub-divided from the whole, consideration should be given to the inputs and the outputs. Many of the complex electronic systems on a vehicle lend themselves to this form of analysis. Considering the ECU of the system as the control element and looking at its inputs and outputs is the recommended approach.

2.7.3 Open loop systems

An open loop system is designed to give the required output whenever a given input is applied. A good example of an open loop vehicle system would be the headlights. With the given input of the switch being operated the output required is that the headlights will be illuminated. This can be taken further by saying that an input is also required from the battery and a further input of say the dip switch. The feature which determines that a system is open loop is that no feedback is required for it to operate. Figure 2.8 shows this example in block diagram form.

2.7.4 Closed loop systems

A closed loop system is identified by a feedback loop. It can be described as a system where there is a possibility of applying corrective measures if the output is not quite what is wanted. A good example of this in a vehicle is an automatic temperature control system. The interior temperature of the vehicle is determined by the output from the heater which is switched on or off in response to a signal from a temperature sensor inside the cabin. The feedback loop is the fact that the output from the system, temperature, is also an input to the system. This is represented by Figure 2.9.

The feedback loop in any closed loop system can be in many forms. The driver of a car with a conventional heating system can form a feedback loop by turning the heater down when he/she is too hot and turning it back up when cold. The feedback to a voltage regulator in an alternator is an electrical signal using a simple wire.

2.7.5 Block diagrams

Another secret to good diagnostics is the 'block diagram' approach. Most systems can be considered as consisting of 'inputs to a control which has outputs'. This technique means that complex systems can be considered in manageable 'chunks'.

Many complex vehicle electronic systems can be represented as block diagrams. In this way several inputs can be shown supplying information to an ECU that in turn controls the system outputs. As an example of this, consider the operation of a vehicle alarm system (Figure 2.10). In its simplest form the inputs are the 'sensors' (such as door switches) and the 'outputs' are the actuators (such as the siren). The 'control' section is the alarm ECU.

The diagnostic approach is that if all the sensors are providing the correct information to the control and the actuators respond when tested,

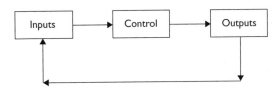

Figure 2.9 Closed loop system

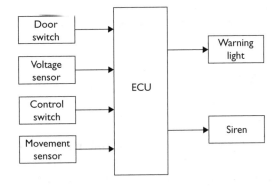

Figure 2.10 Block diagram

Figure 2.8 Open loop system

then the fault must be the control unit. If a sensor does not produce the required information then the fault is equally evident.

2.8 On- and off-board diagnostics

2.8.1 On-board diagnostics

On-board diagnostics refers to the systems on the vehicle carrying out some form of self-monitoring. The more complex automobiles become, the greater the number of electronic systems and the more difficult it is to register the actual condition in case of a defect.

Many connecting cables and adapters are required to achieve this. Data about the different systems and their working together is needed to allow a system specific diagnosis. Modern electronics with self-diagnosis supports the technician by registering actual values, comparing them with the nominal values, and diagnosing faults that are stored for repair purposes.

Internal to an ECU, a checksum of the program memory is calculated. Then a read and write test of the random access memory (RAM) is performed. Other elements such as A/D (analogue/digital) converters are also checked within this test cycle.

During the operating time of the vehicle, the ECUs are constantly checking the sensors they are connected to. The ECUs are then able to determine whether a sensor has a short circuit to ground or battery voltage, or if a cable to the sensor is open circuit. By comparing the measured values and the stored data, an ECU is able to determine whether the measured values exceed or are still within the tolerance required. Combining information provided by other sensors allows the ECU to monitor for plausibility of the sensor signals.

Measuring the current normally taken by their circuits is used to carry out a check on actuators. Powering the actuator and observing the reaction of the system can test the function of an actuator in some cases.

If discrepancies to the nominal values are diagnosed, the information is stored in an internal fault memory together with other parameters, such as engine temperature or speed. In this way, defects that appear intermittent or only under certain conditions can be diagnosed. If a fault occurs only once during a set period of time, it is deleted. The fault memory can be read later in the workshop and provides valuable information for the technician.

When a defective sensor is detected, the measured values are replaced by a nominal value, or an alternative value is calculated using the information from other sensors to provide a limp-home function. With the help of an appropriate code reader or scanner, a technician can communicate with the ECUs, read the fault memory and the measured values, and send signals to the actuators.

Another task of self-diagnosis is to indicate a defect to the driver. A warning light on the dashboard is the most common method used to do this. Regulations concerning exhaust emissions mean an extension of self-diagnosis is desirable. The control units will soon have to be able to control all exhaust gas functions and components and to clearly indicate a defective function or the exceeding of the permissible exhaust limits. Chapter 5 covers this subject in detail.

2.8.2 Off-board diagnostics

The continual increase in the use of electronics within vehicles represents a major challenge for customer service and workshop operations. Modern diagnosis and information systems must cope with this challenge and manufacturers of test equipment must provide instruments that are flexible and easy to handle. Quick and reliable fault diagnosis in modern vehicles requires extensive technical knowledge, detailed vehicle information, up-to-date testing systems and the skill to be able to apply all of these.

The test equipment on the market can be subdivided into two main categories:

- hand-held or portable instruments;
- stationary equipment.

Hand-held instruments are commonly used for the control of engine functions like ignition or fuel injection and the request of error codes from the ECUs.

Stationary test equipment may be able to cover the whole range of function and performance checks of the engine, gear, brakes, chassis, and exhaust monitoring. Most of the common testers are used for diagnosing engine faults.

For repair, service, and maintenance, many different manuals and microfiches are used in workshops. It is difficult to collect all the necessary information, especially when vehicles of different makes have to be repaired. It is, however, becoming common to supply material on CD/DVD. Workshops equipped with appropriate data systems will be able to receive updates via telephone line or by

Diagnostic techniques **21**

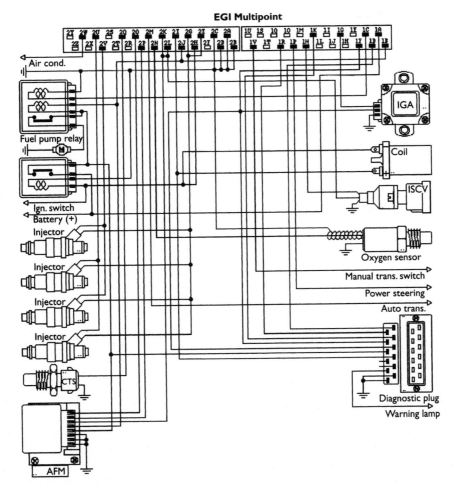

Figure 2.11 Example fuel and ignition circuit diagram

periodic receipt of updated CDs. A committee of the Society of Automotive Engineers (SAE) has prepared rules for the standardisation of manuals.

2.9 Data sources

2.9.1 Introduction

Data is available from a number of sources; clearly the best being direct from the manufacturer. However, for most 'general' repair workshops other sources have to be found.

Examples of the type of data necessary for diagnostic and other work are as follows:

- Component specification (resistance, voltage output etc.)
- Diagnostics charts
- Circuit diagrams (Figure 2.11)
- Adjustment data
- Timing belt fitting data
- Component location (Figure 2.12)

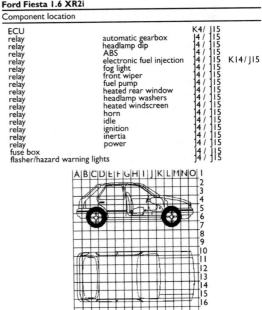

Figure 2.12 Component location information

22 Advanced automotive fault diagnosis

PORSCHE		Technical Data				Autodata
		1	2	3	4	5
1	Vehicle Identification Ref. No.	1064	2811	2810	2812	1073
2	Model	944S	944 Turbo	944 S2	928 GT	928 S4
3						
4						
5	Engine specially tuned for		R-Cat	R-Cat	R-Cat	R-Cat
6	Year	1986-89	1989-93	1989-93	1989-92	1989-94
7	Engine Code	M44/04	M44/52	M44/41	M28/47	M28/41/42
8	No. of cylinders/Type	4/OHC	4/OHC	4/OHC	8/OHC	8/OHC
9	Capacity cm³	2479	2479	2990	4957	4957
10	Output kW (DIN hp) rpm	140 (190) 6000	184 (250) 6000	155 (211) 5800	243 (330) 6200	235 (320) 6000
11	Minimum octane rating RON	95	95	95	95	95
12	Ignition system Description	Map-h	Map-h	Map-h	Map-i	Map-i
13	Trigger location	Crankshaft	Crankshaft	Crankshaft	Crankshaft	Crankshaft
14	Fuel system Make	Bosch	Bosch	Bosch	Bosch	Bosch
15	Type	Motronic	Motronic	Motronic	LH-Jetronic	LH-Jetronic
16	Description	MFI-i	MFI-i	MFI-i	MFI-i	MFI-i
17	Air metering Type	Flow	Flow	Flow	Mass	Mass
18	Combined ignition and fuel ECU	Yes	Yes	Yes	No	No
19	Diagnostic socket	Yes	Yes	Yes	Yes	Yes
20	**Tuning and emissions**			■		■
21	Ignition coil supply voltage V	12.0	12.0	12.0	11.0	12.0
22	Primary resistance Ω	0.4-0.6	0.4-0.6	0.4-0.6	0.4-0.6	0.4-0.6
23	Secondary resistance Ω	5000-7200	5000-7200	5000-7200	5000-7000	5000-7200
24	Firing order	1-3-4-2	1-3-4-2	1-3-4-2	1-3-7-2-6-5-4-8	1-3-7-2-6-5-4-8
25	Ignition distributor (ECU) no.	(0 261 200 080)	(0 261 200 088)	(0 261 200 195)	(0 227 400 164)	(0 227 400 034)
26	Ignition timing BTBC °Engine/rpm	10±3/840	5±3/840	10±3/840	10±2/775	10±2/675
27	alternative °Engine/rpm	–	–	–	–	–
28	o without + with vacuum	o	–	–	–	o
29	Ignition advance checks °Engine/rpm	ECU controlled	ECU controlled	ECU controlled	ECU controlled	ECU controlled
30	a = without vacuum and basic timing °Engine/rpm	–	–	–	–	–
31	b = without vacuum with basic timing °Engine/rpm	–	–	–	–	–
	c = with vacuum and basic timing					
32	Vacuum advance range °Engine	–	–	–	–	–
33	Idle speed rpm	840±40	840±40	840±40	775±25	675±25
34	alternative rpm	–	–	–	–	–
35	Oil temperature for CO test °C	90	90	90	90	90
36	CO content at idle - tail pipe Vol.%	1.0±0.5	0.5 Max	0.5 Max	0.5 Max	0.5 Max
37	- sample pipe Vol.%	–	0.4-0.8	0.4-0.8	0.4-1.2	0.4-1.2
38	CO₂/O₂ content at idle speed Vol.%	13-16/0.5-2.0	14.5-16/0.1-05	14.5-16/0.1-0.5	14.5-16/0.1-0.5	14.5-16/0.1-0.5
39	HC content at idle speed ppm	300	100	100	100	100
40	Increased idle speed for CO test rpm	–	2500-2800	2500-2800	2500-2800	2500-2800
41	CO content at increased idle speed Vol.%	–	0.3	0.3	0.3	0.3
42	Lambda at increased idle speed λ	–	0.97-1.03	0.97-1.03	0.97-1.03	0.97-10.3
43	**Service checks and adjustments**					
44	Spark plugs Make	Bosch	Bosch	Bosch	Bosch	Bosch
45	(also see Spark Plugs list) Type	WR5DC	WR7DC	WR5DC	WR7DC	WR7DC
46	Electrode gap mm	0.7	0.7	0.7	0.7	0.6-0.8
47	Valve clearance - inlet mm	Hydraulic	Hydraulic	Hydraulic	Hydraulic	Hydraulic
48	- exhaust mm	Hydraulic	Hydraulic	Hydraulic	Hydraulic	Hydraulic
49	Compression pressure bar	–	–	–	–	–
50	Oil pressure bar / rpm	3.5/6000	3.5/6000	3.5/6000	5/4000	5/5000
51	**Lubricants and capacities**					
52	Engine oil grade SAE (API)	15W/40 (SF)	15W/40 (SF)	15W/40 (SF)	15W/40 (SF)	15W/40 (SF)
53	Engine with filter litres	6.5	7.0	7.0	7.5	7.5
54	Gearbox oil grade SAE	75W/90	75W/90	75W/90	75W/90	75W/90
55	4/5 speed litres	2.0	2.0	7.0	4.5	4.5
56	Automatic transmission fluid Type	Dexron II D	–	–	–	Dexron II D
57	refill litres	6.0	–	–	–	7.3
58	Differential oil grade SAE	90W	–	–	–	90W
59	front/rear litres	1.0 (AT)	–	–	–	30 (AT)
POR 3	■ = refer to Technical Information at end of this manufacturer				Δ = setting not adjustable	

Figure 2.13 Example data book example (*Source*: Autodata)

- Repair times
- Service schedules

2.9.2 Autodata

One of the best known and respected companies for supplying automotive data is Autodata, both in the UK and the USA.

This range of books and CDs (on subscription) is well known and well respected. Very comprehensive manuals are available ranging from the standard 'Data book' to full vehicle circuit diagrams and engine management diagnostic tests data (Figure 2.13).

Information about testing procedures is available as shown in Figure 2.14. These sheets include

TOYOTA — Autodata

☐ Carina E 2,0 GLi   1992-

2.7 Injector valves

Technical Data
Resistance between terminals   13,4–14,2 Ω (approx.)

Injector spray pattern and leak rate — refer to General Test Procedures.

Resistance - 12
- Ensure ignition switched OFF.
- Disconnect injector valve multi-plug.
- Connect ohmmeter across injector valve terminals.
- Compare resistance indicated with that specified.

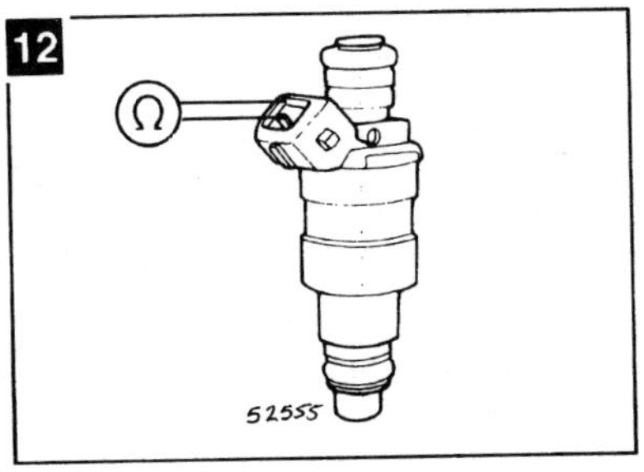

2.8 Lambda sensor

Self-diagnosis code: 21

Technical Data
Heater resistance at 20°C   5100–6300 Ω

Checking sensor - 13
- Bridge terminals TE1 and E1 of diagnostic socket.
- Run engine at 2500 rev/min for two minutes to heat up Lambda sensor.
- Connect voltmeter between terminals VF1 and E1 of diagnostic socket.
- Hold engine speed at 2500 rev/min
- Check that voltmeter needle fluctuates more than 6 times in 10 seconds.
- If less than 6 times, disconnect bridge between terminals TE1 and E1.
- Engine at 2500 rev/min, check voltage between terminals VF1 and E1.
- If more than 0 V, replace sensor.

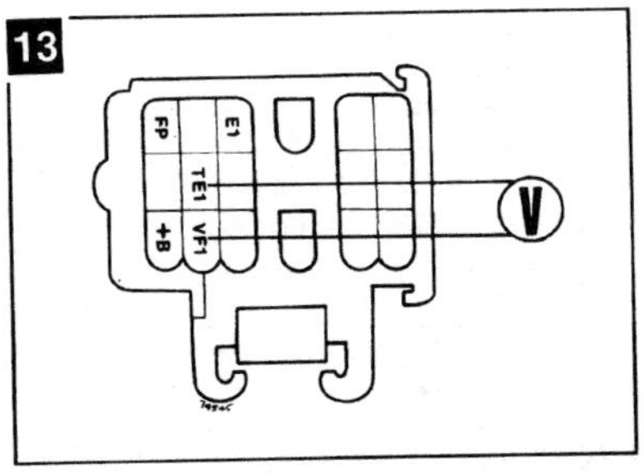

Checking sensor heater - 14
- Ensure ignition switched OFF.
- Disconnect sensor multi-plug.
- Connect ohmmeter across terminals +B and HT of sensor connector.
- Compare resistance indicated with that specified.

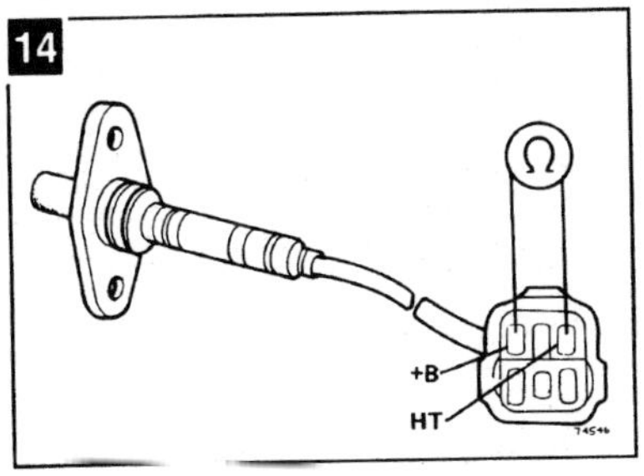

2.9 Fuel pump relay

Checking - 15
- Ensure ignition switched OFF.
- Remove relay located in LH fascia.
- Check for continuity with ohmmeter between terminals STA and E1 and +B and FC.
- Check for open circuit with ohmmeter connected between terminals +B and FP.
- Connect battery voltage between terminals STA and E1.
- Check for continuity between terminals +B and FP.
- Connect battery voltage between terminals +B and FC and check for continuity between terminals +B and FP.

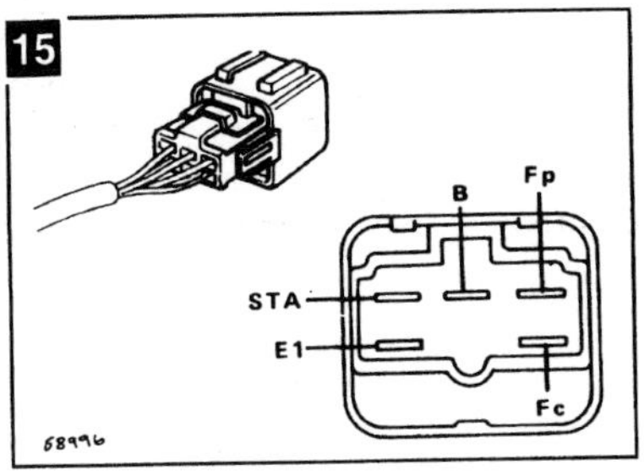

Figure 2.14 Fuel injection testing example (*Source*: Autodata)

24 Advanced automotive fault diagnosis

Figure 2.15 ESI[tronic] data (*Source*: Bosch Press)

test data as well as test procedures related to specific vehicles or systems.

Bosch ESI[tronic]

There are already over 10 million cars in the UK and over 60 million in the USA! Most of these now have engine management systems.

These engines need sophisticated test equipment to diagnose faults and system failures. Ineffective diagnostic work inevitably leads to vehicle problems, dissatisfied customers and labour costs which far exceed a realistic invoice value for the workshop.

Clearly this is where good data comes in. The Bosch system runs from a DVD and as well as information about test procedures and test results, much more is included as well (Figure 2.15).

Summary

Both of the previously mentioned sources of data are excellent – and essential. It is possible to carry out diagnostic work without data, but much more difficult and less reliable. The money will be well spent.

Knowledge check questions

To use these questions, you should first try to answer them without help but if necessary, refer back to the content of the chapter. Use notes, lists and sketches as appropriate to answer them. It is not necessary to write pages and pages of text!

1. List the six-stage diagnostic process in your own words.
2. Explain how the six-stage process is used by giving a simple example.
3. State the length of a standard piece of string and explain why this is relevant to diagnostics!
4. Describe how to carry out tests for an electrical short circuit.
5. Explain using a sketch, what is meant by 'black box' faultfinding.

3
Tools and equipment

3.1 Basic equipment

3.1.1 Introduction

As the complexity of the modern vehicle continues to increase, developments in suitable test equipment must follow. Many mechanical and electronic systems now have ECUs, that contain self-diagnosis circuits. This is represented by Figure 3.1. On earlier systems this was done by activating the blink code output to access the information held in the ECU memory. This was done in some cases by connecting two wires and then switching on the ignition. It is now usual to read out the fault codes on a scanner.

Diagnostic techniques are very much linked to the use of test equipment. In other words you must be able to interpret the results of tests. In most cases this involves comparing the result of a test to the reading given in a data book or other source of information. By way of an introduction, the following table lists some of the basic words and descriptions relating to tools and equipment.

Hand tools	Spanners and hammers and screwdrivers and all the other basic bits!
Special tools	A collective term for items not held as part of a normal tool kit. Or items required for just one specific job
Test equipment	In general, this means measuring equipment. Most tests involve measuring something and comparing the result of that measurement to data. The devices can range from a simple ruler to an engine analyser
Dedicated test equipment	Some equipment will only test one specific type of system. The large manufacturers supply equipment dedicated to their vehicles. For example, a diagnostic device which plugs in to a certain type of fuel injection ECU
Accuracy	Careful and exact, free from mistakes or errors and adhering closely to a standard

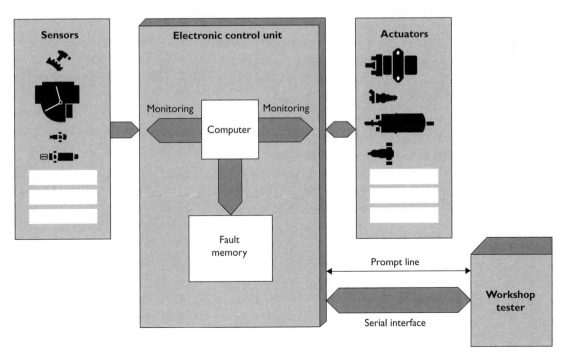

Figure 3.1 Engine control with self-diagnosis

26 *Advanced automotive fault diagnosis*

Calibration	Checking the accuracy of a measuring instrument
Serial port	A connection to an ECU, a diagnostic tester or computer for example. Serial means the information is passed in a 'digital' string like pushing black and white balls through a pipe in a certain order
Code reader or scanner	This device reads the 'black and white balls' mentioned above, or the on-off electrical signals, and converts them to language we can understand
Combined diagnostic and information system	Usually now PC based, these systems can be used to carry out tests on vehicle systems and they also contain an electronic workshop manual. Test sequences guided by the computer can also be carried out
Oscilloscope	The main part of 'scope' is the display, which is like a TV or computer screen. A scope is a voltmeter but instead of readings in numbers it shows the voltage levels by a trace or mark on the screen. The marks on the screen can move and change very fast allowing us to see the way voltages change

3.1.2 Basic hand tools

You cannot learn to use tools from a book; it is clearly a very practical skill. However, you can follow the recommendations made here and of course by the manufacturers. Even the range of basic hand tools is now quite daunting and very expensive.

It is worth repeating the general advice and instructions for the use of hand tools.

- Only use a tool for its intended purpose.
- Always use the correct size tool for the job you are doing.
- Pull a wrench rather than pushing whenever possible.
- Do not use a file or similar, without a handle.
- Keep all tools clean and replace them in a suitable box or cabinet.
- Do not use a screwdriver as a pry bar.
- Always follow manufacturers' recommendations (you cannot remember everything).
- Look after your tools and they will look after you!

3.1.3 Accuracy of test equipment

Accuracy can mean a number of slightly different things:

- careful and exact
- free from mistakes or errors; precise
- adhering closely to a standard

Consider measuring a length of wire with a steel rule. How accurately could you measure it to the nearest 0.5 mm? This raises a number of issues. Firstly, you could make an error reading the ruler. Secondly, why do we need to know the length of a bit of wire to the nearest 0.5 mm? Thirdly the ruler may have stretched and not give the correct reading!

The first and second of these issues can be dispensed with by knowing how to read the test equipment correctly and also knowing the appropriate level of accuracy required. A micrometer for a plug gap? A ruler for valve clearances? I think you get the idea. The accuracy of the equipment itself is another issue.

Accuracy is a term meaning how close the measured value of something is to its actual value. For example, if a length of about 30 cm is measured with an ordinary wooden ruler, then the error may be up to 1 mm too high or too low. This is quoted as an accuracy of ± 1 mm. This may also be given as a percentage which in this case would be 0.33%.

Resolution, or in other words the 'fineness' with which a measurement can be made, is related to accuracy. If a steel ruler was made to a very high standard but only had markings of one per centimetre it would have a very low resolution even though the graduations were very accurate. In other words the equipment is accurate but your reading will not be!

To ensure instruments are, and remain, accurate there are just two simple guidelines.

- Look after the equipment; a micrometer thrown on the floor will not be accurate.
- Ensure that instruments are calibrated regularly – this means being checked against known good equipment.

Here is a summary of the steps to ensure a measurement is accurate:

Step	Example
Decide on the level of accuracy required	Do we need to know that the battery voltage is 12.6 V or 12.635 V?
Choose the correct instrument for the job	A micrometer to measure the thickness of a shim
Ensure that the instrument has been looked after and calibrated when necessary	Most instruments will go out of adjustment after a time. You should arrange for adjustment at regular intervals. Most tool suppliers will offer the service or in some cases you can compare older equipment to new stock

Study the instructions for the instrument in use and take the reading with care. Ask yourself if the reading is about what you expected	Is the piston diameter 70.75 mm or 170.75 mm?
Make a note if you are taking several readings	Don't take a chance, write it down

3.1.4 Multimeters

An essential tool for working on vehicle electrical and electronic systems is a good digital multimeter. Digital meters are most suitable for accuracy of reading as well as available facilities. The following list of functions, broadly in order from essential to desirable, should be considered.

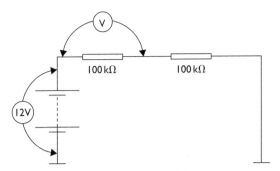

Figure 3.2 Loading effect of a test meter

Function	Range	Accuracy
DC voltage	500 V	0.3%
DC current	10 A	1.0%
Resistance	0 to 10 MΩ	0.5%
AC voltage	500 V	2.5%
AC current	10 A	2.5%
Dwell	3,4,5,6,8 cylinders	2.0%
RPM	10 000 rev/min	0.2%
Duty cycle	% on/off	0.2% (kHz)
Frequency	100 kHz	0.01%
Temperature	>900°C	0.3% + 3°C
High current clamp	1000 A (DC)	Depends on conditions
Pressure	3 bar	10.0% of standard scale

A way of determining the quality of a meter, as well as by the facilities provided, is to consider the following:

- accuracy;
- loading effect of the meter;
- protection circuits.

The loading effect is a consideration for any form of measurement. With a multimeter this relates to the internal resistance of the meter. It is recommended that the internal resistance of a meter should be a minimum of 10 MΩ. This not only ensures greater accuracy but also prevents the meter from damaging sensitive circuits.

Figure 3.2 shows two equal resistors connected in series across a 12 V supply. The voltage across each resistor should be 6 V. However, the internal resistance of the meter will affect the circuit conditions and change the voltage reading. If the resistor values were 100 kΩ the effect of meter internal resistance would be as follows.

Figure 3.3 Multimeter and accessories

Meter resistance 1 MΩ

With a parallel combined value of 1 MΩ and 100 kΩ = 91 kΩ the voltage drop in the circuit across this would be:

$$91/(100 + 91) \times 12 = 5.71 \text{ V}$$

This is an error of about 5%.

Meter resistance 10 MΩ

With parallel combined value of 10 MΩ and 100 kΩ = 99 kΩ the voltage drop in the circuit across this would be:

$$99/(100 + 99) \times 12 = 5.97 \text{ V}$$

This is an error of about 0.5%.

> Note: This 'invasive measurement' error is in addition to the basic accuracy of the meter.

Figure 3.3 shows a digital multimeter. Now that this is recognised, only two further skills are required – where to put the probes and what the reading you get actually means!

3.1.5 Logic probe

This device is a useful way of testing logic circuits but it is also useful for testing some types of

28 Advanced automotive fault diagnosis

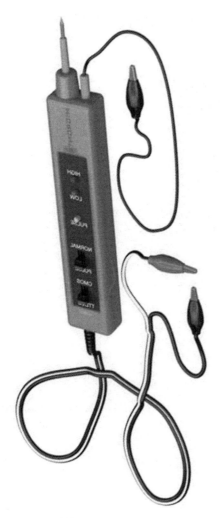

Figure 3.4 Logic probe

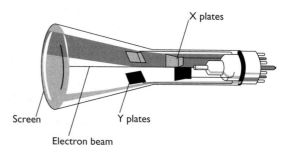

Figure 3.5 Analogue oscilloscope principle

sensor. Figure 3.4 shows a typical logic probe. Most types consist of two power supply wires and a metal 'probe'. The display consists of two or three LEDs labelled 'high', 'low' and on some devices 'pulse'. These LEDs light up, together with an audible signal in some cases, when the probe touches either a high, low or pulsing voltage. Above or below 2.5 V is often used to determine high or low on a 5 V circuit.

3.2 Oscilloscopes

3.2.1 Introduction

Two types of oscilloscope are available: analogue or digital. Figure 3.5 shows the basic operation of an analogue oscilloscope. Heating a wire creates a source of electrons, which are then accelerated by suitable voltages and focused into a beam. This beam is directed towards a fluorescent screen where it causes light to be given off. This is the basic cathode ray tube.

The plates as shown in the figure are known as X and Y plates as they make the electron beam draw a 'graph' of a voltage signal. The X plates are supplied with a saw tooth signal, which causes the beam to move across the screen from left to right and then to 'fly back' and start again. Being attracted towards whichever plate has a positive potential the beam moves simply. The Y plates can now be used to show voltage variations of the signal under test. The frequency of the saw tooth signal, known as the time base, can be adjusted either automatically as is the case with many analysers or manually on a stand alone oscilloscope. The signal from the item under test can either be amplified or attenuated (reduced), much like changing the scale on a voltmeter. The trigger, in other words when the trace across the screen starts, can be caused internally or externally. In the case of the engine analyser triggering is often external, each time an individual spark fires or each time number one spark plug fires.

A digital oscilloscope has much the same end result as the analogue type but the signal can be thought of as being plotted rather than drawn on the screen. The test signal is A/D converted and the time base is a simple timer or counter circuit. Because the signal is plotted digitally on a screen from data in memory, the 'picture' can be saved, frozen or even printed. The speed of data conversion and the sampling rate as well as the resolution of the screen are very important in ensuring accurate results. This technique is becoming the norm as including scales and notes or superimposing two or more traces for comparison can enhance the display.

A very useful piece of equipment becoming very popular is the PC based system as shown in Figure 3.6. This is a digital oscilloscope, which allows data to be displayed on a PC. The Scope

Tools and equipment 29

Figure 3.6 Multiscope

can be used for a large number of vehicle tests. The waveforms used as examples in this book were 'captured' using this device. This type of test equipment is highly recommended.

3.2.2 Waveforms

You will find the words 'waveform', 'pattern' and 'trace' are used in books and workshop manuals but they mean the same thing. When you look at a waveform on a screen you must remember that the height of the scale represents voltage and the width represents time. Both of these axes can have their scales changed. They are called axes because the 'scope' is drawing a graph of the voltage at the test points over a period of time. The time scale can vary from a few µs to several seconds. The voltage scale can vary from a few mV to several kV. For most test measurements only two connections are needed just like a voltmeter. The time scale will operate at intervals pre-set by the user. It is also possible to connect a 'trigger' wire so that, for example, the time scale starts moving across the screen each time the ignition coil fires. This keeps the display in time with the speed of the engine. When you use a full engine analyser, all the necessary connections are made as listed in the table in Section 3.4.1. A hand-held scope has to be connected for each waveform as required.

All the waveforms shown in various parts of this book are from a correctly operating vehicle. The skill you will learn by practice is to note when your own measurements vary from those shown here.

3.2.3 PC based two-channel automotive oscilloscope (PicoScope)

Author's Note: This section will outline the use and features of the PicoScope® automotive oscilloscope. I have chosen this PC based scope as a case study because it provides some very advanced features at a very reasonable price. At the time of writing (early 2006) the price was about £800/$1400 – very reasonable for such a useful piece of equipment. Most (if not all) of the example waveforms in this book were captured using this diagnostics kit. For more information: www.picoscope.com.

The latest PicoScope two-channel PC oscilloscope automotive kit is an ideal diagnostic tool. It is designed with mechanics and automotive technicians in mind, allowing fast and accurate diagnosis of all the electrical and electronic components and circuits in a modern vehicle. The kit includes all of the components necessary to test the engine and other systems for abnormalities. It is supplied with excellent automotive software

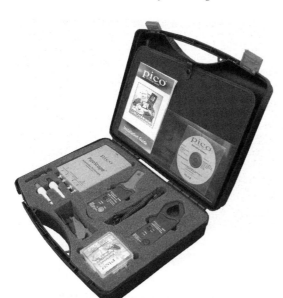

Figure 3.7 Pico Technology two-channel PC automotive oscilloscope

to collect, display, store and analyse readings. The accessories including current clamps, ignition pick ups, test leads and probes, come packaged in a hard-wearing carry case.

The powerful combination of the two-channel oscilloscope diagnostic kit and PicoScope automotive software allows thorough testing of the multitude of sensors and actuators in current vehicles (Figure 3.7). The software contains a technical reference library of tests and tutorials on over fifty topics, illustrated with waveforms. The components and circuits that can be tested include ignition, injectors, ABS, lambda oxygen sensors, relative compression, fuel pumps, CAN Bus and many others. The kit is a cost-effective ignition and engine diagnostic tool for all automotive technicians.

The two-channel automotive kit contains, amongst other things, the following items:

- PicoScope 3223 automotive oscilloscope
- 600 A AC/DC current clamp
- 60 A DC current clamp
- Wide range of general leads, clips and probes for all applications
- Secondary ignition pickup leads
- 2-pin break out lead.

The equipment is connected to, and powered through, the USB port on any modern PC or laptop. This eliminates the need for a power supply or batteries.

The two-channel PC oscilloscope in the new kit has an improved sampling rate of 20 MS/s, a 12 bit resolution, a huge buffer memory of 512 k samples and an excellent 1% level of accuracy. The automotive software is continually updated and free upgrades are available from the Pico Technology web site. The kit even comes with lifetime technical support from the Pico automotive specialists.

The waveforms in Chapter 4 were all captured using this equipment.

3.3 Scanners/fault code readers

3.3.1 Serial port communications

A special interface of the type that is stipulated by ISO 9141 is required to read data. This standard is designed to work with a single or two wire port allowing many vehicle electronic systems to be connected to a central diagnostic plug. The sequence of events to extract data from the ECU is as listed below:

- test unit transmits a code word;
- ECU responds by transmitting a baud rate recognition word;
- test unit adopts the appropriate setting;
- ECU transmits fault codes.

The test unit converts these to suitable output text.

3.3.2 The scanner/fault code reader

Serial communication is an area that is continuing to grow. A special interface is required to read data. This standard is designed to work with a single or two wire port, which connects vehicle electronic systems to a diagnostic plug. Many functions are then possible when a scanner is connected. They include the following.

- Identification of ECU and system to ensure that the test data is appropriate to the system currently under investigation.
- Read out of current live values from sensors so that spurious figures can be easily recognised. Information such as engine speed, temperature air flow and so on can be displayed and checked against test data.

Tools and equipment

Figure 3.8 Snap-on scanner

Figure 3.9 AutoTap scanner and extension cable

Figure 3.10 Diagnostic connector

- System function stimulation allows actuators to be tested by moving them and watching for suitable response.
- Programming of system changes. Basic idle CO or changes in basic timing can be programmed into the system.

Figure 3.8 shows a scanner from snap-on. TAs scanner allows the technician to perform all the necessary operations, such as fault code reading, via a single common connector. The portable hand-held tool has a large graphics display allowing clear instructions and data. Context-sensitive help is available to eliminate the need to refer back to manuals to look up fault code definitions. It has a memory so that data can be reused even after disconnecting power from the tool. This scanner will even connect to a controller area network (CAN) system with a suitable adapter.

3.3.3 OBD scanner (AutoTap)

> Author's Note: This section will outline the use and features of the AutoTap scanner. I have chosen this particular tool as a case study because it provides some very advanced features at a very competitive price. At the time of writing (2006) the price was about £140/$200 – very reasonable for such a powerful tool. The scanner is designed to work with OBD2 systems. However, it worked fine on all the EOBD systems I have used it on so far. For more information: www.autotap.com.

Like any other scan tool or code reader, the AutoTap scan tool connects the special OBD2 (CAN) plug-in-port on or under the dash (Figure 3.10). A USB connection then makes the scanner connection to a computer. The AutoTap scanner translates the signals from the vehicle's computer-controlled sensors to easy-to-read visual displays. It also reads out the diagnostic trouble codes (DTCs) (Figure 3.11).

The software allows the technician to choose which parameters or signals they want to see and whether you want to view them in tables or as graphs, meters or gauges (Figure 3.11).

It is possible to set the ranges and alarms and pick display colours. Once a screen configuration is created it can be saved for future use. Different screen configurations are useful for different vehicles, or perhaps one for major maintenance, one for tuning, one for quick checks at the track.

Lots of data is provided in easy-to-read views with multiple parameters. Graphs can be used to show short-term logs, and gauges for instant readings.

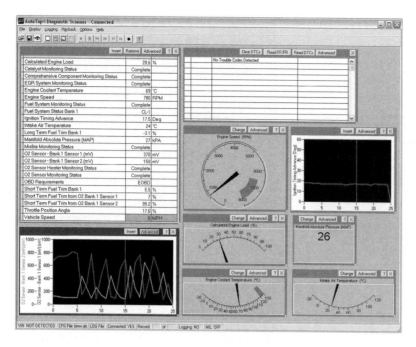

Figure 3.11 Screen grab showing gauges and graphs

DTCs can be checked immediately on connecting the scanner and starting up the software. This gives the critical info needed in the shortest time possible. When repairs are completed the tool can be used to turn off the malfunction indicator light (MIL). This light is also described as the check engine light.

The software will also log data, for example, during a road test. This is particularly useful for diagnosing intermittent faults. The data can be played back after a road or dynamometer test. It can also be exported to a spreadsheet file for later analysis.

Overall, to read live data and get access to powertrain (engine related) system DTCs, this is an excellent piece of equipment.

3.4 Engine analysers

3.4.1 Engine analysers

Some form of engine analyser became an almost essential tool for faultfinding modern vehicle engine systems. However, hand-held equipment is now tending to replace the larger analysers. The latest machines are now generally based around a PC. This allows more facilities that can be added to by simply changing the software. Whilst engine analysers are designed to work specifically with the motor vehicle, it is worth remembering that the machine consists basically of three parts:

- multimeter;
- gas analyser;
- oscilloscope.

This is not intended to imply that other available tests, such as cylinder balance, are less valid, but to show that the analyser is not magic, it is just able to present results of electrical tests in a convenient way to allow diagnosis of faults. The key component of any engine analyser is the oscilloscope facility, which allows the user to 'see' the signal under test.

The following is a description of the facilities available on a typical engine analyser. The new concept in garage equipment design is based on a PC, specially engineered for workshop use and enabling a flexibility of use far exceeding the ability of machines previously available.

Software is used to give the machine its 'personality' as an engine analyser, system tester, wheel aligner or even any of the other uses made of PCs. Either an infrared handset or a standard 'qwerty' keyboard controls the machine. The information is displayed on a super VGA monitor giving high resolution colour graphics. Output can be sent to a standard printer when a hard copy is required for the customer.

A range of external measurement modules and software application programmes is available.

The modules are connected to the host computer by high speed RS422 or RS232 serial communication links. Application software is loaded onto a hard disk. Vehicle specific data can also be stored on disk to allow fast easy access to information but also to allow a guided test procedure.

The modern trend with engine analysers seems to be to allow both guided test procedures with pass/fail recommendations for the less skilled technician, and freedom to test any electrical device using the facilities available in any reasonable way. This is more appropriate for the highly skilled technician. Some of the routines available on modern engine analysers are listed below.

Tune-up

This is a full prompted sequence that assesses each component in turn with results and diagnosis displayed at the end of each component test. Stored data allows pass/fail diagnosis by automatically comparing results of tests with data on the disk. Printouts can be taken to show work completed.

Symptom analysis

This allows direct access to specific tests relating to reported driveability problems.

Waveforms

A comprehensive range of digitised waveforms can be displayed with colour highlights. The display can be frozen or recalled to look for intermittent faults. A standard lab scope mode is available to allow examination of EFI (electronic fuel injection) or ABS traces for example. Printouts can be made from any display. An interesting feature is 'transient capture' which ensures that even the fastest spikes and intermittent signals are captured and displayed for detailed examination.

Adjustments

Selecting specific components from a menu can make simple quick adjustments. Live readings are displayed appropriate to the selection.

MOT emissions (annual UK test)

Full MOT procedure tests are integrated and displayed on the screen with pass/fail diagnosis to the department of transport specifications for both gas analysis and diesel smoke if appropriate options are fitted. The test results include engine rev/min and oil temperature as well as the gas readings. These can all be printed for garage or customer use.

The connections to the vehicle for standard use are much the same for most equipment manufacturers. These are listed as follows.

Connection	Purpose or one example of use
Battery positive	Battery and charging voltages
Battery negative	A common earth connection
Coil positive	To check supply voltage to coil
Coil negative (adapters for DIS – distributorless ignition systems)	To look at dwell, rev/min and primary waveforms
Coil HT lead clamp (adapters are available for DIS)	Secondary waveforms
Number one cylinder plug lead clamp	Timing light and sequence of waveforms
Battery cable amp clamp	Charging and starting current
Oil temperature probe (dip stick hole)	Oil temperature
Vacuum connection	Engine load

Figure 3.12 shows a digital oscilloscope engine analyser. This test equipment has many features as listed previously and others such as the following.

- **High tech test screens** Vacuum waveform, cylinder time balance bar graph, power balance waveform and dual trace lab scope waveform.
- **Scanner interface** This allows the technician to observe all related information at the same time.
- **Expanded memory** This feature allows many screens to be saved at once, then recalled at a later time for evaluation and reference.

The tests are user controlled whereas some machines have pre-programmed sequences. Some of the screens available are as follows.

Primary	Secondary	Diagnostic	Cylinder test
Primary waveform	Secondary waveform	Voltage waveform	Vacuum waveform
Primary parade waveform	Secondary parade waveform	Lab scope waveform	Power balance waveform
Dwell bar graph	kV histogram	Fuel injector waveform	Cylinder time balance bar graph
Duty cycle/dwell bar graph	kV bar graph	Alternator waveform	Cylinder shorting even/odd bar graph
Duty cycle/voltage bar graph	Burn time bar graph		Cranking amps bar graph

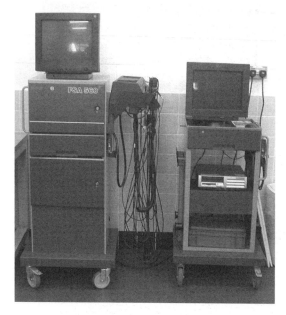

Figure 3.12 Engine analyser

3.4.2 Exhaust gas measurement

It has now become standard to measure four of the main exhaust gases namely:

- carbon monoxide (CO);
- carbon dioxide (CO_2);
- hydrocarbons (HC);
- oxygen (O_2).

The emission test module is often self-contained with its own display but can be linked to the main analyser display. Often lambda value and the air fuel ratio are displayed in addition to the four gasses. The Greek symbol lambda (λ) is used to represent the ideal air fuel ratio (AFR) of 14.7:1 by mass. In other words just the right amount of air to burn up all the fuel. Typical gas, lambda and AFR readings are given in Table 3.1 for a closed loop lambda control system, before (or without) and after the catalytic converter. These are for a modern engine in excellent condition (examples only – always check current data).

The composition of exhaust gas is now a critical measurement and hence a certain degree of accuracy is required. To this end the infrared measurement technique has become the most suitable for CO, CO_2 and HC. Each individual gas absorbs infrared radiation at a specific rate.

Oxygen is measured by electro-chemical means in much the same way as the on vehicle lambda sensor. CO is measured using a beam of infrared light. A similar technique is used for the

Table 3.1

Reading	CO%	Hc (ppm)	CO_2%	O_2%	Lambda (λ)	AFR
Before catalyst	0.6	120	14.7	0.7	1.0	14.7
After catalyst	0.2	12	15.3	0.1	1.0	14.7

measurement of CO_2 and HC. At present it is not possible to measure NOx without more sophisticated laboratory equipment.

Good four-gas emission analysers often have the following features.

- Stand alone unit not dependent on other equipment.
- Graph screen simultaneously displays up to four values as graphs and the graph display order is user selectable. Select from HC, CO, CO_2, O_2 and rev/min for graph display.
- User can create personalised letterhead for screen printouts.
- Uses the non-dispersive infrared (NDIR) method of detection (each individual gas absorbs infrared light at a specific rate).
- Display screens may be frozen or stored in memory for future retrieval.
- Recalibrate at the touch of a button (if calibration gas and a regulator are used).
- Display exhaust gas concentrations in real time numerics or create live exhaust gas data graphs in selectable ranges.
- Calculate and display lambda (λ) (the ideal air fuel ratio of about 14.7:1).
- Display engine rev/min in numeric or graph form and display oil temperature along with current time and date.
- Display engine diagnostic data from a scanner.
- Operate from mains supply or a 12 V battery.

Accurate measurement of exhaust gas is not only required for annual tests but is essential to ensure an engine is correctly tuned. Table 3.1 lists typical values measured from a typical exhaust. Note the toxic emissions are small, but none-the-less dangerous.

3.4.3 Pressure testing

Measuring the fuel pressure on a fuel injection engine is of great value when faultfinding. Many types of pressure testers are available and they

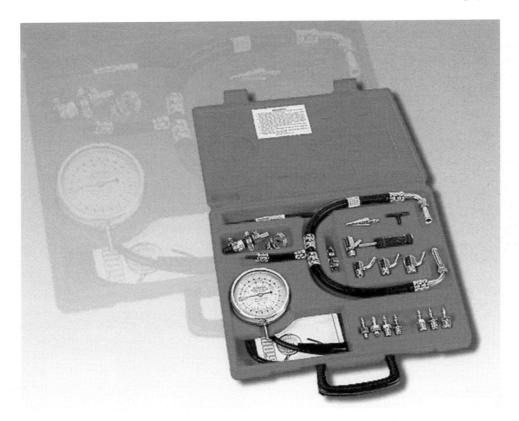

Figure 3.13 Pressure gauge and kit

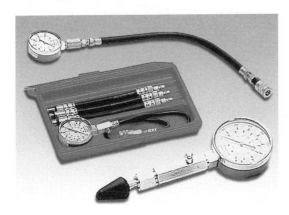

Figure 3.14 Compression testers

often come as part of a kit consisting of various adapters and connections. The principle of the gauges is that they contain a very small tube wound in a spiral. As fuel under pressure is forced into a spiral tube it unwinds causing the needle to move over a graduated scale. Figure 3.13 shows a fuel pressure gauge as part of a kit.

Measuring engine cylinder compression or leakage is a useful test. Figure 3.14 shows an engine compression tester. This device is used more to compare cylinder compressions than to measure actual values.

3.4.4 Bosch KTS diagnostic equipment

Author's Note: This section will outline the use and features of the Bosch KTS 650 diagnostic system. I have chosen this particular tool as a case study because it provides everything that a technician needs to diagnose faults. The system is a combination of a scanner, multimeter, oscilloscope and information system (when used with Esitronic). At the time of writing (2006) the price was about £5000/$8000. For more information: www.bosch.com.

Modern vehicles are being fitted with more and more electronics. That complicates diagnosis and repair, especially as the individual systems are often interlinked. The work of service and repair workshops is being fundamentally changed. Automotive engineers have to continually update their knowledge of vehicle electronics. But this is no longer sufficient on its own. The ever-growing number of electrical and electronic vehicle components is no longer manageable without modern diagnostic technology – such as the latest range of KTS control unit diagnostic testers from Bosch (Figure 3.15). In addition, more and more of the previously purely mechanical interventions on vehicles now require the use of electronic control units – such as the oil change, for example.

36 Advanced automotive fault diagnosis

Figure 3.15 Diagnostic system in use (*Source:* Bosch Press)

Vehicle workshops operate in a very competitive environment and have to be able to carry out demanding repair work efficiently, to a high standard and at a competitive price on a wide range of vehicle makes and models. The Bosch KTS control-unit diagnostic testers, used in conjunction with the comprehensive Esitronic workshop software, offers the best possible basis for efficient diagnosis and repair of electrical and electronic components. The testers are available in different versions, suited to the individual requirements of the particular workshop: The portable KTS 650 with built-in computer and touch-screen can be used anywhere. It has a 20 GB hard drive, a touch-screen and a DVD drive. When being used away from the workshop, the power supply of the KTS 650 comes from the vehicle battery or from rechargeable batteries with one to two hours' service life. For use in the workshop, there is a tough wheeled trolley with a built-in charger unit. As well as having all the necessary adapter cables, the trolley can also carry an inkjet printer and an external keyboard, which can be connected to the KTS 650 via the usual PC interfaces (Figure 3.16).

The Esitronic software package accounts for the in-depth diagnostic capacity of the KTS diagnostic testers. With the new common rail diesel systems, for example, even special functions such as quantitative comparison and compression testing can be carried out. This allows for reliable diagnosis of the faulty part and avoids unnecessary dismantling and re-assembly or the removal and replacement of non-faulty parts.

Modern diagnostic equipment is also indispensable when workshops have to deal with braking

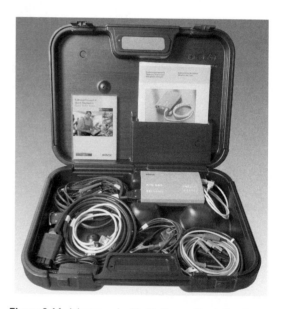

Figure 3.16 Adapter and cable kit (*Source:* Bosch Press)

systems with electronic control systems such as ABS, ASR and ESP. Nowadays, the diagnostic tester may even be needed for bleeding a brake system.

In addition, KTS and Esitronic allow independent workshops to reset the service interval warning; for example, after an oil change or a routine service, or perhaps find the correct default position for the headlamps after one or both of these have been replaced.

As well as ISO norms for European vehicles and SAE norms for American and Japanese vehicles, the KTS testers can also deal with CAN norms for checking modern CAN bus systems, which are coming into use more and more frequently in new vehicles. The testers are connected directly to the diagnostics socket via a serial diagnostics interface by means of an adapter cable.

The system automatically detects the control unit and reads out the actual values, the error memory and other controller-specific data. Thanks to a built-in multiplexer, it is even easier for the user to diagnose the various systems in the vehicle. The multiplexer determines the connection in the diagnostics socket so that communication is established correctly with the selected control unit.

The sequence of images shown below detail a number of steps taken to diagnose a fault, using the KTS, on a vehicle that had poor running symptoms and the MIL was illuminated.

(My thanks to Kevin and Steve at the Bosch training centre for their help with this section.)

The first step in this procedure was to connect the equipment to the car's diagnostic socket. The ignition should be off when the connection is made and then switched on.	 Connect the serial lead to the diagnostic socket
On this system the data for a wide range of vehicles is included on the system. The particular make and engine, etc. can be selected from the menu system.	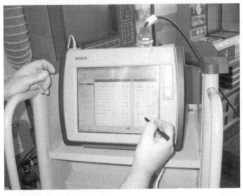 Choose the vehicle type
The standard test for stored diagnostic trouble codes (DTCs) was run and the result suggested that there was a fault with the air flow sensor. The specific fault was that the signal value was too low. No real surprise as we had disconnected the sensor to simulate a fault!	 Take a readout from the control unit memory (DTC display)

This is the connection that was causing the problems. Further information about its pin configuration can be looked up in the Esitronic database.

The system also provides typical readings that should be obtained on different pins. For example, the supply and earth as well as the signal outputs.

Additional tests can be carried out to determine the fault.

Air flow sensor connection

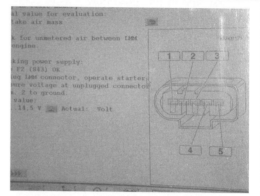
Esitronic information for the air flow sensor

The faulty connection was repaired and general checks carried out to ensure no other components had been disturbed during the testing and repair process.

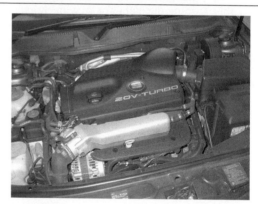
Make repairs

The final task is to clear the fault code memory and turn off the malfunction indicator light (MIL).

Road tests showed that the fault had been rectified.

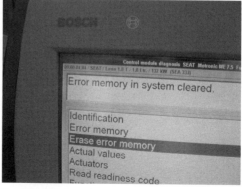
Erase the fault from the memory

Knowledge check questions

To use these questions, you should first try to answer them without help but if necessary, refer back to the content of the chapter. Use notes, lists and sketches to answer them. It is not necessary to write pages and pages of text!

1. Explain why a good multimeter has a high internal resistance.
2. List three advantages of using an oscilloscope for testing signals.
3. Describe how a scanner is connected to a vehicle and what information it can provide.
4. State what is meant by the term 'accuracy'.
5. List five tests carried out on a vehicle using a pressure gauge.

4
Oscilloscope diagnostics

4.1 Introduction

This chapter outlines the methods used and the results of using an oscilloscope to test a variety of systems. It will be a useful reference as all the waveforms shown are from a correctly operating system. The chapter is split into three main sections: sensors, actuators and ignition.

> *Author's Note:* The waveforms in this chapter were captured using the PicoScope® automotive oscilloscope. I am most grateful to the PicoTech team for supplying information and equipment to assist in the production of this chapter (www.picoscope.com).

4.2 Sensors

4.2.1 ABS speed sensor waveform

The anti-lock braking system (ABS) wheel speed sensors have become increasingly smaller and more efficient in the course of time (Figure 4.1). Recent models not only measure the speed and direction of wheel rotation but can be integrated into the wheel bearing as well.

The ABS relies upon information coming in from the sensors to determine what action should

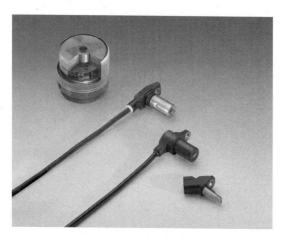

Figure 4.1 ABS wheel speed sensors (*Source:* Bosch Press)

be taken. If, under heavy braking, the ABS electronic control unit (ECU) loses a signal from one of the road wheels, it assumes that the wheel has locked and releases that brake momentarily until it sees the signal return. It is therefore imperative that the sensors are capable of providing a signal to the ABS ECU. If the signal produced from one wheel sensor is at a lower frequency than the others the ECU may also react.

The operation of an ABS sensor is similar to that of a crank angle sensor. A small inductive pick-up is affected by the movement of a toothed wheel, which moves in close proximity. The movement of the wheel next to the sensor, results in a 'sine wave' (Figure 4.2). The sensor, recognisable by its two electrical connections (some may have a coaxial braided outer shield), will produce an output that can be monitored and measured on the oscilloscope.

4.2.2 Air flow meter – air vane waveform

The vane type air flow meter is a simple potentiometer that produces a voltage output that is proportional to the position of a vane (Figure 4.3). The vane in turn positions itself in a position proportional to the amount of air flowing.

The voltage output from the internal track of the air flow meter should be linear to flap movement; this can be measured on an oscilloscope and should look similar to the example shown in Figure 4.4.

The waveform should show approximately 1.0 volt when the engine is at idle. This voltage will rise as the engine is accelerated and will produce an initial peak. This peak is due to the natural inertia of the air vane and drops momentarily before the voltage is seen to rise again to a peak of approximately 4.0 to 4.5 V. This voltage will however depend on how hard the engine is accelerated, so a lower voltage is not necessarily a fault within the air flow meter. On deceleration the voltage will drop sharply as the wiper arm, in

Oscilloscope diagnostics **41**

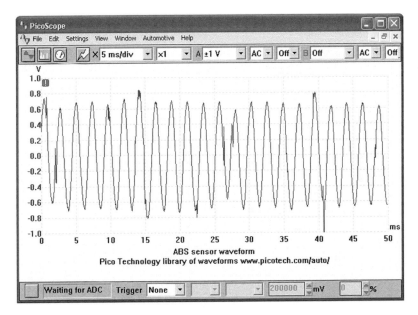

Figure 4.2 ABS speed sensor waveform

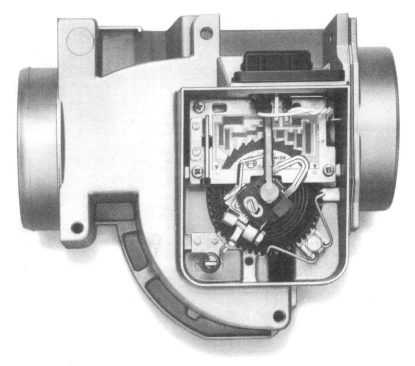

Figure 4.3 Vane or flap type airflow sensor (*Source*: Bosch)

contact with the carbon track, returns back to the idle position. This voltage may in some cases 'dip' below the initial voltage before returning to idle voltage. A gradual drop will be seen on an engine fitted with an idle speed control valve as this will slowly return the engine back to base idle as an anti-stall characteristic.

A time base of approximately 2 seconds plus is used. This enables the movement to be shown on one screen, from idle, through acceleration and back to idle again. The waveform should be clean with no 'drop-out' in the voltage, as this indicates a lack of electrical continuity. This is common on an AFM with a dirty or faulty carbon track. The problem will show as a 'flat spot' or hesitation when the vehicle is driven; this is a typical problem on vehicles with high mileage that have spent the majority of their working life with the throttle in one position.

The 'hash' on the waveform is due to the vacuum change from the induction pulses as the engine is running.

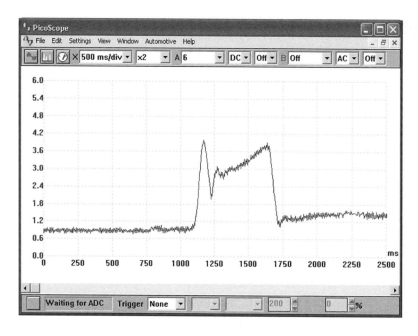

Figure 4.4 Air vane output voltage

4.2.3 Air flow meter – hot wire waveform

Figure 4.5 shows a micro mechanic mass airflow sensor from Bosch. This type has been in use since 1996. As air flows over the hot wire it cools it down and this produces the output signal. The sensor measures air mass because the air temperature is taken into account due to its cooling effect on the wire.

The voltage output should be linear to airflow. This can be measured on an oscilloscope and should look similar to Figure 4.6. The waveform should show approximately 1.0 volt when the engine is at idle. This voltage will rise as the engine is accelerated and air volume is increased producing an initial peak. This peak is due to the initial influx of air and drops momentarily before the voltage is seen to rise again to another peak of approximately 4.0 to 4.5 V. This voltage will however depend on how hard the engine is accelerated; a lower voltage is not necessarily a fault within the meter.

On deceleration the voltage will drop sharply as the throttle butterfly closes, reducing the airflow, and the engine returns back to idle. The final voltage will drop gradually on an engine fitted with idle speed control valve as this will slowly return the engine back to base idle as an anti-stall characteristic. This function normally only affects the engine speed from around 1200 rev/min back to the idle setting.

A time base of approximately 2 seconds plus is used because this allows the output voltage on

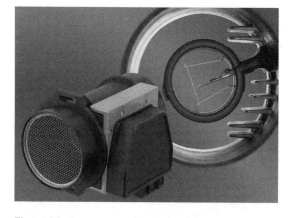

Figure 4.5 Hot wire air mass meter (Source: Bosch Press)

one screen, from idle, through acceleration and back to idle again. The 'hash' on the waveform is due to airflow changes caused by the induction pulses as the engine is running.

4.2.4 Inductive crankshaft and camshaft sensor waveform

The inductive type crank and cam sensors work in the same way. A single tooth, or toothed wheel, induces a voltage into a winding in the sensor. The cam sensor provides engine position information as well as which cylinder is on which stroke (Figure 4.7). The crank sensor provides engine speed. It also provides engine position in many cases by use of a 'missing' tooth.

In this particular waveform (Figure 4.8) we can evaluate the output voltage from the crank

Oscilloscope diagnostics **43**

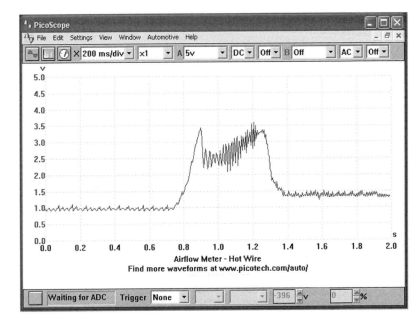

Figure 4.6 Air mass meter waveform

Figure 4.7 Crank sensor in position near the engine flywheel

sensor. The voltage will differ between manufacturers and it also increases with engine speed. The waveform will be an alternating voltage signal.

The gap in the picture is due to the 'missing tooth' in the flywheel or reluctor and is used as a reference for the ECU to determine the engine's position. Some systems use two reference points per revolution.

The camshaft sensor is sometimes referred to as the cylinder identification (CID) sensor or a 'phase' sensor and is used as a reference to time sequential fuel injection.

This particular type of sensor generates its own signal and therefore does not require a voltage

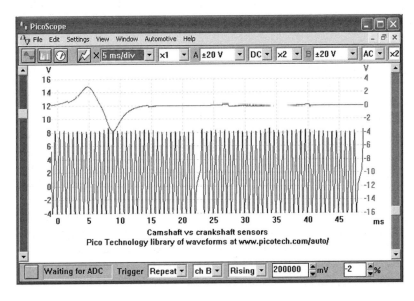

Figure 4.8 Crank and cam sensor output signals

supply to power it. It is recognisable by its two electrical connections, with the occasional addition of a coaxial shielding wire.

The voltage produced by the camshaft sensor will be determined by several factors, these being the engine's speed, the proximity of the metal rotor to the pick-up and the strength of the magnetic field offered by the sensor. The ECU needs to see the signal when the engine is started for its reference; if absent it can alter the point at which the fuel is injected. The driver of the vehicle may not be aware that the vehicle has a problem if the CID sensor fails, as the drivability may not be affected. However, the MIL should illuminate.

The characteristics of a good inductive camshaft sensor waveform is a sinewave that increases in magnitude as the engine speed is increased and usually provides one signal per 720° of crankshaft rotation (360° of camshaft rotation). The voltage will be approximately 0.5 V peak to peak while the engine is cranking, rising to around 2.5 V peak to peak at idle as seen in the example show.

4.2.5 Coolant temperature sensor waveform

Most coolant temperature sensors are NTC thermistors; their resistance decreases as temperature increases (Figure 4.9). This can be measured on most systems as a reducing voltage signal (Figure 4.10).

The coolant temperature sensor (CTS) will usually be a two wire device with a voltage supply of approximately 5 V.

The resistance change will therefore alter the voltage seen at the sensor and can be monitored for any discrepancies across its operational range. By selecting a time scale of 500 seconds and connecting the oscilloscope to the sensor, the output voltage can be monitored. Start the engine and in the majority of cases the voltage will start in the region of 3 to 4 V and fall gradually. The voltage will depend on the temperature of the engine.

The rate of voltage change is usually linear with no sudden changes to the voltage, if the sensor displays a fault at a certain temperature, it will show up in this test.

4.2.6 Hall effect distributor pick-up waveform

Hall sensors are now used in a number of ways. The ignition distributor is very common but they

Figure 4.9 Temperature sensor

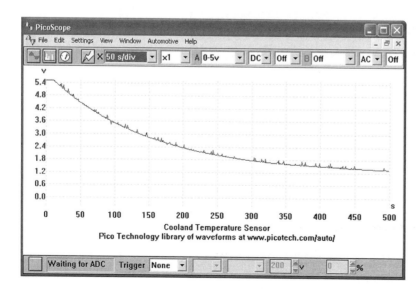

Figure 4.10 Decreasing voltage from the temperature sensor

are also used by ABS for monitoring wheel speed and as transmission speed sensors, for example (Figure 4.11).

This form of trigger device is a simple digital 'on/off switch' which produces a square wave output that is recognised and processed by the ignition control module or engine management ECU.

The trigger has a rotating metal disc with openings that pass between an electromagnet and the semiconductor (Hall chip). This action produces a square wave that is used by the ECU or amplifier (Figure 4.12).

The sensor will usually have three connections which are: (1) a stabilised supply voltage, (2) an earth and (3) the output signal. The square wave when monitored on an oscilloscope may vary in amplitude; this is not usually a problem as it is the frequency that is important, not the height of the voltage. However, in most cases the amplitude/voltage will remain constant.

4.2.7 Inductive distributor pick-up waveform

This particular type of pick-up generates its own signal and therefore does not require a voltage supply to power it. The pick-up is used as a signal to trigger the ignition amplifier or an ECU. The sensor normally has two connections. If a third connection is used it is normally a screen to reduce interference.

As a metal rotor spins, a magnetic field is altered which induces an ac voltage from the pick-up. This type of pick-up could be described as a small alternator because the output voltage rises as the metal rotor approaches the winding, sharply dropping through zero volts as the two components are aligned and producing a voltage in the opposite direction as the rotor passes. The waveform is similar to a sine wave (Figure 4.13); however, the design of the components are such that a more rapid switching is evident.

The voltage produced by the pick-up will be determined by three main factors:

- Engine speed – the voltage produced will rise from as low as 2 to 3 V when cranking, to over 50 V, at higher engine speeds
- The proximity of the metal rotor to the pick-up winding – an average air gap will be in the order of 0.2 to 0.6 mm (8 to 14 thou), a larger air gap will reduce the strength of the magnetic field seen by the winding and the output voltage will be reduced
- The strength of the magnetic field offered by the magnet – the strength of this magnetic field determines the effect it has as it 'cuts' through the windings and the output voltage will be reduced accordingly.

Figure 4.11 Distributors usually contain a Hall effect or inductive pulse generator (*Source:* Bosch press)

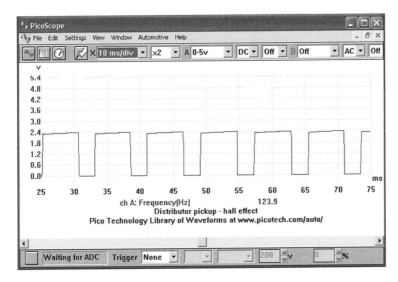

Figure 4.12 Hall output waveform

46 Advanced automotive fault diagnosis

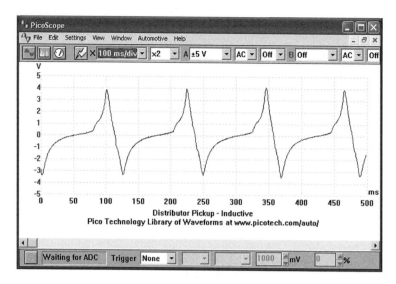

Figure 4.13 Inductive pick-up output signal

Figure 4.14 Knock sensor

A difference between the positive and the negative voltages may also be apparent as the negative side of the sine wave is sometimes attenuated (reduced) when connected to the amplifier circuit, but will produce perfect ac when disconnected and tested under cranking conditions.

4.2.8 Knock sensor waveform

The optimal point at which the spark ignites the air/fuel mixture is just before knocking occurs. However, if the timing is set to this value, under certain conditions knock (detonation) will occur. This can cause serious engine damage as well as increasing emissions and reducing efficiency.

A knock sensor is used by some engine management systems (Figure 4.14). The sensor is a small piezo-electrical crystal that, when coupled with the ECU, can identify when knock occurs and retard the ignition timing accordingly (Figure 4.15).

The frequency of knocking is approximately 15 kHz. As the response of the sensor is very fast an appropriate time scale must be set – in the case of the example waveform 0–500 ms and a 0–5 volt scale. The best way to test a knock sensor is to remove the knock sensor from the engine and to tap it with a small spanner; the resultant waveform should be similar to the example shown.

> *Note:* When refitting the sensor tighten to the correct torque setting as over tightening can damage the sensor and/or cause it to produce incorrect signals.

4.2.9 Oxygen sensor (Titania) waveform

The lambda sensor, also referred to as the oxygen sensor, plays a very important role in the control of exhaust emissions on a catalyst equipped vehicle (Figure 4.16).

The main lambda sensor is fitted into the exhaust pipe before the catalytic converter. The sensor will have four electrical connections. It reacts to the oxygen content in the exhaust system and will produce an oscillating voltage between 0.5 V (lean) to 4.0 V, or above (rich) when running correctly. A second sensor to monitor the catalyst performance may be fitted downstream of the converter.

Titania sensors, unlike Zirconia sensors, require a voltage supply as they do not generate their own voltage. A vehicle equipped with a lambda sensor is said to have 'closed loop', this means that after the fuel has been burnt during the combustion process, the sensor will analyse the emissions and adjust the engine's fuelling accordingly.

Titania sensors have a heater element to assist the sensor reaching its optimum operating

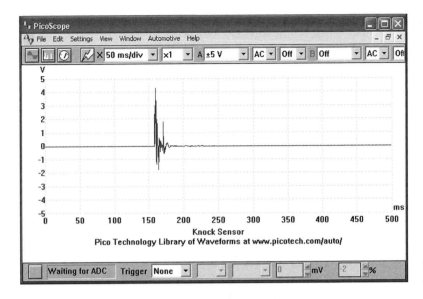

Figure 4.15 Knock sensor output signal

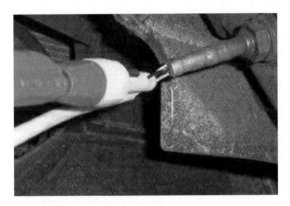

Figure 4.16 Titania knock sensor in position

temperature. The sensor when working correctly will switch approximately once per second (1 Hz) but will only start to switch when at normal operating temperature. This switching can be seen on the oscilloscope, and the waveform should look similar to the one in Figure 4.17.

4.2.10 Oxygen sensor (Zirconia) waveform

The lambda sensor is also referred to as the oxygen sensor or a heated exhaust gas oxygen (HEGO) sensor and plays a very important role in control of exhaust emissions on a catalytic equipped vehicle. The lambda sensor is fitted into the exhaust pipe before the catalytic converter (Figure 4.18). A second sensor to monitor the catalyst performance may be fitted downstream of the converter.

The sensor will have varying electrical connections and may have up to four wires; it reacts to the oxygen content in the exhaust system and will produce a small voltage depending on the Air/Fuel mixture seen at the time. The voltage range seen will, in most cases, vary between 0.2 and 0.8 V. The 0.2 V indicates a lean mixture and a voltage of 0.8 V shows a richer mixture.

A vehicle equipped with a lambda sensor is said to have 'closed loop', this means that after the fuel has been burnt during the combustion process, the sensor will analyse the emissions and adjust the engine's fuelling accordingly.

Lambda sensors can have a heater element to assist the sensor reaching its optimum operating temperature. Zirconia sensors when working correctly will switch approximately once per second (1 Hz) and will only start to switch when at normal operating temperature. This switching can be seen on the oscilloscope, and the waveform should look similar to the one in the example waveform of Figure 4.19.

4.2.11 Throttle position potentiometer waveform

This sensor or potentiometer is able to indicate to the ECU the exact amount of throttle opening due to its linear output (Figure 4.20).

The majority of modern management systems use this type of sensor. It is located on the throttle butterfly spindle. The 'throttle pot' is a three-wire device having a 5 V supply (usually), an earth connection and a variable output from the centre pin. As the output is critical to the vehicle's performance, any 'blind spots' within the internal carbon track's swept area, will cause 'flat spots' and 'hesitations'. This lack of continuity can be seen on an oscilloscope.

48 Advanced automotive fault diagnosis

A good throttle potentiometer should show a small voltage at the throttle closed position, gradually rising in voltage as the throttle is opened and returning back to its initial voltage as the throttle is closed. Although many throttle position sensor voltages will be manufacturer specific, many are non-adjustable and the voltage will be in the region of 0.5 to 1.0 V at idle rising to 4.0 V (or more) with a fully opened throttle. For the full operational range, a time scale around 2 seconds is used.

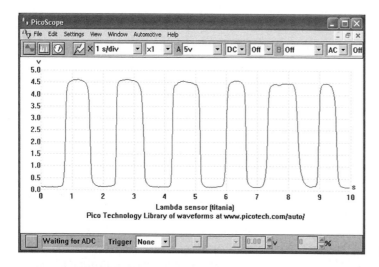

Figure 4.17 Titania lambda sensor output

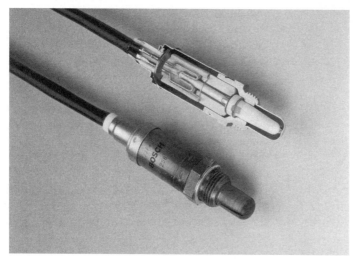

Figure 4.18 Zirconia type oxygen sensor

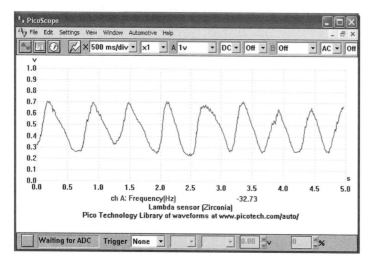

Figure 4.19 Zirconia oxygen sensor output

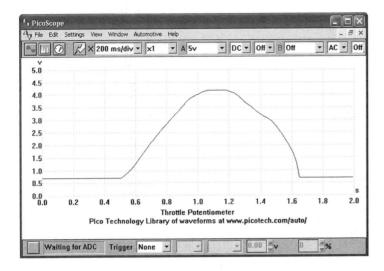

Figure 4.20 Throttle pot output voltage

4.2.12 Road speed sensor (Hall effect)

To measure the output of this sensor, jack up the driven wheels of the vehicle and place on axle stands on firm level ground. Run the engine in gear and then probe each of the three connections (+, − and signal).

As the road speed is increased the frequency of the switching should be seen to increase. This change can also be measured on a multimeter with frequency capabilities. The sensor will be located on either the speedometer drive output from the gearbox or to the rear of the speedometer head if a speedo cable is used. The signal is used by the engine ECU and if appropriate, the transmission ECU.

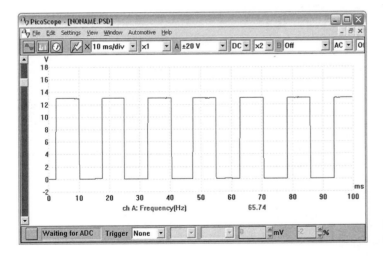

Figure 4.21 Hall effect road speed sensor waveform

4.3 Actuators

4.3.1 Single point injector waveform

Single point injection is also sometimes referred to as throttle body injection (Figure 4.22).

A single injector is used (on larger engines two injectors can be used) in what may have the outward appearance to be a carburettor housing.

The resultant waveform from the single point system shows an initial injection period followed by multi-pulsing of the injector in the remainder of the trace (Figure 4.23). This 'current limiting' section of the waveform is called the supplementary duration and is the part of the injection trace that expands to increase fuel quantity.

4.3.2 Multi-point injector waveform

The injector is an electromechanical device which is fed by a 12 volt supply (Figure 4.24).

50 Advanced automotive fault diagnosis

Figure 4.22 Throttle body with a single injector

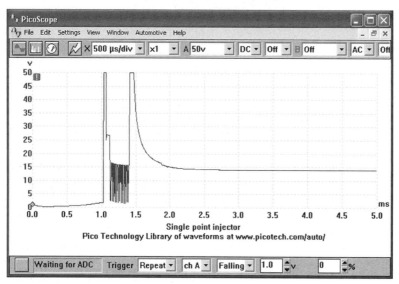

Figure 4.23 Single point injector waveform

Figure 4.24 Multi-point injectors on the rail

The voltage will only be present when the engine is cranking or running because it is controlled by a relay that operates only when a speed signal is available from the engine. Early systems had this feature built into the relay; most modern systems control the relay from the ECU.

The length of time the injector is held open will depend on the input signals seen by the ECU from

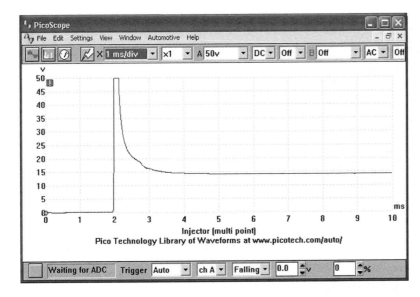

Figure 4.25 Multi-point injector waveform

its various engine sensors. The held open time or 'injector duration' will vary to compensate for cold engine starting and warm-up periods. The duration time will also expand under acceleration. The injector will have a constant voltage supply while the engine is running and the earth path will be switched via the ECU, the result can be seen in the example waveform (Figure 4.25). When the earth is removed a voltage is induced into the injector and a spike approaching 60 V is recorded.

The height of the spike will vary from vehicle to vehicle. If the value is approximately 35 V, it is because a zener diode is used in the ECU to clamp the voltage. Make sure the top of the spike is squared off, indicating the zener dumped the remainder of the spike. If it is not squared, that indicates the spike is not strong enough to make the zener fully dump, meaning there is a problem with a weak injector winding. If a zener diode is not used in the computer, the spike from a good injector will be 60 V or more.

Multi-point injection may be either sequential or simultaneous. A simultaneous system will fire all four injectors at the same time with each cylinder receiving two injection pulses per cycle (720° crankshaft rotation). A sequential system will receive just one injection pulse per cycle, this is timed to coincide with the opening of the inlet valve.

As a very rough guide the injector durations for an engine at normal operating temperature, at idle speed are:

- 2.5 ms – simultaneous
- 3.5 ms – sequential.

Monitoring the injector waveform using both voltage and amperage, allows display of the 'correct' time that the injector is physically open. The current waveform (the one starting on the zero line) shows that the waveform is 'split' into two defined areas.

The first part of the current waveform is responsible for the electromagnetic force lifting the pintle; in Figure 4.26 the time taken is approximately 1.5 ms; this is often referred to as the solenoid reaction time. The remaining 2 ms is the actual time the injector is fully open. This, when taken as a comparison against the injector voltage duration, is different to the 3.5 ms shown. The secret is to make sure you compare like with like!

4.3.3 Bosch common rail diesel injector waveform

Common rail diesel systems are becoming more common (Figure 4.27)!

It can be clearly seen from the example waveform (Figure 4.28) that there are two distinctive points of injection, the first being the 'pre-injection' phase, with the second pulse being the 'main' injection phase.

As the throttle is opened, and the engine is accelerated, the 'main' injection pulse expands in a similar way to a petrol injector. As the throttle is released, the 'main' injection pulse disappears until such time as the engine returns to just above idle.

Under certain engine conditions a third phase may be seen, this is called the 'post injection'

52 *Advanced automotive fault diagnosis*

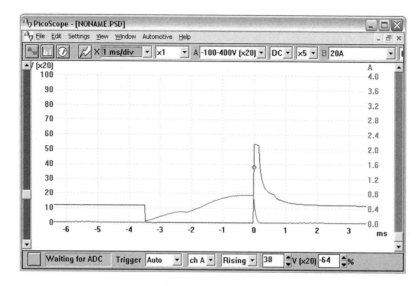

Figure 4.26 Injector voltage and current measurements

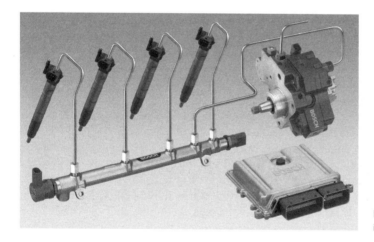

Figure 4.27 Common rail diesel pump, rail, injectors and ECU (*Source*: Bosch Press)

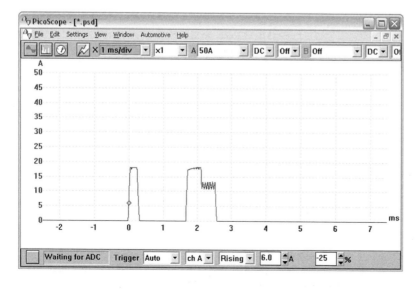

Figure 4.28 CR injector (current) waveform showing pre and main injection pulses

phase and is predominantly concerned with controlling the exhaust emissions.

4.3.4 Electromagnetic idle speed control valve waveform

This device contains a winding, plunger and spring (Figure 4.29). When energised the port opens and when not it closes.

The electromagnetic idle speed control valve (ISCV) will have two electrical connections; usually a voltage supply at battery voltage and a switched earth.

The rate at which the device is switched is determined by the ECU to maintain a prerequisite speed according to its programming. The valve will form an air bypass around the throttle butterfly. If the engine has an adjustable air by-pass and an ISCV, it may require a specific routine to balance the two air paths. The position of the valve tends to take up an average position determined by the supplied signal.

As the example waveform in Figure 4.30 shows, the earth path is switched and the resultant picture is produced. Probing onto the supply side will produce a straight line at system voltage. When the earth circuit is monitored a 'saw tooth' waveform will be seen.

4.3.5 Rotary idle speed control valve waveform

The rotary idle speed control valve (ISCV) will have two or three electrical connections, with a voltage supply at battery voltage and either a single or a double switched earth path (Figure 4.31). The device is like a motor that only ever rotates about half a turn in each direction!

Figure 4.29 Electromagnetic idle speed control valve

Figure 4.31 Rotary idle speed control valve

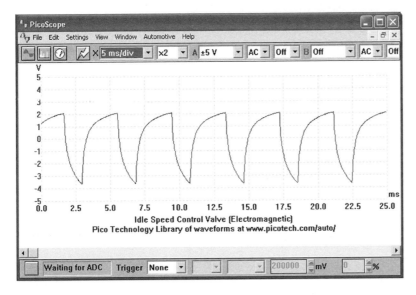

Figure 4.30 Signal produced by an electromagnetic idle speed control valve

The rate at which the earth path is switched is determined by the ECU to maintain a prerequisite idle speed according to its programming.

The valve will form an air bypass past the throttle butterfly, to form a controlled air bleed within the induction tract. The rotary valve will have the choice of either single or twin earth paths, the single being pulled one way electrically and returned to its closed position via a spring; the double switched earth system will switch the valve in both directions. This can be monitored on a dual trace oscilloscope. As the example waveform in Figure 4.32 shows, the earth path is switched and the resultant picture is produced. The idle control device takes up a position determined by the on/off ratio (duty cycle) of the supplied signal.

Probing onto the supply side will produce a straight line at system voltage and when the earth circuit is monitored a square wave will be seen. The frequency can also be measured as can the on/off ratio.

4.3.6 Stepper motor waveform

The stepper or stepper motor is a small electro-mechanical device that allows either an air bypass circuit or a throttle opening to alter in position depending on the amounts that the stepper is indexed (moved in known steps) (Figure 4.33).

Stepper motors are used to control the idle speed when an idle speed control valve is not employed. The stepper may control an 'air bypass' circuit by having four or five connections back to the ECU. The earths enable the control unit to move the motor in a series of 'steps' as the contacts are earthed to ground. These devices may also be used to control the position of control flaps, for example, as part of a heating and ventilation system.

The individual earth paths can be checked using the oscilloscope. The waveforms should be similar on each path. Variations to the example shown in Figure 4.34 may be seen between different systems.

4.4 Ignition system

4.4.1 Ignition primary waveform

The ignition primary waveform is a measurement of the voltage on the negative side of the ignition coil. The earth path of the coil can produce over 350 V. Different types of ignition coils

Figure 4.33 Stepper motor and throttle potentiometer on a throttle body

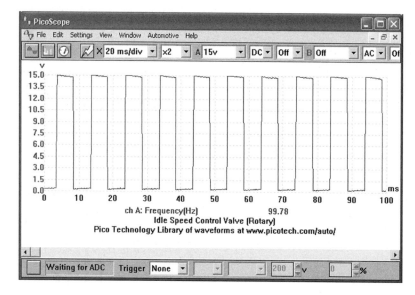

Figure 4.32 Signal supplied to a rotary idle control valve

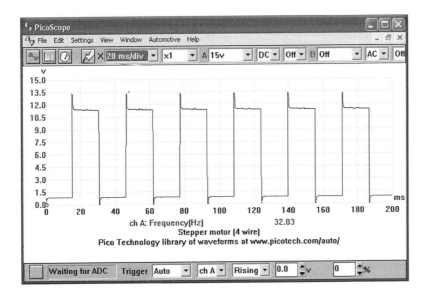

Figure 4.34 Stepper motor signals

Figure 4.35 Direct ignition coils in position

produce slightly different traces but the fundamental parts of the trace and principles are the same (Figure 4.35).

In the waveform shown in Figure 4.36, the horizontal voltage line at the centre of the oscilloscope is at fairly constant voltage of approximately 40 V, which then drops sharply into what is referred to as the coil oscillation. The length of the horizontal voltage line is the 'spark duration' or 'burn time', which in this particular case is about 1 ms. The coil oscillation period should display a minimum of four to five peaks (both upper and lower). A loss of peaks would indicate a coil problem.

There is no current in the coil's primary circuit until the dwell period. This starts when the coil is earthed and the voltage drops to zero. The dwell period is controlled by the ignition amplifier or ECU and the length of the dwell is determined by the time it takes to build up to about 8 A. When this predetermined current has been reached, the amplifier stops increasing the primary current and it is maintained until the earth is removed from the coil. This is the precise moment of ignition.

The vertical line at the centre of the trace is in excess of 200 V, this is called the 'induced voltage'. The induced voltage is produced by magnetic inductance. At the point of ignition, the coil's earth circuit is removed and the magnetic flux collapses across the coil's windings. This induces a voltage between 150 and 350 V. The coil's high tension output will be proportional to this induced voltage. The height of the induced voltage is sometimes referred to as the primary peak volts.

From the example current waveform (in Figure 4.37), the limiting circuit can be seen in operation. The current switches on as the dwell period starts and rises until the required value is achieved (usually about 8 A). At this point the current is maintained until it is released at the point of ignition.

The dwell will expand as the engine revs are increased to maintain a constant coil saturation time. This gives rise to the term 'constant energy'. The coil saturation time can be measured and this will remain the same regardless of engine speed. Figure 4.37 shows a charge time of about 3.5 ms.

4.4.2 Ignition secondary waveform

The ignition secondary waveform is a measurement of the HT output voltage from the ignition coil. Some coils can produce over 50,000 V. Different types of ignition coils produce slightly different traces but the fundamental parts of the trace and principles are the same (Figure 4.38).

56 *Advanced automotive fault diagnosis*

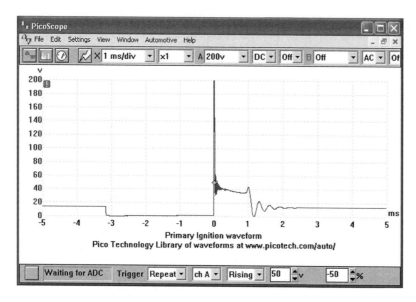

Figure 4.36 Primary ignition voltage trace

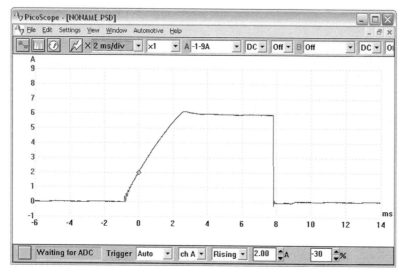

Figure 4.37 Primary ignition current trace

The ignition secondary picture shown in Figure 4.39 waveform is from an engine fitted with electronic ignition. In this case, the waveform has been taken from the main coil lead (king lead). Suitable connection methods mean that similar traces can be seen for other types of ignition system.

The secondary waveform shows the length of time that the HT is flowing across the spark plug electrode after its initial voltage, which is required to initially jump the plug gap. This time is referred to as either the 'burn time' or the 'spark duration'. In the trace shown it can be seen that the horizontal voltage line in the centre of the oscilloscope is at fairly constant voltage of approximately 3 or 4 kV, which then drops sharply into the 'coil oscillation' period.

The coil oscillation period should display a minimum of four or five peaks (both upper and lower). A loss of peaks indicates that the coil may

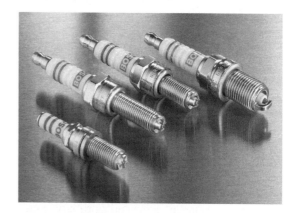

Figure 4.38 Spark plugs (*Source*: Bosch Press)

be faulty. The period between the coil oscillation and the next 'drop down' is when the coil is at rest and there is no voltage in the secondary circuit. The 'drop down' is referred to as the 'polarity peak', and produces a small oscillation in the opposite direction to the plug firing voltage. This is due to the initial switching on of the coil's primary current.

The plug firing voltage is the voltage required to jump and bridge the gap at the plug's electrode, commonly known as the 'plug kV'. In this example the plug firing voltage is about 12 or 13 kV.

When the plug kVs are recorded on a DIS or coil per cylinder ignition system, the voltage seen on the waveform (Figure 4.40) should be in the 'upright position'. If the trace is inverted it would suggest that either the wrong polarity has been selected from the menu or in the case of DIS, the inappropriate lead has been chosen. The plug voltage, while the engine is running, is continuously fluctuating and the display will be seen to move up and down. The maximum voltage at the spark plug, can be seen as the 'Ch A: Maximum (kV)' reading at the bottom of the screen.

It is a useful test to snap the throttle and observe the voltage requirements when the engine is under load. This is the only time that the plugs are placed under any strain and is a fair assessment of how they will perform on the road.

The second part of the waveform after the vertical line is known as the spark line voltage. This second voltage is the voltage required to keep the plug running after its initial spark to jump the gap. This voltage will be proportional to the resistance within the secondary circuit. The length of the line can be seen to run for approximately 2 ms.

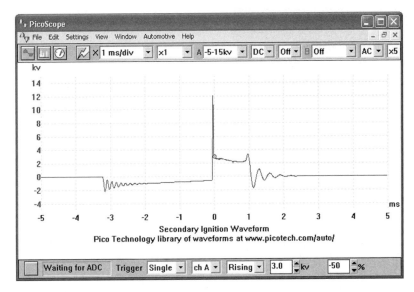

Figure 4.39 Ignition secondary trace

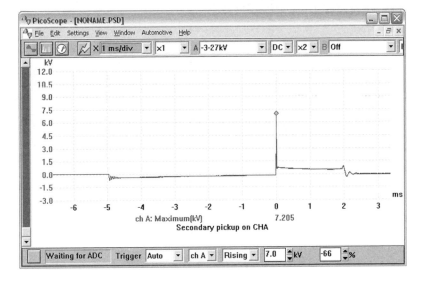

Figure 4.40 Distributorless ignition

4.5 Other components

4.5.1 Alternator waveform

Checking the ripple voltage produced by an alternator is a very good way of assessing its condition (Figure 4.41).

The example waveform shown in Figure 4.42 illustrates the rectified output from the alternator. The output shown is correct and that there is no fault within the phase windings or the diodes (rectifier pack).

The three phases from the alternator have been rectified to dc from its original ac and the waveform shows that the three phases are all functioning.

If the alternator is suffering from a diode fault, long downward 'tails' appear from the trace at regular intervals and 33% of the total current output will be lost. A fault within one of the three phases will show a similar picture to the one illustrated but is three or four times the height, with the base to peak voltage in excess of 1 V.

The voltage scale at the side of the oscilloscope is not representative of the charging voltage, but is used to show the upper and lower limits of the ripple. The 'amplitude' (voltage/height) of the waveform will vary under different conditions. A fully charged battery will show a 'flatter' picture, while a discharged battery will show an exaggerated amplitude until the battery is charged. Variations in the average voltage of

Figure 4.41 Alternator

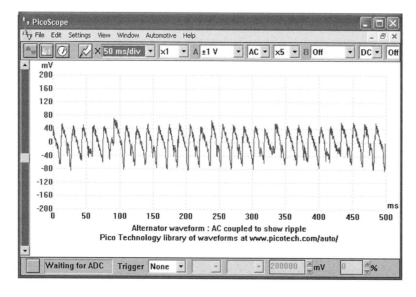

Figure 4.42 Alternator ripple voltage

the waveform are due to the action of the voltage regulator.

4.5.2 Relative compression petrol waveform

Measuring the current drawn by the starter motor (Figure 4.43) is useful to determine starter condition but it is also useful as an indicator of engine condition.

The purpose of this particular waveform (Figure 4.44) is therefore to measure the current required to crank the engine and to evaluate the relative compressions.

The amperage required to crank the engine depends on many factors, such as: the capacity of the engine, the number of cylinders, the viscosity of the oil, the condition of the starter motor, the condition of the starter's wiring circuit and the compressions in the cylinders. To evaluate the compressions therefore, it is essential that the battery is charged, and the starter and associated circuit are in good condition.

The current for a typical 4 cylinder petrol/gasoline engine is in the region of 100 to 200 A.

In the waveform shown, the initial peak of current (approximately 400 A) is the current required to overcome the initial friction and inertia to rotate the engine. Once the engine is rotating, the current will drop. It is also worth mentioning the small step before the initial peak, which is being caused by the switching of the starter solenoid.

The compressions can be compared against each other by monitoring the current required to push each cylinder up on its compression stroke. The better the compression the higher the current demand and vice versa. It is therefore important that the current draw on each cylinder is equal.

4.5.3 CAN-H and CAN-L waveform

Controller area network (CAN) is a protocol used to send information around a vehicle on data bus. It is made up of voltage pulses that represent ones and zeros, in other words, binary signals. The data is applied to two wires known as CAN-high and CAN-low (Figure 4.45).

In this display, it is possible to verify that data is being continuously exchanged along the CAN bus. It is also possible to check that the peak to peak voltage levels are correct and that a signal is present on both CAN lines. CAN uses a differential signal, and the signal on one line should be a coincident mirror image (the signals should line up) of the data on the other line (Figure 4.46).

The usual reason for examining the CAN signals is where a CAN fault has been indicated by

Figure 4.43 Starter and ring gear

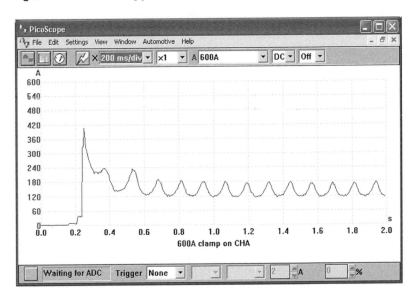

Figure 4.44 Spark ignition engine cranking amps

OBD, or to check the CAN connection to a suspected faulty CAN node. The vehicle manufacturers' manual should be referred to for precise waveform parameters.

The signal shown is captured on a fast timebase and allows the individual state changes to be viewed. This enables the mirror image nature of the signals, and the coincidence of the edges to be verified.

4.6 Summary

'Scope' diagnostics is now an essential skill for the technician to develop. As with all diagnostic techniques that use test equipment, it is necessary for the user to know how:

- the vehicle system operates;
- to connect the equipment;
- readings should be interpreted.

Remember that an oscilloscope is really just a voltmeter or ammeter but that it draws a picture of the readings over a set period of time. Learn what good waveforms look like, and then you will be able to make good judgements about what is wrong when they are not so good!

Figure 4.45 OBD socket – pin 6 is CAN-high and pin 14 is CAN-low

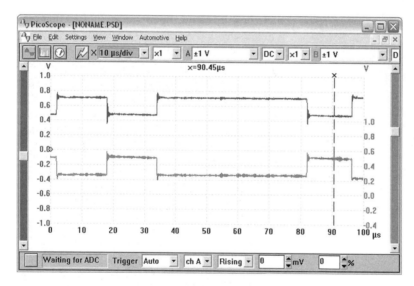

Figure 4.46 CAN-high and low signals on a dual trace scope

Knowledge check questions

To use these questions, you should first try to answer them without help but if necessary, refer back to the content of the chapter. Use notes, lists and sketches to answer them. It is not necessary to write pages and pages of text!

1. Explain the terms 'timebase', 'amplitude' and 'voltage scale'.
2. Make a sketch of ignition primary and secondary waveforms. Label each part and state which aspects indicate that no faults are present.
3. Describe how to connect an oscilloscope to examine the signal supplied to a single point (throttle body) injector.
4. State the typical output voltage (peak to peak) of an inductive crankshaft sensor at cranking, idle and 3000 rev/min. Sketch the waveform to show the aspects that indicate engine speed and engine position.
5. Explain with the aid of a sketch, why current limiting is used on the primary circuit of an ignition system.

5
On-board diagnostics

5.1 On-board diagnostics – a first perspective

> *Authors Note: I am most grateful to Alan Malby (Ford Motor Company) for his excellent contribution to this chapter.*

5.1.1 Introduction

Originating in the USA, and subsequently followed by Europe, Asia and many others, governments around the globe have augmented vehicle emissions control legislation. This includes a requirement that all vehicles sold within their territories must support an on-board diagnostic system that can be operated to determine the serviceability of the vehicle's emission control systems, sub-systems and components.

Enabled by the increasing advances in electronics and microprocessor software development, this system, now commonly termed as on-board diagnostics (OBD), has been developed over recent years and is now implemented by all major motor vehicle manufacturers. Furthermore, this has been extended to allow diagnosis of none-emission related vehicle systems.

Later in this chapter, and in the final chapter of the book, there is more information about the development of on-board diagnostics and the potential for further improvements. Some aspects are repeated for reasons of clarity.

5.1.2 Vehicle emissions and environmental health

From as early as 1930, the subject of vehicle engine emissions influencing environmental health was very topical in the State of California. Already with a population of 2 million vehicles, scores of people have died and thousands have become sick due to air pollution related illnesses.

Following the outbreak of WW2, the population of California had risen to some 7 million people in 1943, with 2.8 million vehicles travelling over a total of 24 billion miles (Figure 5.1). Already, smog was apparent and people suffered with stinging eyes, sore throats and breathing difficulties. Local Government initiated a study into the cause of the problem, and scientists at CALTEC and the University of California investigated the problem of smog (Figure 5.2).

In 1945, after the conclusion of the war, Los Angeles began its air pollution control program and established the Bureau of Smoke Control. On June 10, 1947 the then California Governor, Earl Warren, signed the Air Pollution Control Act. By

Figure 5.1 Early traffic jam!

Figure 5.2 Smog over Los Angeles

1950, California's population had reached 11 million people. Total registered vehicles in California exceeded 4.5 million and Vehicle Miles Travelled (VMT) was 44.5 billion. The search for the root cause of smog production went on. Reports of deaths in other countries became apparent. For example, in the early 1950s, thousands of people died in London of a 'mystery fog'.

In 1952, Dr. Arie Haagen-Smit determined the root cause of smog production. He surmised that engine pollutants, Carbon Monoxide (CO), Hydrocarbons (HC) and various Oxides of Nitrogen or 'NOx' combine to generate the smog, which consists of Ozone and Carbon Dioxide.

Carbon Dioxide is a pollutant, which is now said to contribute to global warming and climate change. Ozone, occupying a region of the lower atmosphere, is now known to cause respiratory ill health, lung disease and is also thought to make a much greater contribution to the green house effect than even Carbon Dioxide.

The State became a centre for environmental activism. Naturally, amidst a public outcry to preserve the local environment, the State began to legislate for controls on motor vehicle emissions. So began an initiative that would span over 50 years, one that would drive change in a world industry and lead the world in the fight for clean air.

5.1.3 History of the emissions control legislation

In 1960, The Motor Vehicle Pollution Control Board was established with a mandate to certify devices proposed to be fitted on cars for sale in California. In addition, The Federal Motor Vehicle Act of 1960 was enacted, requiring Federal research to combat motor vehicle engine pollution. Manufacturers made technology improvements and during this period, California's population reached 16 million people. Total registered vehicles approached 8 million and Vehicle Miles Travelled (VMT) was 71 billion.

In 1961, in an effort to control hydrocarbon crankcase emissions, the first piece of vehicle emissions control legislation mandating the use of specific hardware was issued. Positive Crankcase Ventilation (PCV) controls hydrocarbon crankcase emissions by extracting gases from the crankcase and re-circulating them back into the fresh air/fuel charge in the cylinders.

A key turning point in history, in 1966 the California Motor Vehicle Pollution Control Board pioneered the adoption of vehicle tailpipe emissions standards for hydrocarbons (HC) and carbon monoxide (CO) and the California Highway Patrol began random roadside inspections of the smog control devices fitted to vehicles.

The following year the Governor of California, Ronald Reagan, signed the Mulford-Carrell Air Resources Act. This effectively allowed the state of California to set its own emissions standards. The same year saw the formation of the California Air Resources Board (CARB), which was created from the amalgamation of the Motor Vehicle Pollution Control Board and the Bureau of Air Sanitation.

In 1969 the first California State Ambient Air Quality Standards were extended by California for photochemical oxidants, suspended particulates, sulphur dioxide (SO_2), nitrogen dioxide (NO_2), and carbon monoxide (CO). California's population reached 20 million people. Total registered vehicles exceeded 12 million and VMT is 110 billion.

Total cumulative California vehicle emissions for HC and NOx are estimated at 1.6 million tons per year. In 1970, the US Environmental Protection Agency (EPA) was created; its primary directive was, and still is, to protect all aspects of the environment. The next seven years witnessed further development of emissions control legislation and increasing employment of vehicle emissions control technology.

In 1971 CARB adopted the first vehicle NOx standards. The EPA announced National Ambient Air Quality Standards for particulates, hydrocarbons (HC), carbon monoxide (CO), nitrogen dioxide (NO_2), photochemical oxidants (including ozone) and sulphur dioxide (SO_2).

The first Two-Way Catalytic Converters came into use in 1975 as part of CARB's Motor Vehicle Emission Control Program followed by an announcement that CARB will limit lead in gasoline.

Table A California State-wide emissions 1969

State-wide average emissions (per vehicle)	
NOx (g/mile)	HC (g/mile)
5.3	8.6

Table B California State-wide emissions 1980

State-wide average emissions (per vehicle)	
NOx (g/mile)	HC (g/mile)
4.8	5.5

In 1977, Volvo introduced a vehicle marketed as 'Smog-Free'. This vehicle supported the first Three-Way Catalytic (TWC) Converter to control hydrocarbon (HC), nitrogen oxides (NOx), and carbon monoxide (CO) emissions.

In 1980, the California population reached 24 million people. Total registered vehicles were in the region of 17 million and VMT was 155 billion.

Total cumulative California vehicle emissions for NOx and HCs remain at 1970 levels of 1.6 million tons/year despite a rise of 45 billion in VMT over those ten years.

The legislative controls had clearly begun to have a positive effect. Spurred on by this victory, CARB began a program of compliance testing on 'in use' vehicles in order to determine whether they continue to comply with emission standards as vehicle mileage increases. Vehicle manufacturers commissioned the development of more durable emission control systems.

1984 saw the introduction of the biennial California Smog Check Program, the aim of which was to identify vehicles in need of maintenance and to confirm the effectiveness of their emissions control systems.

The mid-term period of emissions control legislation ended in 1988 with a key announcement, which saw the beginning of on-board diagnostics. The California Clean Air Act was signed and CARB adopted regulations that required that all 1994 and beyond model year cars were fitted with 'On-board Diagnostic' systems. The task of these systems was, as it is now:

> To monitor the vehicle emissions control systems performance and alert owners when here is a malfunction that results in the lack of function of an emissions control system/sub-system or component.

5.1.4 Introduction of vehicle emissions control strategies

To meet the ever increasing but justifiable and 'wanted' need of vehicle emissions control legislation, vehicle manufacturers were forced to invest heavily in the research and development of Vehicle Emission Control Strategies. Building upon the foundation laid by PCV, the two-way and three-way catalyst, manufacturers further developed emissions control hardware. Such systems included exhaust gas re-circulation, secondary air injection, fuel tank canister purge, spark timing adjustment, air/fuel ratio control biasing, fuel shut off under negative torque conditions (overrun or cruise down), to name but a few.

This development continued and expanded meaning that these systems demanded an ever-increasing array of sensors and actuators. The resolution of measurement, control of air/fuel ratio, actuator displacement rates and accuracy of displacement, etc. was way beyond that which could be provided by traditional existing mechanical technologies.

At about this time, an enabler was provided in the form of recent advances in microprocessor technology. The path was clear, the drivers for on-board diagnostic system monitoring were in force and the enablers were available.

On-board diagnostics was born.

5.2 What is on-board diagnostics?

Fundamentally, a contemporary microprocessor based on-board diagnostics or OBD system is intended to self diagnose and report when the performance of the vehicle's emissions control systems or components have degraded. This is to the extent that the tailpipe emissions have exceeded legislated levels or are likely to be exceeded in the long term.

When an issue occurs the OBD system illuminates a warning lamp known as the malfunction indicator lamp (MIL) or Malfunction Indicator (MI) on the instrument cluster. In the United States this symbol often appears with the phrase 'Check Engine', 'Check' or 'Service Engine Soon' contained within it (Figure 5.3). European vehicles

Figure 5.3 Malfunction indicator lamp symbols

tend to use the engine symbol on an orange background.

When the fault occurs the system stores a diagnostic trouble code (DTC) that can be used to trace and identify the fault. The system will also store important information that pertains to the operating conditions of the vehicle when the fault was set. A service technician is able to connect a diagnostic scan tool or a code reader that will communicate with the microprocessor and retrieve this information. This allows the technician to diagnose and rectify the fault, make a repair/replacement, reset the OBD system and restore the vehicle emissions control system to a serviceable status.

As vehicles and their systems become more complex, the functionality of OBD is being extended to cover vehicle systems and components that do not have anything to do with vehicle emissions control. Vehicle body, chassis and accessories such as air conditioning or door modules can now also be interrogated to determine their serviceability as an aid to fault diagnosis.

5.2.1 OBD scenario example

Whilst driving, a vehicle owner observes that the vehicle's engine 'lacks power' and 'jumps sometimes'. This is a problem often faced by technicians in that customers often have no engineering or automotive knowledge and use lay terms to describe what is happening with a very complex system. The driver does however report that the MIL has been illuminated.

The technician connects a scan tool that can communicate using an industry standard communications protocol. The OBD code memory is checked and data is presented in a way that also conforms to a standard. DTC P1101 with the description 'MAF sensor out of self test range' is stored in memory, which means that the OBD System Component Monitor has identified the mass airflow (MAF) sensor circuit voltage as outside an acceptable range (Figure 5.4).

Upon confirming the fault the system was smart – it defaulted to a 'safe' value of mass airflow, a concept know as failure mode effects management (FMEM) to allow the driver to take the vehicle to a place of repair. Whilst this FMEM value was a good short-term solution it is not a sufficient substitute for the full functionality of a serviceable MAF sensor.

Since the MAF sensor determines the mass airflow going into the engine intake it will be impossible for the system to run at the optimum air/fuel ratio for efficient burning of the air/fuel charge

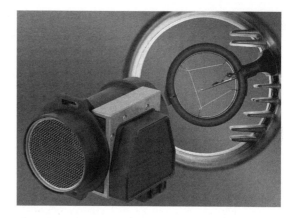

Figure 5.4 New mass airflow sensor (*Source*: Bosch)

within the cylinder. It may be that tailpipe emissions are likely to rise beyond legislated limits.

Also the MAF sensor is used by other emissions control systems on the vehicle – now that its input is unreliable it follows that those systems are no longer working at their optimum levels and may not work at all. This is the reason for the malfunction indication lamp (MIL) illumination, which says, in as many words:

> An emissions control system/sub system or component has become unserviceable!

Visual inspection of the MAF sensor reveals that it has become damaged beyond repair and needs replacing. This is carried out, the technician clears the DTC from the OBD system memory, resets the system, and a short test drive later the diagnostic scan tool confirms that the DTC is no longer present. The road test also confirms that the previous drive issue is no longer apparent.

5.2.2 Origins of OBD in the USA

The previous example relates to the current situation, but when OBD was first introduced standards and practices were less well defined. Manufacturers developed and applied their own systems and code descriptions. This state of affairs was obviously undesirable since none franchised service, and repair centres had to understand the various subtleties of each system; this meant having different scan tools, as well as a multitude of leads, manuals and connectors. This made diagnostics unwieldy and expensive. This stage became known as 'OBD1', the first stage of OBD introduction.

In the late 1980s the Society of Automotive Engineers (SAE) defined a list of standard practices and recommended these to the

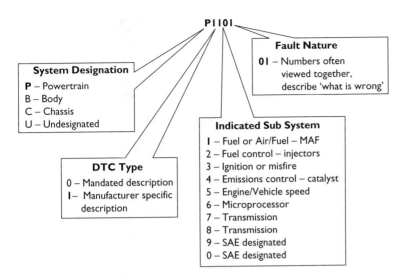

Figure 5.5 P Code composition

Environmental Protection Agency (EPA). The EPA acknowledged the benefits of these standards and recommendations, and adopted them. In combination, they changed the shape and application of OBD. The recommendations included having a standard diagnostic connector, a standard scan tool and a communications protocol that the standard scan tool could use to interface with the vehicle of any manufacturer.

The standard also included mandatory structures and descriptions for certain emission control system/component defects. These were called 'P0' Codes. Manufacturers were still free to generate their own 'manufacturer specific code descriptions' known as 'P1' Codes. This phase of implementation became known as OBD2, and was adopted for implementation by 1st January 1996.

5.2.3 P Code composition

The diagnostic trouble code (DTC) is displayed as a 5-character alphanumeric code. The first character is a letter that defines which vehicle system set the code be it Powertrain, Body, or Chassis.

- **P** – means Powertrain System set the code
- **B** – means Body System set the code
- **C** – means Chassis System set the code
- **U** – is currently unused but has been 'stolen' to represent communication errors

P Codes are requested by the microprocessor controlling the powertrain or transmission and refer to the emissions control systems and their components.

B Codes are requested by the microprocessor controlling the body control systems. Collectively these are grouped as lighting, air conditioning, instrumentation, or even in-car entertainment or telematics.

C Codes are requested by the microprocessor controlling the chassis systems that control vehicle dynamics such as ride height adjustment, traction control, etc.

The four numbers that follow the letter detail information pertaining to what sub-system declared the code. Using the example from before, see Figure 5.5.

5.2.4 European on-board diagnostics and global adoption

Europe was not immune to the environmental issues associated with smog. As mentioned earlier, a major smog episode occurred in London in late 1952; this lasted for five days and resulted in about 4000 deaths. In 1956, the UK government passed its first Clean Air Act, which aimed to control domestic sources of smoke pollution.

In 1970 the then European Community adopted directive 70/220 EEC – 'Measures to be taken against Air Pollution by Emissions from Motor Vehicles'. This basically set the foundation for future legislation to curb motor vehicle pollution in Europe. This Directive was amended over the next three decades when in October 1998 the amendment 98/69/EC 'On-Board Diagnostics (OBD) for Motor Vehicles' was adopted, which added Annex XI to the original 70/220 document. Annex XI details the functional aspects of OBD for motor vehicles in Europe and across the Globe. This became known as EOBD.

5.2.5 Summary

A major contributing factor to environmental health issues in the Unites States was found to be motor vehicle emissions pollution. Scientific studies by government sponsored academic establishments and vehicle manufacturers then took place over several years. Legislative bodies were formed which later developed and enacted the vehicle emissions control legislation, which forced vehicle manufacturers to develop control strategies and incorporate them within their production vehicles.

As microprocessor technologies became more advanced and commercially viable, the legislation was augmented to include a self-diagnosing onboard diagnostic system, which would report when the emissions control system was unserviceable. First attempts by manufacturers to use such a system were applied unilaterally which resulted in confusion, regenerative work and a poor reception of the OBD (now termed OBD1) concept. A revision of the legislation adopted SAE recommended standards, which resulted in the OBD (now termed OBD2) system becoming largely generic and applicable across the whole range of vehicle manufacturers.

As environmental activism spread across to Europe, vehicle manufacturers realised they had to support a philosophy of sustainable growth. Similar legislation was adopted and European OBD (EOBD) manifested itself in a form very similar to that observed in the United States.

5.3 Petrol/gasoline on-board diagnostic monitors

5.3.1 Introduction

This section will cover the fundamentals of some of the on-board diagnostic systems employed on mainstream petrol/gasoline vehicles. The concept of how the OBD system is divided into a series of software based serviceability indicators known as 'OBD monitors', is also covered.

5.3.2 Legislative drivers

In Europe – The European Directive 70/220 EEC was supplemented by European Directive 98/69/EC (Year: 1998. OJ Series: L – OJ Number: 350/1). This introduced legislation mandating the use of on-board diagnostic systems in passenger vehicles manufactured and sold after January 1st 2001.

In the USA – Legislation was first introduced in California by the California Air Requirements Board (CARB) in 1988 and later federally by the Clean Air Act Amendments of 1990. This meant that the enforcing body, the environmental protection agency, requires that States have to develop state implementation plans (SIPs) that explain how each state will implement a plan to clean up pollution from sources including motor vehicles. One aspect of the requirement is the performance of on-board diagnostic (OBD) system checks as part of the required periodic inspection.

In order to be compliant with legislation and still sell vehicles, manufacturers needed to engineer 'early warning' monitoring sub-systems that would determine when emission control systems had malfunctioned to the extent that tailpipe emissions had (or were likely to in the long-term) exceed a legislated level. On-board diagnostic 'monitors' were derived for this purpose.

5.3.3 Component monitoring

The emission control systems integral with the vehicle employ many sensors and actuators. A software program housed within a microprocessor defines their actions.

The 'component monitor' is responsible for determining the serviceability of these sensors and actuators. Intelligent component drivers linked to the microprocessor have the ability to enable/disable sensors/actuators, and to receive signals. The analogue inputs from the sensors are converted to digital values within the microprocessor.

In combination with these component drivers, the microprocessor possesses the functionality to detect circuit faults on the links between microprocessor and component. In addition, rationality tests can be performed to determine whether the sensor is operating out of range of its specification.

5.3.4 Rationality testing

Rationality tests can be performed on such sensors as the mass airflow (MAF) sensor and throttle body. For example, the MAF is tested by observing its output value in comparison to a 'mapped' value normalised by throttle position and engine speed. The map or table contains expected MAF output values for the engine speed/throttle set point.

Should the MAF output lie outside of an acceptable range (threshold) of values for that engine speed/throttle set point, then a fault is reported.

5.3.5 Circuit testing

The component monitor is capable of monitoring for circuit faults. Open circuits, short circuits to ground or voltage can be detected. Many manufacturers also include logic to detect intermittent errors.

5.3.6 Catalyst monitor

The purpose of the catalyst is to reduce tailpipe/exhaust emissions. The 'catalyst monitor' is responsible for determining the efficiency of the catalyst by inferring its ability to store oxygen. The method favoured most by the majority of manufacturers is to fit an oxygen sensor before and after the catalyst (Figure 5.6).

As the catalyst's ability to store oxygen (and hence perform three-way catalysis) deteriorates, the oxygen sensor downstream of the sensor will respond to the oxygen in the exhaust gas stream and its signal response will exhibit a characteristic similar to the upstream oxygen sensor.

An algorithm within the microprocessor analyses this signal and determines whether the efficiency of the catalyst has degraded beyond the point where the vehicle tailpipe emissions exceed legislated levels (Figure 5.7). If the microprocessor determines that this has occurred then a malfunction and a DTC is reported. Repeat detections

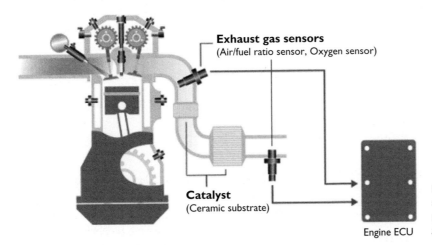

Figure 5.6 Exhaust gas oxygen sensors positioned pre and post catalyst (*Source*: www.globaldensoproducts.com)

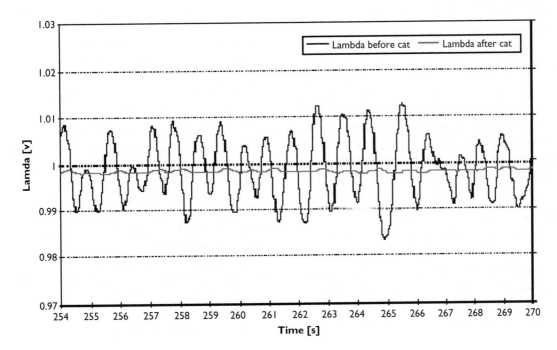

Figure 5.7 Upstream and downstream exhaust gas sensor activity – good catalyst (*Source*: SAE 2001-01-0933 New Cat Preparation Procedure for OBD2 Monitoring Requirements)

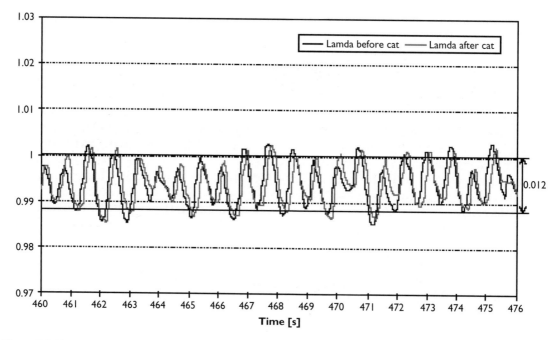

Figure 5.8 Upstream and downstream exhaust gas sensor activity – failed catalyst (*Source*: SAE 2001-01-0933 New Cat Preparation Procedure for OBD2 Monitoring Requirements)

of a failed catalyst will result in MIL illumination (Figure 5.8).

5.3.7 Evaporative system monitor

The purpose of the evaporative (EVAP) emissions control system is to store and subsequently dispose of unburned hydrocarbon emissions thus preventing them from entering the atmosphere. This is achieved by applying a vacuum across the fuel tank. The vacuum then causes fuel vapour to be drawn through a carbon canister in which the hydrocarbon vapours are collected and stored.

During certain closed loop fuel control conditions the microprocessor activates a solenoid controlled 'vapour management valve'. This allows the manifold vacuum to draw vapour from the carbon canister along vapour lines, which terminate in the intake manifold. The fuel vapour is then combined and combusted with the standard air /fuel charge, the closed loop fuel control system caters for the additional air fuel ratio enrichment to ensure that stoichiometric fuelling continues.

The evaporative system monitor is responsible for determining the serviceability of the EVAP system components and to detect leaks in the vapour lines. Most manufacturers check for fuel vapour leaks by employing a diagnostic that utilises a pressure or vacuum test on the fuel system.

European legislation dictates that these checks are not required. However, vehicles manufactured in the USA after 1996 and before 1999 generally employ a system that uses a pressure or vacuum system. This must be able to detect a leak in a hose or filler cap that is equivalent to that generated by a hole, which is 1 mm (0.40″) in diameter. Vehicles manufactured after 2000 must support diagnostics that are capable of detecting a 0.5 mm (0.20″) hole.

5.3.8 Fuel system monitoring

As vehicles accumulate mileage then so also do the components, sensors and actuators of the emissions control systems. Mass airflow sensors become dirty and their response slows with age. Exhaust gas oxygen sensors also respond slower as they are subject to the in-field failure modes such as oil and fuel contamination, thermal stress and general ageing. Fuel pressure regulators perform outside of their optimum capacity, fuel injectors become slower in their response, and partial blockages mean that they deliver less and sometimes more fuel than requested.

If this component ageing were not compensated for it would mean that the fuel system would not be able to maintain normal fuelling around stoichiometric air fuel ratio as shown in Figure 5.9. The end result would be the potential to

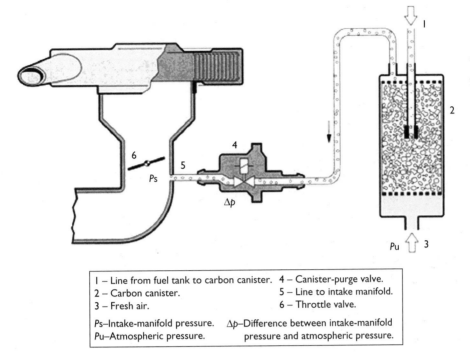

1 – Line from fuel tank to carbon canister.	4 – Canister-purge valve.
2 – Carbon canister.	5 – Line to intake manifold.
3 – Fresh air.	6 – Throttle valve.
P_s–Intake-manifold pressure.	Δp–Difference between intake-manifold pressure and atmospheric pressure.
P_u–Atmospheric pressure.	

Figure 5.9 Evaporative emissions control system (*Source*: Bosch)

exceed emission limits. A lambda value of one is required in order for the three-way catalyst to work. In addition to this, more severe fuelling errors would cause noticeable effects in the performance of the vehicle leading to customer complaints and potential to damage the manufacturer's brand image.

This compensation strategy is known as adaptive learning. A dedicated piece of software contained within the ECU learns these deviations away from stoichiometry, while the fuel system is in closed loop control. They are stored in a memory that is only reset when commanded by a technician and which is also robust to battery changes.

These memory-stored corrections are often termed 'long-term' fuel corrections. They are often stored in memory as a function of air mass, engine speed or engine load.

An exhaust gas oxygen sensor detects the amount of oxygen in the catalyst feed gas and the sensor produces a voltage, which is fed back to the microprocessor. This is then processed to determine the instantaneous or 'short-term' fuel correction to be applied. This is done in order to vary the fuel around stoichiometry and allow three-way catalysis to occur.

The microprocessor then calculates the amount of fuel required using an equation, which is shown in its most basic form below.

$$\text{Fuel mass} = \frac{\text{Air mass} \times \text{Long-term fuel trim}}{\text{Short-term fuel trim} \times 14.64}$$

Referring to Figures 5.10 and 5.11, it can be seen that when there is a component malfunction, which causes the AFR in the exhaust stream to be rich, then there is a need to adapt to this to bring the AFR back into the region of stoichiometry. The value of the long-term fuel trim correction must decrease because less fuel is required. Should the situation continue and the problem causing rich AFR becomes slowly worse, then the error adaption will continue with an ever-decreasing value for the long-term fuel trim being applied, learned and stored in memory.

The purpose of the fuel monitor is to determine when the amount of long-term adaptive correction has reached the point where the system can no longer cope. This is also where long-term fuel trim values reach a pre-defined or 'calibrated' limit at which no further adaption to error is allowed. This limit is calibrated to coincide with exhaust tailpipe emissions exceeding legislated levels. At this point, and when a short-term fuelling error exceeds another 'calibrated' limit, a DTC is stored and after consecutive drives the MI is illuminated. The opposite occurs, with extra fuel being added, via the long-term fuel trim parameter, should an error occur that causes the AFR at the exhaust gas oxygen sensor to be lean.

5.3.9 Misfire monitor

When an engine endures a period of misfire, at best tailpipe emissions will increase and at worst

Voltage characteristic of the Lambda-sensor signal

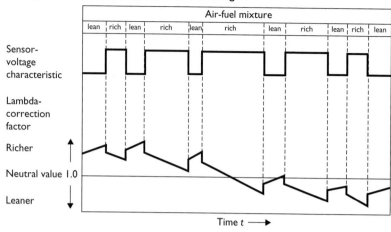

Figure 5.10 Rich AFR lambda sensor signal fuelling error (*Source*: Bosch)

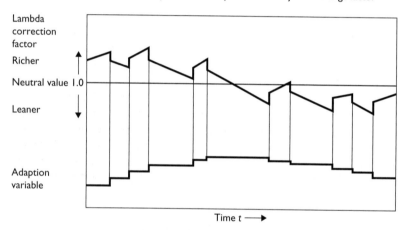

Figure 5.11 Adaptive fuel strategy in operation (*Source*: Bosch)

catalyst damage and even destruction can occur. When misfire occurs, the unburned fuel and air is discharged direct to the exhaust system where it passes directly through the catalyst.

Subsequent normal combustion events can combust this air/fuel charge in something akin to a bellows effect, which causes catalyst temperatures to rise considerably. Catalyst damage failure thresholds are package specific but are in the region of 1000°C. The catalyst itself is a very expensive service item whether replaced by the customer or the manufacturer under warranty.

The misfire monitor is responsible for determining when misfire has occurred, calculating the rate of engine misfire and then initiating some kind of protective action in order to prevent catalyst damage. The misfire monitor is in operation continuously within a 'calibrateable' engine speed/load window defined by the legislation (Figure 5.12). The USA requires misfire monitoring

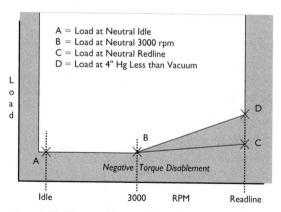

Figure 5.12 Misfire enablement window (Ford Motor Company)

throughout the revs range but European legislation requires monitoring only up to 4500 rev/min.

The crankshaft sensor generates a signal as the wheel rotates and the microprocessor processes

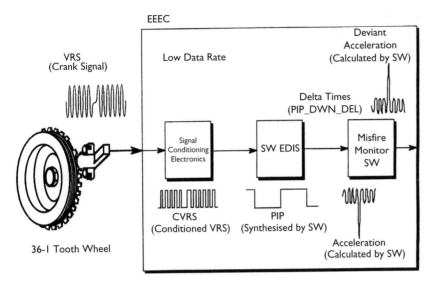

Figure 5.13 Crankshaft mounted wheel and sensor source of angular acceleration (*Source*: Ford Motor Company)

this signal to determine the angular acceleration of the crankshaft produced by each engine cylinder when a firing event occurs. When a misfire occurs the crankshaft decelerates and a cam position sensor identifies the cylinder that misfired.

Processing of the signal from the crank position sensor is not straightforward. A considerable amount of post-processing takes place to filter the signal and disable monitoring in unfavourable conditions. The misfire monitor must learn and cater for the differences in manufacturing tolerances of the crankshaft wheel and so has a specific sub-algorithm for learning these differences and allowing for them when calculating the angular acceleration of the crankshaft (Figure 5.13). These correction factors are calculated during deceleration, with the injectors switched off. They should be re-learned following driveline component changes such as flywheel, torque converter, crankshaft sensor, etc.

The misfire monitor must be able to detect two types of misfire:

- Type A misfire;
- Type B misfire.

A type A misfire is defined as that rate of misfire, which causes catalyst damage. When this occurs the MI will flash at a rate of 1 Hz and is allowed to stop flashing should the misfire disappear. The MI will stay on steady state should the misfire re-occur on a subsequent drive and the engine operating conditions are 'similar' i.e. engine speed is within 375 rev/min, engine load is within 20%, and the engine's warm-up status is the same as that under which the malfunction was first detected (and no new malfunctions have been detected).

The rate of misfire that will cause catalyst damage varies as a function of engine speed and load. Misfire rates in the region of 45% are required to damage a catalyst at neutral idle whilst at 80% engine load and 4000 rev/min misfire rates in the region of only 5% are needed (Figure 5.14).

A type B misfire is defined as that rate of misfire that will cause the tailpipe emissions to exceed legislated levels. This varies from vehicle to vehicle and is dependent upon catalyst package. MI operation is the same as for standard MI DTCs.

5.3.10 Exhaust gas recirculation monitor

As combustion takes place within the engine cylinders, nitrogen and oxygen combine to form various oxides of nitrogen; collectively these are termed NOx. NOx emissions can be reduced up to a certain point by enriching the air/fuel ratio beyond the point at which hydrocarbon (HC) and carbon monoxide (CO) emissions begin to increase. NOx emissions are generated as a function of combustion temperature, so another way to reduce these is to decrease the compression ratio which leads to other inefficiencies like poor fuel economy.

Most manufacturers employ an emissions control sub-system known as exhaust gas recirculation (EGR). This by definition recirculates some of the exhaust gases back into the normal intake charge. These 'combusted' gases cannot be burnt again so they act to dilute the intake charge. As a result, in-cylinder temperatures are reduced along with NOx emissions (Figure 5.15).

72 Advanced automotive fault diagnosis

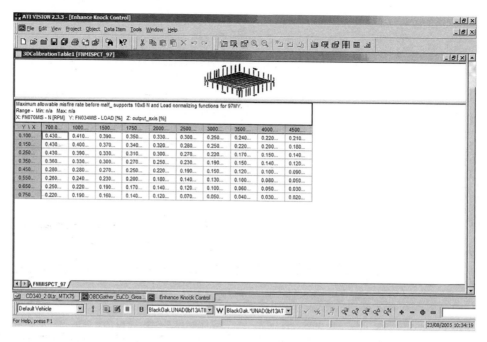

Figure 5.14 System development screen showing type A misfire rates normalised by engine speed and load (*Source*: Ford Motor Company)

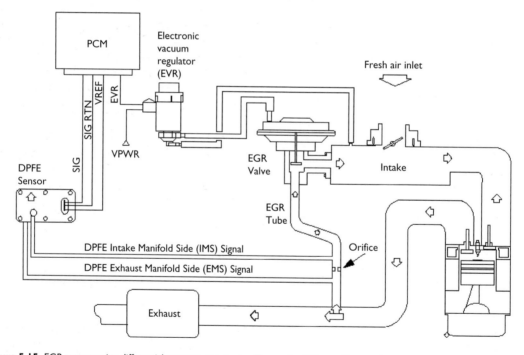

Figure 5.15 EGR system using differential pressure monitoring (*Source*: Ford Motor Company)

The exhaust gas recirculation system monitor is responsible for determining the serviceability of the sensors, hoses, valves and actuators that belong to the EGR system. Manufacturers employ systems that can verify that the requested amount of exhaust gas is flowing back into the engine intake. Methods can be both intrusive and non-intrusive, such as a change in manifold pressure as EGR is flowed and then shut off.

One method monitors AFR excursions after the EGR valve is opened and then closed as the AFR becomes lean. Another system employs a differential pressure scheme that determines the pressure both upstream and downstream of the

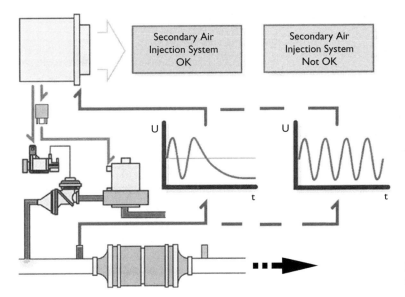

Figure 5.16 Secondary airflow diagnostic monitoring

exhaust to determine whether the requested flow rate is in effect. Yet another system employs a temperature sensor, which reports the change in temperature as EGR gases flow past the sensor. The temperature change will be mapped against the amount of EGR flowing so when an amount of EGR is requested then the flow rate is inferred by measuring the change in temperature.

5.3.11 Secondary air monitor

The exhaust system catalyst is not immediately operative following a start where the engine and exhaust system is cool. Temperature thresholds above which the catalyst is working, and three-way catalysis is occurring, vary as a function of the exhaust gas system package. Typically, this 'light off' point occurs at temperatures of about 260°C/ 500°F. Some manufacturers employ electrically heated catalysts to reach this temperature rapidly but these are expensive to manufacture and replace.

Most manufacturers rely on the exhaust gases as a source of heat in order to bring the catalyst up to light off temperature. When the vehicle is started from cold the AFR is rich; this is required to ensure a stable engine start for cold pull-away. From an emissions perspective, the impact is observed in the production of HC and CO in the exhaust stream because the exhaust system catalyst has not reached light off.

The secondary air system uses a pump, which adds more air into the exhaust stream at a point before the catalyst following a cold start. The secondary air combusts the HC in the catalyst, generating heat, which in turn, promotes light off and further emissions reduction.

Older systems support a belt driven mechanical pump with a bypass valve for when secondary airflow is not required. Modern vehicles employ an electric air pump operated by the engine management ECU (powertrain control module [PCM]) via relays.

The secondary air monitor is responsible for determining the serviceability of the secondary air system components (Figure 5.16). Most strategies monitor the electrical components and ensure the system pumps air when requested by the ECU. To check the airflow, the ECU observes the response of the exhaust gas oxygen sensor after it commands the fuel control system to enter open loop control and force the AFR to become rich. The secondary air pump is then commanded on and the ECU determines the time taken for the exhaust gas oxygen sensor to indicate a lean AFR. If this time exceeds a calibrated threshold then a diagnostic trouble code is stored.

5.3.12 Exhaust gas oxygen sensor monitor

The outside surface of the ceramic measuring tube protrudes into the exhaust gas flow, and the inner surface is in contact with the outside air (Figure 5.17).

At high temperatures a zirconia type sensor allows the passage of oxygen ions between the two platinum electrodes and a voltage is generated. The voltage is proportional to the relationship between the residual oxygen in the exhaust gas and that of the surrounding air.

When the AFR of the exhaust gas is rich there is less free oxygen at the outer electrode with more

74 Advanced automotive fault diagnosis

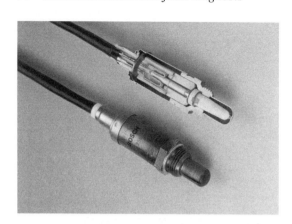

Figure 5.17 Exhaust gas oxygen sensor – zirconia type (*Source*: Bosch Press)

available at the inner electrode. This results in a positive voltage of between 0.6 V and 0.8 V being generated. When the AFR of the exhaust gas is lean there is little difference in the oxygen concentrations and a low voltage is produced. This voltage is processed by the microprocessor as part of a closed loop fuel control system which in-turn ensures three-way catalysis. The exhaust gas oxygen sensor monitor is responsible for determining the serviceability of all of the oxygen sensors and their heater elements. Manufacturers employ an algorithm similar to the component monitor to detect open circuits and other common shorts.

Additional diagnostics exist for when the sensor is 'stuck' lean or rich. The monitor waits for the sensor to 'switch' as it normally would and if this does not occur within a calibrated timeframe then a DTC is reported and MI illumination if the fault is apparent on consecutive drives.

Diagnostics also exist for when a sensor response is slowed. As the sensor ages it continues to switch but with a much reduced amplitude and frequency. When this occurs it induces fuelling errors. Manufacturers are obliged to report when the performance of the sensor has been degraded to the point where exhaust emissions exceed legislated limits.

5.4 On-board diagnostics – a second perspective

> *Authors note: The section on OBD repeats some of the information supplied previously but because this is such an important subject I decided that looking at it from two different view points would be useful. I am most grateful to Dave Rogers (www.autoelex.co.uk) for his excellent contribution to this chapter.*

5.4.1 Introduction

In the mid-1980s the US environmental protection agency (EPA) introduced a policy which made the use of on-board diagnostics (OBD) compulsory for vehicles in the United States. This was followed by similar requirements introduced by the California Air Resources Board (CARB), which promoted the development of technology such that the diagnostic technician would have access to information stored within the engine electronic control unit (ECU). This information, relating to faults that have occurred and that have been logged and stored in the ECU memory, significantly assists in fault diagnosis and rectification. The main reason for this legislation is the requirements to reduce exhaust emissions. The basic objectives are:

- to improve emissions compliance by alerting the driver immediately to a fault condition whilst driving;
- to assist repair/diagnostics technicians in identifying system faults and faulty components in the emission control system.

On-board diagnostic monitoring applies to systems which are most likely to cause an increase in harmful exhaust emission, namely:

- All main engine sensors
- Fuel system
- Ignition system
- Exhaust gas recirculation (EGR) system

The system uses information from sensors to judge the performance of the emission controls but these sensors do not directly measure the vehicle emissions.

An important part of the system, and the main driver information interface, is the 'check engine' warning light. Also known as the malfunction indicator light (MIL). This is the main source of feedback to the driver to indicate if an engine problem has occurred or is present. When a malfunction or fault occurs the warning light illuminates to alert the driver. Additionally the fault is stored in ECU memory. If normal condition is re-instated, the light extinguishes but the fault remains logged to aid diagnostics. Circuits are monitored for open or short circuits as well as plausibility. When a malfunction is detected, information about the malfunctioning component is stored.

An additional benefit allows the diagnostic technician to be able to access fault information and monitor engine performance via data streamed

1943: first smog alarm in LA
1950: 4,5 Mio. vehicles in california
1952: Dr. Arie Haagen-Smit analyses the reasons for smog development
1960: 8 Mio. vehicles in California
1961: introduction of crankcase ventilation (PCV)
1966: Federal Clean Air Act
1967: foundation of CARB, Chairman: Haagen-Smit
1970: foundation of EPA
1980: 17 Mio. vehicles in California
1988: CARB decides OBD II for 1994 MY
1990: number of smog days goes down, CARB decides LEV and ZEV – program
1995: 26 Mio. vehicles in California
1996: Ozon-pollution 59% below 1965, number of smog days 94% below 1975

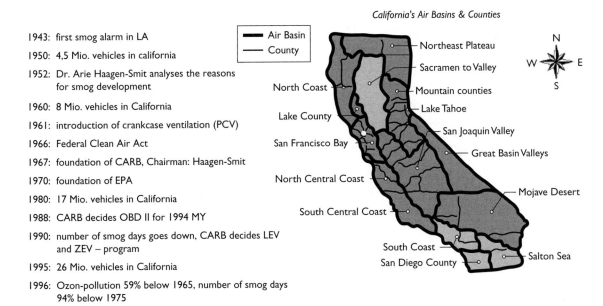

Figure 5.18 History of CARB Emission legislation activity

directly from the ECU whilst the engine is running (on certain vehicles). This information can be accessed via various scan tools available on the market and is communicated in a standardised format so one tool (more or less!) works with all vehicles. The data is transmitted in a digital form via this serial interface. Thus data values are transmitted as data words and the protocol used for this data stream has to be known in order to evaluate the information properly.

The benefits of having an OBD system are that it:

- encourages vehicle and engine manufacturers to have a responsible attitude to reducing harmful emissions from their engines via the development of reliable and durable emission control systems;
- aids diagnosis and repair of complex electronic engine and vehicle control systems;
- reduces global emissions by identifying and highlighting immediately to the driver or user emission control systems in need of repair;
- provides 'whole life' emission control of the engine;
- on-board diagnostics, or OBD, was the name given to the early emission control and engine-management systems introduced in cars. There was no single standard – each manufacturer often using quite different systems (even between individual car models). OBD systems have been developed and enhanced, in line with United States government requirements, into the current OBD2 standard. The OBD2 requirement applies to all cars sold in the United States from 1996. EOBD is the European equivalent of the American OBD2 standard, which applies to petrol cars sold in Europe from 2001 (and diesel cars three years later).

OBD2 (also OBDII) was developed to address the shortcomings of OBD1 and make the system more user friendly for service and repair technicians.

5.4.2 OBD2

Even though new vehicles sold today are cleaner than they have ever been, the millions of cars on the road and the ever increasing miles they travel each day make them our single greatest source of harmful emissions. While a new vehicle may start out with very low emissions, infrequent maintenance or failure of components can cause the vehicle emission levels to increase at an undesirable rate. OBD2 works to ensure that the vehicles remain as clean as possible over their entire life (Figure 5.19). The main features of OBD2 are, therefore, as follows.

- Malfunction of emission relevant components to be detected when emission threshold values are exceeded.
- Storage of failures and boundary conditions in the vehicle's fault memory.
- Diagnostic light (MIL – Malfunction Indicator Light) to be activated in case of failures.
- Read out of failures with generic scan tool.

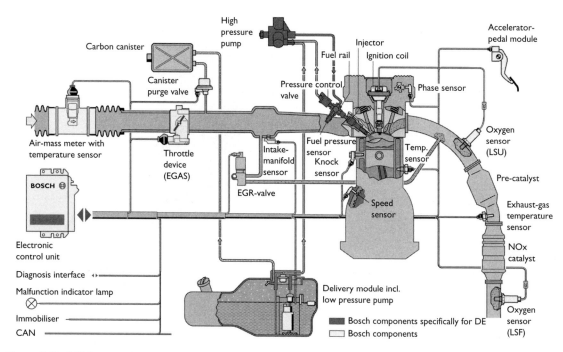

Figure 5.19 OBD2 system showing the main components of a gasoline direct injection system (*Source*: Bosch Press)

The increased power of micro controllers (CPUs) in ECUs has meant that a number of important developments could be added with the introduction of OBD2. These include Catalyst efficiency monitoring, misfire detection, canister purge and EGR flow rate monitoring. An additional benefit is the standardisation of diagnostic equipment interfaces.

For OBD1 each manufacturer applied specific protocols. With the introduction of OBD2 a standardised interface was developed with a standard connector for all vehicles, and a standardised theory for fault codes relating to the engine and powertrain (more about this later). This meant that generic scan tools could be developed and used in the repair industry by diagnostic technicians to aid troubleshooting of vehicle problems.

Another feature of OBD2 is that the prescribed thresholds at which a fault is deemed to have occurred are in relation to regulated emission limits. The basic monitor function is:

- monitoring of catalyst efficiency, engine misfire and oxygen sensors function such that crossing a threshold of 1.5 times the emission limit will record a fault;
- monitoring of the evaporation control system such that a leak greater than the equivalent leak from a 0.04 inch hole will record a fault.

The main features of an OBD2 compliant system (as compared to OBD1) are:

- pre and post-catalyst oxygen sensors to monitor conversion efficiency;
- much more powerful ECU with 32bit processor;
- ECU Map data held on EEPROMS such that they can be accessed and manipulated via an external link. No need to remove ECU from vehicle for software updates or tuning;
- more sophisticated EVAP system, can detect minute losses of fuel vapour;
- EGR systems with feedback of position/flow rate;
- sequential fuel injection with MAP (Manifold Air Pressure) and MAF (Mass Air Flow) sensing for engine load.

5.4.3 EOBD

EOBD is an abbreviation of European on-board diagnostics. All petrol/gasoline cars sold in Europe since 1 January 2001, and diesel cars manufactured from 2003, must have on-board diagnostic systems to monitor engine emissions. These systems were introduced in line with European directives to monitor and reduce emissions from cars. All such cars must also have a standard EOBD diagnostic socket that provides access to this system. The EOBD standard is similar to the US OBD2

Table C Emission limits table for comparison

Legislation	OBD Malfunction Limit (grams/km)			
	HC	CO	NOx	Pm
EPA	⩾1.5 times the applicable federal standard			
EPA – Method	Multiplicative relative to limits			
CARB 1 and 2	⩾1.5 times the relevant CARB emission limits			
CARB 1 and 2 – Method	Multiplicative relative to limits			
EOBD Positive Ign. 2000	0.40	3.20	0.60	–
EOBD Diesel 2003	0.40	3.20	1.20	0.18
EOBD Positive Ign. 2005	0.20	1.40	0.30	–
EOBD Diesel 2008 (for indication only)	0.30	2.40	0.90	0.14
EOBD – Method	Absolute limits			

Table D Cars not exceeding 2.5 tonnes laden

	Number of seats	Fuel	Directive	Limit values (gm/km)					Implementation dates	
				CO	HC	NOx	HC + NOx	PM	Type approval	In-use
Euro III	up to 9	P	98/69/EC	2.30	0.20	0.15	–	–	01/01/00	01/01/01
	up to 9	D	98/69/EC	0.64	–	0.50	0.56	0.05	01/01/00	01/01/01
	Note (i)	D	98/69/EC	0.80	–	0.65	0.72	0.07	01/01/01	01/01/02
	Note (ii)	D	98/69/EC	0.95	–	0.78	0.86	0.10	01/01/01	01/01/02
Euro IV	up to 9	P	98/69/EC	1.00	0.10	0.08	–	–	01/01/05	01/01/06
	up to 9	D	98/69/EC	0.50	–	0.25	0.30	0.025	01/01/05	01/01/06

Key: P – Petrol, D – Diesel, di – direct injection, CO – Carbon Monoxide, HC – Hydrocarbons, NOx – Oxides of Nitrogen, PM – Particulate mass
Note (i) – Temporary concession for diesel cars over 2.0 tonnes laden weight which are off-road or more than six seats (unladen weight from 1206 to 1660 kg). Concession ceased on 31/12/02.
Note (ii) – Temporary concession for diesel cars over 2.0 tonnes laden weight which are off-road or more than six seats (unladen weight over 1660 kg). Concession ceased on 31/12/02.
Source: Vehicle Certification Agency manual, May 2005.

Table E Heavy motor car (more than 2.5 tonnes laden or more than six seats). Unladen weight between 1151 and 1600 kg

	Number of seats	Fuel	Directive	Limit values (gm/km)					Implementation dates	
				CO	HC	NOx	HC + NOx	PM	Type approval	In-use
Euro III	up to 9	P	98/69/EC	4.17	0.25	0.18	–	–	01/01/01	01/01/02
	up to 9	D	98/69/EC	0.80	–	0.65	0.72	0.07	01/01/01	01/01/02
Euro IV	up to 9	P	98/69/EC	1.81	0.13	0.10	–	–	01/01/06	01/01/07
	up to 9	D	98/69/EC	0.63	–	0.33	0.39	0.04	01/01/06	01/01/07

Key: P – Petrol, D – Diesel, di – direct injection, CO – Carbon Monoxide, HC – Hydrocarbons, NOx – Oxides of Nitrogen, PM – Particulate mass

standard. In Japan, the JOBD system is used. The implementation plan for EOBD was as follows:

- January 2000 OBD for all new Petrol/Gasoline vehicle models;
- January 2001 OBD for all new Petrol/Gasoline vehicles;
- January 2003 OBD for all new Diesel vehicle models PC/LDV;
- January 2004 OBD for all new Diesel vehicles PC/LDV;
- January 2005 OBD for all new Diesel vehicles HDV.

The EOBD system is designed, constructed and installed in a vehicle such as to enable it to identify types of deterioration or malfunction over the entire life of the vehicle. The system must be

Table F Heavy motor car (more than 2.5 tonnes fully laden or more than six seats). Unladen weight over 600 kg

	Number of seats	Fuel	Directive	Limit values (gm/km)					Implementation Dates	
				CO	HC	NOx	HC + NOx	PM	Type Approval	In-use
Euro III	up to 9	P	98/69/EC	5.22	0.29	0.21	–	–	01/01/01	01/01/02
	up to 9	D	98/69/EC	0.95	–	0.78	0.86	0.10	01/01/01	01/01/02
Euro IV	up to 9	P	98/69/EC	2.27	0.16	0.11	–	–	01/01/06	01/01/07
	up to 9	D	98/69/EC	0.74	–	0.39	0.46	0.06	01/01/06	01/01/07

Key: P – Petrol, D – Diesel, di – direct injection, CO – Carbon Monoxide, HC – Hydrocarbons, NOx – Oxides of Nitrogen, PM – Particulate mass

Note: The test procedure for Euro III and Euro IV is more severe than that for Euro I and Euro II. This results in some emission levels having an apparent increase when in fact they are more tightly controlled.

designed, constructed and installed in a vehicle to enable it to comply with the requirements during conditions of normal use.

In addition, EOBD and OBD2 allow access to manufacturer-specific features available on some OBD2/EOBD compliant scan tools. This allows additional parameters or information to be extracted from the vehicle systems. These are in addition to the normal parameters and information available within the EOBD/OBD2 standard. These enhanced functions are highly specific and vary widely between manufacturers.

EOBD monitoring requirements on the vehicle

The monitoring capabilities of the EOBD system are defined for petrol/gasoline (spark ignition) and diesel (compression ignition) engines. The following is an outline.

Spark ignition engines

- Detection of the reduction in the efficiency of the catalytic converter with respect to emissions of HC only.
- The presence of engine misfires in the engine operation region within the following boundary conditions.
- Oxygen sensor deterioration.
- Other emission control system components or systems, or emission-related powertrain components or systems which are connected to a computer, the failure of which may result in tailpipe emission exceeding the specified limits.
- Any other emission-related powertrain component connected to a computer must be monitored for circuit continuity.
- The electronic evaporative emission purge control must, at a minimum, be monitored for circuit continuity.

Compression ignition engines

- Where fitted, reduction in the efficiency of the catalytic converter.
- Where fitted, the functionality and integrity of the particulate trap.
- The fuel injection system electronic fuel quantity and timing actuator(s) is/are monitored for circuit continuity and total function failure.
- Other emission control system components or systems, or emission-related powertrain components or systems which are connected to a computer, the failure of which may result in tailpipe emission exceeding the specified limits given. Examples of such systems or components are those for monitoring and control of air massflow, air volumetric flow (and temperature), boost pressure and inlet manifold pressure (and relevant sensors to enable these functions to be carried out).
- Any other emission-related powertrain component connected to a computer must be monitored for circuit continuity.

5.4.4 Features and technology of current systems

To avoid false detection, the legislation allows verification and healing strategies. These are outlined as follows.

MIL activation logic for detected malfunctions

To avoid wrong detections the legislation allows verification of the detected failure. The failure is stored in the fault memory as a pending code immediately after the first recognition but the MIL is not activated. The MIL will be illuminated in the third driving cycle, in which the failure has been detected; the failure is then recognised as a confirmed fault.

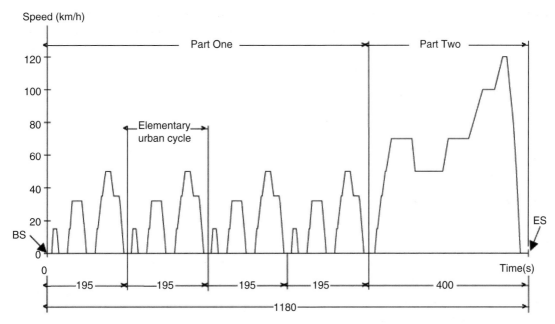

Figure 5.20 New European driving cycle (NEDC)

MIL healing

The MIL may be de-activated after three subsequent sequential driving cycles during which the monitoring system responsible for activating the MIL ceases to detect the malfunction and if no other malfunction has been identified that would independently activate the MIL.

Healing of the fault memory

The OBD system may erase a fault code, distance travelled and freeze-frame information if the same fault is not re-registered in at least forty engine warm-up cycles.

Freeze frame

This is a feature that can assist in the diagnosis of intermittent faults. Upon determination of the first malfunction of any component or system, 'freeze frame' engine conditions present at the time must be stored in the computer memory. Stored engine conditions must include, but are not limited to:

- calculated/derived load value;
- engine speed;
- fuel trim values (if available);
- fuel pressure (if available);
- vehicle speed (if available);
- coolant temperature;
- intake manifold pressure (if available);
- closed or open loop operation (if available);
- the fault code which caused the data to be stored.

5.4.5 OBD cycles

Directive 98/69/EC defines three different types of driving cycle:

1. Type 1 test (the NEDC, (Figure 5.20)) defined in Annex III, Appendix 1 of the directive;
2. A 'driving cycle consists of engine start-up, a driving mode where a malfunction would be detected if present, and engine shut off;
3. A 'warm-up cycle' means sufficient vehicle operation such that the coolant temperature has risen by at least 22°C from engine starting and reaches a minimum temperature of 70°C.

5.4.6 Monitors and readiness flags

An important part of the OBD system is the system monitors and associated readiness flags (Figure 5.21). These readiness flags indicate when a monitor is active. Certain monitors are continuous, for example, misfire and fuel system monitors.

Monitor status (Ready/Not Ready) indicates if a monitor has completed its self evaluation sequence. System monitors are set to 'Not Ready'

Insert | Remove | Advanced | ? | X

Calculated Engine Load	28.6	%
Catalyst Monitoring Status	Complete	
Comprehensive Component Monitoring Status	Complete	
EGR System Monitoring Status	Complete	
Engine Coolant Temperature	69	°C
Engine Speed	780	RPM
Fuel System Monitoring Status	Complete	
Fuel System Status Bank 1	CL-1	
Ignition Timing Advance	17.5	Deg
Intake Air Temperature	24	°C
Long Term Fuel Trim Bank 1	-3.1	%
Manifold Absolute Pressure (MAP)	27	kPA
Misfire Monitoring Status	Complete	
O2 Sensor - Bank 1 Sensor 1 (mV)	370	mV
O2 Sensor - Bank 1 Sensor 2 (mV)	150	mV
O2 Sensor Heater Monitoring Status	Complete	
O2 Sensor Monitoring Status	Complete	
OBD Requirements	EOBD	
Short Term Fuel Trim Bank 1	5.5	%
Short Term Fuel Trim from O2 Bank 1 Sensor 1	7	%
Short Term Fuel Trim from O2 Bank 1 Sensor 2	99.2	%
Throttle Position Angle	17.6	%
Vehicle Speed	0	MPH

Figure 5.21 System monitors (marked as 'Complete') and live data shown in scan tool

if cleared by scan tool and/or the battery is disconnected. Some of the monitors must test their components under specific, appropriate preconditions.

- The evaporative system monitor has temperature and fuel fill level constraints.
- The misfire monitor may ignore input on rough road surfaces to prevent false triggers.
- The oxygen sensor heater must monitor from a cold start.

Most other system monitors are not continuous and are only active under certain conditions. If these conditions are not fulfilled then the readiness flag for that monitor is set to 'not ready'. Until the readiness flags are set appropriately. It is not possible to perform a test of the OBD system and its associated components.

There is no universal drive cycle that is guaranteed to set all the system monitors appropriately for a test of the OBD system. Most manufacturers and even cars have their own specific requirements and irrespective of this, there are still some specific vehicles that have known issues when trying to set readiness flag status. As mentioned above, some allowance has been made for vehicles of model year 1996 to 2000 which can show two readiness flags to be 'not ready'. After this, 2001 onwards, one readiness flag is allowed to be 'not ready' prior to a test.

5.4.7 Fault codes

An integral feature of the OBD system is its ability to store fault codes relating to problems that

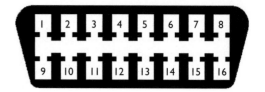

Figure 5.22 Sixteen Pin DLC OBD2/EOBD connector

occur with the engine electronic control system; particularly faults that could affect the emission control system and this is one of its primary functions. For the diagnostic technician this is a powerful feature which can clearly assist in locating and rectifying problems on the vehicle when they occur.

The diagnostic socket used by systems conforming to EOBD/OBD2 standards (Figure 5.22) should have the following pin configuration:

1. Ignition positive supply
2. Bus + Line, SAE J1850 (PWM)
3. Manufacturer's discretion
4. Chassis ground
5. Signal ground
6. CAN bus H
7. K Line
8. Manufacturer's discretion
9. Manufacturer's discretion
10. Bus – Line (PWM)
11. Manufacturer's discretion
12. Manufacturer's discretion
13. Manufacturer's discretion
14. CAN bus L
15. L line or second K line
16. Vehicle battery positive.

With the introduction of OBD2 and EOBD this feature was made even more powerful by making it more accessible. Standardisation of the interface connector known as the diagnostic link connector (DLC) and communication protocol allowed the development of generic scan tools, which could be used on any OBD compliant vehicle.

Most systems in use involve two types of codes. The generic codes associated with the OBD standard and manufacturer specific codes. The generic codes have a specific structure that allows easy identification and are common across all manufacturers (Figure 5.23). That is, a certain fault code means the same thing on any vehicle, irrespective of the manufacturer (Figure 5.24). All basic scan tools are capable of reading these generic codes.

The second type of code available is a manufacturer specific code. These are enhanced codes which can give additional information about the engine, vehicle and powertrain electronic

On-board diagnostics

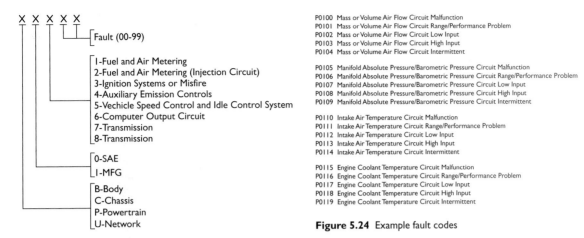

Figure 5.23 Structure of generic codes

Figure 5.24 Example fault codes

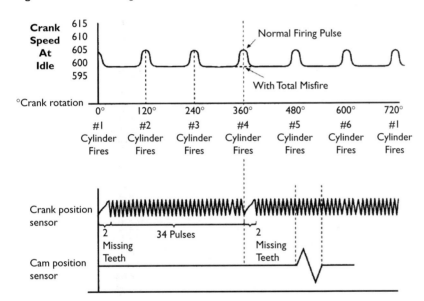

Figure 5.25 Misfire detection via crank sensor

systems and vary widely between manufacturers. Typically they will cover non-emission related components; for example, ABS, SRS, etc. In order to be able to use these enhanced codes it is necessary to have the relevant information available from the manufacturer (as to identification of the codes). In addition, the relevant software or firmware must be available on the scan device such that it can read these additional codes.

5.4.8 Misfire detection

A small engine misfire can raise emissions significantly. If engine misfire occurs in excess of about 17% of the time, permanent catalyst damage will occur. This clearly shows the importance of identifying misfire to keep engine emissions low (both short and medium term). The OBD system monitors individual cylinder misfires, counts these events, and integrates them over time to determine if this is a recurring problem which would cause an emission violation or just a single event. If the problem occurs over two consecutive trips, the MIL will illuminate to alert the driver.

A number of strategies for achieving this are available.

Crank speed fluctuation

A misfire event in a cylinder results in a lost power stroke. The gap in the torque output of the engine and a consequential momentary deceleration of the crankshaft can be detected using the crankshaft position sensor (Figure 5.25). By closely monitoring the speed and acceleration of the crankshaft misfiring cylinders can be detected. This technology is very commonly used in OBD systems to detect non-firing cylinders that can cause harmful emissions and catalyst damage.

There are a number of technical challenges that have to be overcome with this technique; the accuracy achieved and reliability of the system is very dependent on the algorithms used for signal processing and analysis. Under certain conditions, misfire detection can be difficult; particularly at light load with high engine speed. Under these conditions the damping of firing pulses is low due to the light engine load and this creates high momentary accelerations and decelerations of the crankshaft. This causes speed variation which can be mistakenly taken by the OBD system as a misfire. With this method of misfire detection, careful calibration of the OBD system is necessary to avoid false detection. Another vehicle operation mode that can cause problems is operation of the vehicle on rough or poorly made roads. This also causes rapid crankshaft oscillation that could activate false triggers – under these conditions the misfire detection must be disabled.

Ionising current monitoring

An ionisation current sensing ignition system consists of one ignition coil per cylinder, normally mounted directly above the spark plug (Figure 5.26). Eliminating the distributor and high-voltage leads helps promote maximum energy transfer to the spark plug to ignite the mixture. In this system the spark plug is not only used as a device to ignite the air/fuel mixture, but is also used as an in-cylinder sensor to monitor the combustion process. The operating principle used in this technology is that an electrical current flow in an ionised gas is proportional to the flame electrical conductivity. By placing a direct current bias across the spark plug electrodes, the conductivity can be measured (Figure 5.27). The spark current is used to create this bias voltage and this eliminates the requirement for any additional voltage source.

The ion current is monitored and if no ion generating flame is produced by the spark, no current flows through the measurement circuit during the working part of the cycle. The ion current vs. time trace is very different from that of a cycle when normal combustion occurs and this information can be used as a differentiator to detect misfire from normal combustion. This method has proven to be very effective at monitoring for misfires under test conditions and also in practice.

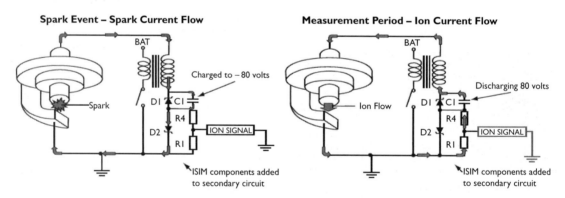

Figure 5.26 Ion sensing circuit in direct ignition system

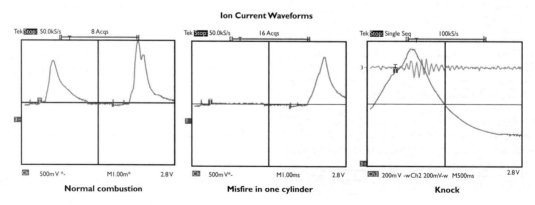

Figure 5.27 Resulting waveforms from the ion sensing system

The signal the system produces contains misfire information and in addition, can provide objective knock or detonation information. This can be used for engine control systems where knowledge of the actual combustion process is required (as mentioned above).

Cylinder pressure sensing

This technology has great potential not just for OBD applications but also for additional feedback to the engine management system about the combustion process due to the direct measurement technique. This additional control dimension can be utilised to improve engine performance and reduce emissions further. With respect to misfire detection, this method provides reliable detection of a positive combustion event and can easily detect misfire with utmost reliability (Figure 5.28).

The major drawback is the availability of suitable sensors that could be installed into the engine at production and would be durable enough to last the life of the engine and provide the required performance expected of sensors in an OBD system. For certain engine applications sensors are available, and currently combustion sensor technology is under rapid development such that this technical hurdle will soon be overcome.

Figure 5.28 Cylinder pressure sensor mounted in the engine

Exhaust pressure analysis

This solution involves using a pressure sensor in exhaust manifold combined with a Fourier analysis as the first stage of the signal processing. Using a sensor to analyse the gas pulses in the exhaust manifold, it is possible to detect single misfires. It is also possible to identify which cylinder is misfiring. This method is less intrusive than the above and could potentially be retrofitted at the production stage. A sensor in the exhaust can detect misfiring cylinders but cannot give useful, qualitative information about the combustion process. This technique has been demonstrated as capable of detecting all misfires at engine speeds up to 6000 rpm, for all engine configurations, loads, and fuels. Generally, a ceramic capacitive type sensor is been employed which has a short response time and good durability.

5.4.9 Testing vehicles for compliance

The manufacturer must demonstrate the correct function of the system to the appropriate authority. For EOBD compliance this requires three complete emission cycle runs (NEDC). This is known as a demonstration test.

A faulty component is installed or simulated which causes a violation of the emission limits; two preconditioning cycles are run and then one complete cycle to show that the error has been recorded and highlighted via illumination of the MIL. These phases are defined in EOBD legislation as:

- simulation of malfunction of a component of the engine management or emission control system;
- preconditioning of the vehicle with a simulated malfunction;
- driving the vehicle with a simulated malfunction over the type 1 test cycle (NEDC) and measuring the emissions of the vehicle;
- determining whether the OBD system reacts to the simulated malfunction and indicates malfunction in an appropriate manner to the vehicle driver.

Typical failure modes induced to be detected are:

- *Petrol/Gasoline Engines*
  - Replacement of the catalyst with a deteriorated or defective catalyst or electronic simulation of such a failure

- Engine misfire conditions according to the conditions for misfire monitoring given in
- Replacement of the oxygen sensor with a deteriorated or defective oxygen sensor or electronic simulation of such a failure
- Electrical disconnection of any other emission-related component connected to a powertrain management computer
- Electrical disconnection of the electronic evaporative purge control device (if equipped). For this specific failure mode, the type 1 test must not be performed

- *Diesel Engines*
  - Where fitted, replacement of the catalyst with a deteriorated or defective catalyst or electronic simulation of this condition
  - Where fitted, total removal of the particulate trap or, where sensors are an integral part of the trap, a defective trap assembly
  - Electrical disconnection of any fuelling system electronic fuel quantity and timing actuator
  - Electrical disconnection of any other emission related component connected to a powertrain management computer

Conditioning Run After Fault Rectification

If an error has occurred with a component and this error has been recorded by the OBD system, then (after the problem has been rectified) it is necessary to clear the fault code memory and test or condition the vehicle to ensure that:

- the fault has really been fixed and does not reoccur;
- the system is set up ready for correct future detection of any faults.

This can be done by putting the vehicle through drive cycle. A typical manufacturer defined drive cycle would consist of the following.

1. A cold start (coolant temperature less than 50°C, coolant and air temp within 11°C of each other).
2. Switch on ignition to allow oxygen sensor heating and diagnostics.
3. Idle engine for 2 minutes with typical electrical loads on (air conditioning and rear screen heater).
4. Turn off loads and accelerate to cruise at half throttle. The OBD system will check for misfire, fuel trim and EVAP (canister purge) systems.
5. Hold speed steady at cruise for 3 minutes. The OBD system monitors EGR, secondary air system, oxygen sensors and EVAP system.
6. Overrun/coast down to low speed (i.e. 20 mph) without using the brake or clutch. The OBD systems check EGR and EVAP systems.
7. Accelerate back up to cruise for 5 minutes at three quarter throttle. OBD checks misfire, fuel trim and EVAP.
8. Hold steady speed of cruise for 5 minutes. OBD monitors catalytic converter efficiency, misfire, fuel trim, oxygen sensors and EVAP systems.
9. Slow down to a stop without braking, OBD checks EGR and EVAP.

The system is now fully reset and ready for detection of new faults. The necessary drive cycle to guarantee reset of the whole system is manufacturer specific and should be checked appropriately.

Roadside test

An official in-service OBD2 emission test, as carried out in the USA by inspectors from the regulatory authority, consists of the following three parts (a likely European development therefore).

1. Check MIL function at ignition switch on.
2. Plug in OBD scanner, check monitor readiness. If monitors are not all showing as ready, the vehicle is rejected and further road testing is to be done in order to activate all the readiness flags. At this stage the scanner will also download any fault codes that are present.
3. An additional test, scanner command illumination of MIL via ECU to verify the correct function of the OBD system.

5.5 Summary

Clearly OBD is here to stay – and be developed. It should be seen as a useful tool for the technician as well as a key driver towards cleaner vehicles. The creating of generic standards has helped those of us at the 'sharp end' of diagnostics significantly.

OBD has a number of key emission related systems to 'monitor'. It saves faults in these systems in a standard form that can be accessed using a scan tool.

In the final chapter of this book there is a short discussion on OBD3 and ways in which it may be implemented in the future.

Knowledge check questions

To use these questions, you should first try to answer them without help but if necessary, refer back to the content of the chapter. Use notes, lists and sketches to answer them. It is not necessary to write pages and pages of text!

1. State the main reasons why OBD was developed.
2. Explain what is meant by OBD monitors and list the most common.
3. Describe how the P-codes are used to indicate faults.
4. Explain with the aid of a sketch, how the 'before and after cat' lambda sensor signals are used by the OBD system to monitor catalyst operation.
5. Explain what is meant by 'healing of the fault memory'.

6
Sensors and actuators

6.1 Introduction

Sensors and actuators have a chapter all to themselves because they are so important! And also because the issues and diagnostic techniques are common to many systems. For example, the testing procedure for an inductive engine speed sensor on a fuel injection system, is the same as for an inductive speed sensor on an ABS system.

Testing sensors to diagnose faults is usually a matter of measuring their output signal. In some cases the sensor will produce this on its own (an inductive sensor for example). In other cases, it will be necessary to supply the correct voltage to the device to make it work (Hall sensor for example).

> *Note:* It is normal to check that the vehicle circuit is supplying the voltage before proceeding to test the sensor.

After the description of most sensors, and actuators a table is included listing the sensor, equipment necessary, test method(s), results of the tests and a scope waveform. A waveform will be reproduced where appropriate, as this is often the recommended method of testing (see Chapter 4 for more details). The waveform shown will either be the output of a sensor or the signal supplied to an actuator.

> *Note:* Any figures given are average or typical values. Refer to a good reference source such as a workshop manual, or 'Autodata' for specific values.
> Sensor and actuator waveforms are shown in Chapter 7.

6.2 Sensors

6.2.1 Thermistors

Thermistors are the most common device used for temperature measurement on the motor vehicle. The principle of measurement is that a change in temperature will cause a change in resistance of the thermistor and hence an electrical signal proportional to the temperature being measured. Figure 6.1 shows a typical thermistor used as an engine coolant temperature sensor.

Most thermistors in common use are of the negative temperature coefficient (NTC) type. The actual response of the thermistors can vary but typical values for those used on motor vehicles will vary from several kΩ at 0°C to few hundred Ω at 100°C. The large change in resistance for a small change in temperature makes the thermistor ideal for most vehicles uses. It can also be easily tested with simple equipment.

Thermistors are constructed of semiconductor materials. The change in resistance with a change in temperature is due to the electrons being able to break free more easily at higher temperatures.

Sensor	Equipment	Method(s)	Results
Thermistor Coolant sensor Air intake temperature sensor Ambient temperature sensor etc.	Ohmmeter	Connect across the two terminals or, if only one, from this to earth	Most thermistors have a negative temperature coefficient (NTC). This means the resistance falls as temperature rises. A resistance check should give readings broadly as follows: 0°C = 4500 Ω 20°C = 1200 Ω 100°C = 200 Ω

6.2.2 Inductive sensors

Inductive type sensors are used mostly for measuring the speed and position of a rotating component.

They work on the very basic principle of electrical induction (a changing magnetic flux will induce an electromotive force in a winding). Figure 6.2

Sensors and actuators 87

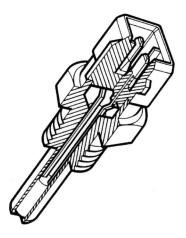

Figure 6.1 Coolant temperature sensor

shows a typical device used as a crankshaft speed and position sensor.

The output voltage of most inductive type sensors approximates to a sine wave. The amplitude of this signal depends on the rate of change of flux. This is determined mostly by the original design as in number of turns, magnet strength and the gap between the sensor and the rotating component. Once in use though the output voltage increases with the speed of rotation. In the majority of applications, it is the frequency of the signal that is used.

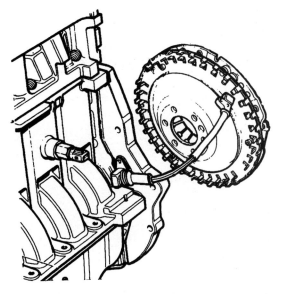

Figure 6.2 Inductive crankshaft sensor and reluctor wheel

6.2.3 Hall effect sensors

If a conductor is carrying a current in a transverse magnetic field then a voltage will be produced at right angles to the supply current. This voltage is proportional to the supply current and to the magnetic field strength.

Sensor	Equipment	Method(s)	Results
Inductive Crankshaft speed and position	Ohmmeter	A resistance test with the sensor disconnected	Values vary from about 200 to 400 Ω on some vehicles to 800 to 1200 Ω on others.
ABS wheel speed Camshaft position	AC voltmeter	AC voltage output with the engine cranking	The 'sine wave' output should be about 5 V (less depending on engine speed)

Many distributors employ Hall effect sensors. The output of this sensor is almost a square wave with constant amplitude. The Hall effect can also be used to detect current flowing in a cable, the magnetic field produced round the cable being proportional to the current flowing.

Hall effect sensors are becoming increasingly popular. This is partly due to their reliability but also to the fact that they directly produce a constant amplitude square wave in speed measurement applications and a varying DC voltage for either position sensing or current sensing.

Hall effect sensors are being used in place of inductive sensors for applications such as engine speed and wheel speed. The two main advantages are that measurement of lower (or even zero) speed is possible and that the voltage output of the sensors is independent of speed. Figure 6.3 shows a selection of Hall effect distributor.

Sensor	Equipment	Method(s)	Results
Hall effect Ignition distributor Engine speed Transmission speed Wheel speed Current flow in a wire (ammeter amp clamp)	DC voltmeter Logic probe Do *not* use an ohmmeter as this will damage the Hall chip	The voltage output is measured as the engine or component is rotated slowly The sensor is normally supplied with 10 to 12 V	This switches between 0 and about 8 V as the Hall chip is magnetised – or not A logic probe will read high and low as the sensor output switches

Figure 6.3 Hall effect distributor

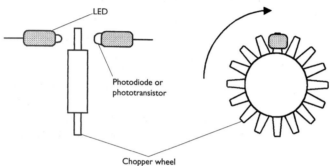

Figure 6.4 Optical sensor

6.2.4 Optical sensors

An optical sensor for rotational position is a relatively simple device. The optical rotation sensor and circuit shown in Figure 6.4 consists of a phototransistor as a detector and a light emitting diode light source. If the light is focused to a very narrow beam then the output of the circuit shown will be a square wave with frequency proportional to speed.

Sensor	Equipment	Method(s)	Results
Optical Ignition distributor Rotational speed	DC voltmeter	The device will normally be supplied with a stabilised voltage. Check the output wire signal as the device is rotated slowly	Clear switching between low and high voltage

6.2.5 Variable resistance

The two best examples of vehicle applications for variable resistance sensors are the throttle position sensor and the flap type airflow sensor. Whereas variable capacitance sensors are used to measure small changes, variable resistance sensors generally measure larger changes in position. This is due to lack of sensitivity inherent in the construction of the resistive track.

The throttle position sensor as shown in Figure 6.5 is a potentiometer in which, when supplied with a stable voltage, often 5 V, the voltage from the wiper contact will be proportional to throttle position. In many cases now the throttle potentiometer is used to indicate rate of change of throttle position. This information is used when implementing acceleration enrichment or inversely over run fuel cut off. The output voltage of a rotary potentiometer is proportional to its position.

The airflow sensor shown as Figure 6.6 works on the principle of measuring the force exerted on the flap by the air passing through it. A calibrated coil spring exerts a counter force on the flap such that the movement of the flap is proportional to the volume of air passing through the sensor. To reduce the fluctuations caused by individual induction strokes a compensation flap is connected to the sensor flap. The fluctuations therefore affect both flaps and are cancelled out. Any damage due to back firing is also minimised due to this design. The resistive material used for the track is a ceramic metal mixture, which is burnt into a ceramic plate at very high temperature. The slider potentiometer is calibrated such that the output voltage is proportional to the quantity of inducted air.

Sensor	Equipment	Method(s)	Results
Variable resistance Throttle potentiometer Flap type airflow sensor Position sensor	DC voltmeter	This sensor is a variable resistor. If the supply is left connected then check the output on a DC voltmeter	The voltage should change *smoothly* from about 0 to the supply voltage (often 5 V)
	Ohmmeter	With the supply disconnected, check the resistance	Resistance should change smoothly

6.2.6 Manifold absolute pressure (MAP) sensor (strain gauges)

When a strain gauge is stretched its resistance will increase and when it is compressed its resistance decreases. Most strain gauges consist of a thin layer of film that is fixed to a flexible backing sheet. This in turn is bonded to the part where strain is to be measured. Most resistance strain gauges have a resistance of about $100\,\Omega$.

Strain gauges are often used indirectly to measure engine manifold pressure. Figure 6.7 shows an arrangement of four strain gauges on a diaphragm

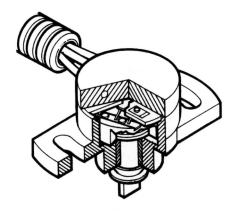

Figure 6.5 Throttle potentiometer

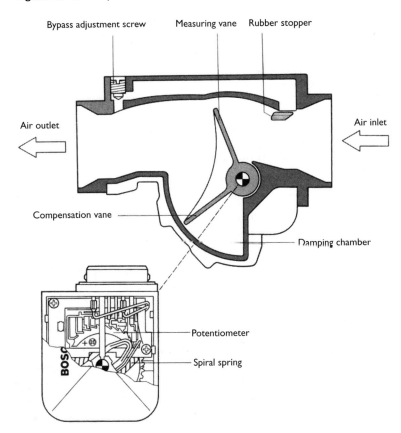

Figure 6.6 Air flow meter (vane type)

90 Advanced automotive fault diagnosis

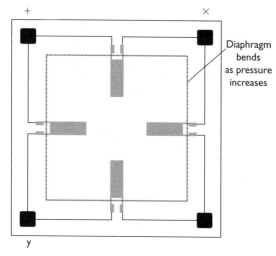

Figure 6.7 Strain gauge pressure sensor

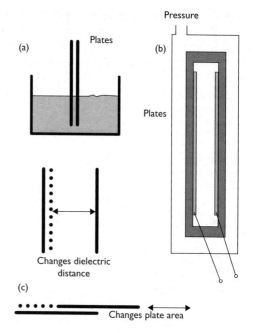

Figure 6.8 Variable capacitance sensors (a) liquid level, (b) pressure, (c) position

forming part of an aneroid chamber used to measure pressure. When changes in manifold pressure act on the diaphragm the gauges detect the strain. The output of the circuit is via a differential amplifier as shown, which must have a very high input resistance so as not to affect the bridge balance. The actual size of this sensor may be only a few millimetres in diameter. Changes in temperature are compensated for by using four gauges which when affected in a similar way cancel out any changes.

Sensor	Equipment	Method(s)	Results
Strain gauges MAP sensor Torque stress	DC voltmeter	The normal supply to an externally mounted MAP sensor is 5 V. Check the output as manifold pressure changes either by snapping the throttle open, road testing or using a vacuum pump on the sensor pipe	The output should change between about 0 and 5 V as the manifold pressure changes – as a general guide 2.5 V at idle speed

6.2.7 Variable capacitance

The value of a capacitor is determined by:

- surface area of its plates;
- distance between the plates;
- the dielectric (insulation between the plates).

Sensors can be constructed to take advantage of these properties. Three sensors each using the variable capacitance technique are shown in Figure 6.8. These are:

(a) **Liquid level sensor** the change in liquid level changes the dielectric value.

(b) **Pressure sensor** similar to the strain gauge pressure sensor but this time the distance between capacitor plates changes.

(c) **Position sensor** detects changes in the area of the plates.

An interesting sensor used to monitor oil quality is now available. The type shown in Figure 6.9 works by monitoring changes in the dielectric constant of the oil. This value increases as antioxidant additives in the oil deplete. The value rapidly increases if coolant contaminates the oil. The sensor output increases as the dielectric constant increases.

Sensor	Equipment	Method(s)	Results
Variable capacitance	DC voltmeter	Measure the voltage at the sensor	Small changes as the input to the sensor is varied – note this is difficult to assess because of very low capacitance values

Sensors and actuators **91**

Figure 6.9 Bosch oil quality sensor

Figure 6.10 Knock sensor

6.2.8 Knock sensors

A piezo-electric accelerometer is a seismic mass accelerometer using a piezo-electric crystal to convert the force on the mass due to acceleration into an electrical output signal. The crystal not only acts as the transducer but as the suspension spring for the mass. Figure 6.10 shows a typical accelerometer or knock sensor in use on a spark ignition vehicle.

The crystal is sandwiched between the body of the sensor and the seismic mass and is kept under compression by the bolt. Acceleration forces acting on the seismic mass cause variations in the amount of crystal compression and hence generate the piezo-electric voltage. The sensor, when used as an engine knock sensor, will also detect other engine vibrations. These are kept to a minimum by only looking for 'knock' a few degrees before and after top dead centre (TDC). Unwanted signals are also filtered out electrically.

Sensor	Equipment	Method(s)	Results
Accelerometer Knock sensors	Scope	Tap the engine block lightly (13 mm spanner) near the sensor	Oscillating output that drops back to zero. If the *whole* system is operating the engine will slow down if at idle speed

6.2.9 Hot wire airflow sensor

The advantage of this sensor is that it measures air mass flow. The basic principle is that as air passes over a hot wire it tries to cool the wire down. If a circuit is created such as to increase the current through the wire then this current will be proportional to the airflow. A resistor is also incorporated to compensate for temperature variations. The 'hot wire' is made of platinum and is only a few millimetres long and about 70 μm thick. Because of its small size the time constant of the sensor is very short, in fact in the order of a few milliseconds. This is a great advantage as any pulsations of the airflow will be detected and reacted to in a control unit accordingly. The output of the circuit involved with the hot wire sensor is a voltage across a precision resistor. Figure 6.11 shows a hot wire air mass sensor.

The resistance of the hot wire and the precision resistor are such that the current to heat the wire varies between 0.5 A and 1.2 A with different air mass flow rates. High resistance resistors are used in the other arm of the bridge and so current flow is very small. The temperature compensating resistor has a resistance of about 500 Ω which must remain constant other than by way of temperature change. A platinum film resistor is used for these reasons. The compensation resistor can cause the system to react to temperature changes within about three seconds.

The output of this device can change if the hot wire becomes dirty. Heating the wire to a very high temperature for one second every time the engine is switched off prevents this by burning off any contamination. In some air mass sensors a variable resistor is provided to set idle mixture.

The nickel film airflow sensor is similar to the hot wire system. Instead of a hot platinum wire a thin film of nickel is used. The response time of this system is even shorter than the hot wire. Figure 6.12 shows a nickel film airflow sensor.

The advantage which makes a nickel thick-film thermistor ideal for inlet air temperature sensing is its very short time constant. In other words its resistance varies very quickly with a change in air temperature.

Sensor	Equipment	Method(s)	Results
Hot wire Air flow	DC voltmeter or duty cycle meter	This sensor includes electronic circuits to condition the signal from the hot wire. The normal supply is either 5 or 12 V. Measure the output voltage as engine speed/load is varied	The output should change between about 0 and 5 V as the air flow changes. 0.4 to 1 V at idle is typical. Or depending on the system in use the output may be digital

6.2.10 Oxygen sensors

The vehicle application for an oxygen sensor is to provide a closed loop feedback system for engine management control of the air fuel ratio. The amount of oxygen sensed in the exhaust is directly related to the mixture strength or air fuel ratio. The ideal air fuel ratio of 14.7:1 by mass is known as a lambda (λ) value of one. Exhaust gas oxygen (EGO) sensors are placed in the exhaust pipe near to the manifold to ensure adequate heating. The sensors operate reliably at temperatures over 300°C. In some cases, a heating element is incorporated to ensure that this temperature is reached quickly. This type of sensor is known as a heated exhaust gas oxygen sensor or HEGO for short! The heating element (which consumes about 10 W) does not operate all the time to ensure that the sensor does not exceed 850°C at which temperature damage may occur to the sensor. This is why the sensors are not fitted directly in the exhaust manifold. Figure 6.13 shows an EGO sensor in position.

The main active component of most types of oxygen sensors is zirconium dioxide (ZrO_2). This ceramic is housed in gas permeable electrodes of platinum. A further ceramic coating is applied to the side of the sensor exposed to the exhaust gas as a protection against residue from the combustion process. The principle of operation is that at temperatures in excess of 300°C the zirconium dioxide will conduct the negative oxygen ions. The sensor is designed to be responsive very close to a lambda value of one. As one electrode of the sensor is open to a reference value of atmospheric air a greater quantity of oxygen ions will be present on this side. Due to electrolytic action these ions permeate the electrode and migrate through the electrolyte (ZrO_2). This builds up a charge rather like a battery. The size of the charge is dependent on the oxygen percentage in the exhaust.

The closely monitored closed loop feedback of a system using lambda sensing allows very accurate control of engine fuelling. Close control of emissions is therefore possible.

Figure 6.11 Hot wire mass airflow sensor

Figure 6.12 Hot film mass airflow sensor

Sensors and actuators 93

Sensor	Equipment	Method(s)	Results
Oxygen Lambda sensor EGO sensor HEGO sensor Lambdasonde	DC voltmeter	The lambda sensor produces its own voltage a bit like a battery. This can be measured with the sensor connected to the system	A voltage of about 450 mV (0.45 V) is the normal figure produced at lambda value of one. The voltage output, however, should vary smoothly between 0.2 and 0.8 V as the mixture is controlled by the ECU

6.2.11 Dynamic position sensors

A dynamic position or movement of crash sensor can take a number of forms; these can be described as mechanical or electronic. The mechanical system works by a spring holding a roller in a set position until an impact (acceleration/deceleration) above a predetermined limit provides enough force to overcome the spring and the roller moves, triggering a micro switch. The switch is normally open with a resistor in parallel to allow the system to be monitored. Two switches similar to this may be used to ensure that an air bag is deployed only in the case of sufficient frontal impact. Figure 6.14 is a further type of dynamic position sensor. Described as an accelerometer it is based on strain gauges.

There are two types of piezo-electric crystal accelerometer, one much like an engine knock sensor and the other using spring elements. A severe change in speed of the vehicle will cause an output from these sensors as the seismic mass moves or the springs bend. See Section 6.2.8 on knock sensors for further details.

> **Warning**
> For safety reasons, it is *not* recommended to test a sensor associated with an air bag circuit without specialist knowledge and equipment.

Sensor	Equipment	Method(s)	Results
Acceleration switch Dynamic position	DC voltmeter	Measure the supply and output as the sensor is subjected to the required acceleration	A clear switching between say 0 and 12 V

6.2.12 Rain sensor

Rain sensors are used to switch on wipers automatically. Most work on the principle of reflected light. The device is fitted inside the windscreen and light from an LED is reflected back from the outer surface of the glass. The amount of light reflected changes if the screen is wet, even with a few drops of rain. Figure 6.15 shows a typical rain sensor.

Sensor	Equipment	Method(s)	Results
Rain!	DC voltmeter	Locate output wire – by trial and error if necessary and measure dry/wet output (splash water on the screen with the sensor correctly fitted in position)	A clear switching between voltage levels

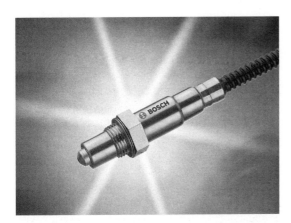

Figure 6.13 Wide band lambda sensor (*Source*: Bosch Press)

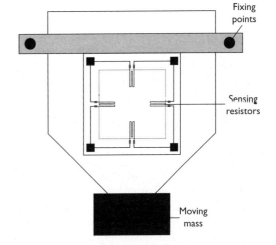

Figure 6.14 Strain gauges accelerometer

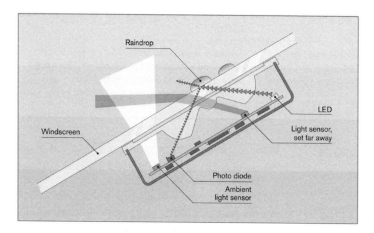

Figure 6.15 Rain sensor (*Source*: Bosch)

6.3 Actuators

6.3.1 Introduction

There are many ways of providing control over variables in and around the vehicle. 'Actuators' is a general term used here to describe a control mechanism. When controlled electrically they will either work by the thermal or magnetic effect. In this section the term actuator will be used to mean a device which converts electrical signals into mechanical movement.

6.3.2 Testing actuators

Testing actuators is simple as many are operated by windings. The resistance can be measured with an ohmmeter. Injectors, for example, often have a resistance of about 16 Ω. A good tip is that where an actuator has more than one winding (stepper motor for example), the resistance of each should be about the same. Even if the expected value is not known, it is likely that if the windings all read the same then the device is in order.

With some actuators, it is possible to power them up from the vehicle battery. A fuel injector should click for example and a rotary air bypass device should rotate about half a turn. Be careful with this method as some actuators could be damaged. At the very least use a fused supply (jumper) wire.

6.3.3 Solenoid actuators

The basic operation of solenoid actuators is very simple. The term 'solenoid' means: 'many coils of wire wound onto a hollow tube'. This is often misused but has become so entrenched that terms like 'starter solenoid', when really it is a starter relay, are in common use.

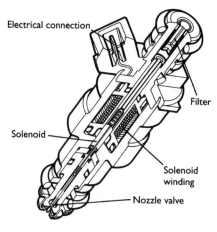

Figure 6.16 Fuel injector

A good example of a solenoid actuator is a fuel injector. Figure 6.16 shows a typical example. When the windings are energised the armature is attracted due to magnetism and compresses the spring. In the case of a fuel injector, the movement is restricted to about 0.1 mm. The period that an injector remains open is very small; under various operating conditions between 1.5 and 10 ms is typical. The time it takes an injector to open and close is also critical for accurate fuel metering. Some systems use ballast resistors in series with the fuel injectors. This allows lower inductance and resistance operating windings to be used, thus speeding up reaction time.

Other types of solenoid actuators, for example door lock actuators, have less critical reaction times. However, the basic principle remains the same.

6.3.4 Motorised and solenoid actuators

Permanent magnet electric motors are used in many applications and are very versatile. The output of a

Actuator	Equipment	Method(s)	Results
Solenoid Fuel injector Lock actuator	Ohmmeter	Disconnect the component and measure its resistance	The resistance of commonly used injectors is about 16 Ω. Lock and other actuators may have two windings (e.g. lock and unlock). The resistance values are very likely to be the same

motor is of course rotation, and this can be used in many ways. If the motor drives a rotating 'nut' through which a plunger is fitted on which there is a screw thread, the rotary action can easily be converted to linear movement. In most vehicle applications the output of the motor has to be geared down, this is to reduce speed and increase torque. Permanent magnet motors are almost universally used now in place of older and less practical motors with field windings. Some typical examples of the use of these motors are listed as follows:

- windscreen wipers;
- windscreen washers;
- headlight lift;
- electric windows;
- electric sun roof;
- electric aerial operation;
- seat adjustment;
- mirror adjustment;
- headlight washers;
- headlight wipers;
- fuel pumps;
- ventilation fans.

One disadvantage of simple motor actuators is that no direct feedback of position is possible. This is not required in many applications; however, in cases such as seat adjustment when a 'memory' of the position may be needed, a variable resistor type sensor can be fitted to provide feedback. Two typical motor actuators are shown in Figure 6.17. They are used as window lift motors. Some of these use Hall effect sensors or an extra brush as a feedback device.

A rotary idle actuator is a type of motor actuator. This device is used to control idle speed by controlling air bypass. There are two basic types in common use. These are single winding types, which have two terminals, and double winding types, which have three terminals. Under ECU control the motor is caused to open and close a shutter controlling air bypass. These actuators only rotate about 90° to open and close the valve. As these are permanent magnet motors the 'single or double windings' refers to the armature.

The single winding type is fed with a square wave signal causing it to open against a spring and then close again, under spring tension. The on/off ratio or duty cycle of the square wave will determine the average valve open time and hence idle speed.

With the double winding type the same square wave signal is sent to one winding but the inverse signal is sent to the other. As the windings are wound in opposition to each other if the duty cycle is 50% then no movement will take place. Altering the ratio will now cause the shutter to move in one direction or the other.

A further type is an on/off solenoid where a port is opened or closed as the solenoid is operated. This type is shown in Figure 6.18.

Actuator	Equipment	Method(s)	Results
Motor See previous list	Battery supply (fused) Ammeter	Most 'motor' type actuators can be run from a battery supply after they are disconnected from the circuit. If necessary the current draw can be measured	Normal operation with current draw appropriate to the 'work' done by the device. For example, a fuel pump motor may draw up to 10 A, but an idle actuator will only draw 1 or 2 A

Actuator	Equipment	Method(s)	Results
Solenoid actuator (idle speed control)	Duty cycle meter	Most types are supplied with a variable ratio square wave	The duty cycle will vary as a change is required

Figure 6.17 Window lift motor and wiper motors

6.3.5 Stepper motors

Stepper motors are becoming increasingly popular as actuators in the motor vehicle. This is mainly because of the ease with which they can be controlled by electronic systems.

Stepper motors fall into three distinct groups, the basic principles of which are shown in Figure 6.19.

- variable reluctance motors;
- permanent magnet (PM) motors;
- hybrid motors.

The underlying principle is the same for each type. All of them have been and are being used in various vehicle applications.

The basic design for a PM stepper motor comprises two double stators. The rotor is often made of barium-ferrite in the form of a sintered annular magnet. As the windings are energised in one direction then the other the motor will rotate in 90° steps. Half step can be achieved by switching on two windings. This will cause the rotor to line up with the two stator poles and implement a half step of 45°. The direction of rotation is determined by the order in which the windings are switched on or off or reversed.

The main advantages of a stepper motor are that feedback of position is not required. This is because the motor can be indexed to a known starting point and then a calculated number of steps will move the motor to any suitable position.

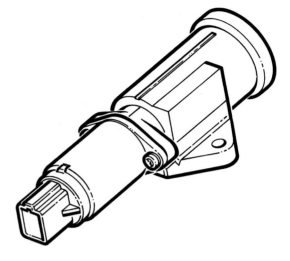

Figure 6.18 Solenoid idle actuator

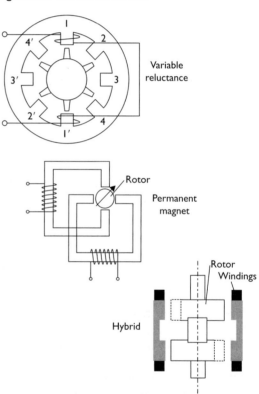

Figure 6.19 Basic principle of variable reluctance permanent magnet and hybrid stepper motor

Actuator	Equipment	Method(s)	Results
Stepper motor Idle speed air bypass Carburettor choke control Speedometer drivers	Ohmmeter	Test the resistance of each winding with the motor disconnected from the circuit	Winding resistances should be the same. Values in the region of 10 to 20 Ω are typical

6.3.6 Thermal actuators

An example of a thermal actuator is the movement of a traditional type fuel or temperature gauge needle. A further example is an auxiliary air device used on many earlier fuel injection systems. The principle of the gauge is shown in

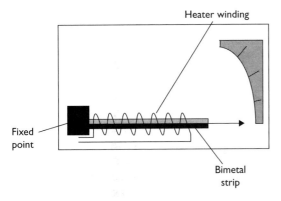

Figure 6.20 Thermal actuator principle

Figure 6.20. When current is supplied to the terminals, a heating element operates and causes a bimetallic strip to bend, which moves the pointer.

The main advantage of this type of actuator, when used as an auxiliary device, apart from its simplicity, is that if it is placed in a suitable position its reaction time will vary with the temperature

Figure 6.21 Rotary valve EGR–Delphi

of its surroundings. This is ideal for applications such as fast idle or cold starting control where, once the engine is hot, no action is required from the actuator.

Actuator	Equipment	Method(s)	Results
Thermal Auxiliary air device Instrument display	Ohmmeter Fused battery supply	Check the winding for continuity. If OK power up the device and note its operation (for instruments, power these but use a resistor in place of the sender unit)	Continuity and slow movement (several seconds to a few minutes) to close the valve or move as required

6.3.7 Exhaust gas recirculation (EGR) valve

Various types of EGR valve are in use based on simple solenoid operation. One development in actuator technology is the rotary electric exhaust gas recirculation (EEGR) valve for use in diesel engine applications (Lucas Varity). This device is shown in Figure 6.21. It has a self-cleaning action, accurate gas flow control and a fast reaction speed. The waveform shown in the table is not necessarily from this type of valve.

Actuator	Equipment	Method(s)	Results
EGR valve	Ohmmeter Fused battery supply	Check the winding(s) for continuity. If OK power up the device and note its operation	Continuity and rapid movement to close the valve

> *Knowledge check questions*
>
> To use these questions, you should first try to answer them without help but if necessary, refer back to the content of the chapter. Use notes, lists and sketches to answer them. It is not necessary to write pages and pages of text!
>
> 1. Explain how a knock sensor operates and why it is used.
> 2. Describe how to test the operation of a Hall sensor using multimeter.
> 3. List in a logical sequence, how to diagnose a fault with one fuel injector on a V6 multipoint system.
> 4. Outline two methods of testing the operation of a sensor that uses a variable resistor (throttle pot or vane type airflow sensors for example).
> 5. Explain with the aid of a sketch, what is meant by 'duty cycle' in connection with an idle speed control valve.

7
Engine systems

7.1 Introduction

The main sections in this chapter that relate to an area of the vehicle start with an explanation of the particular system. The sections then conclude with appropriate diagnostic techniques and symptom charts. Extra tests and methods are explained where necessary.

7.2 Engine operation

7.2.1 Four stroke cycle

Figure 7.1 shows a modern vehicle engine. Engines like this can seem very complex at first but keep in mind when carrying out diagnostic work that, with very few exceptions, all engines operate on the four stroke principle. The complexity is in the systems around the engine to make it operate to its maximum efficiency or best performance. With this in mind then, back to basics!

The engine components are combined to use the power of expanding gas to drive the engine.

When the term 'stroke' is used it means the movement of a piston from top dead centre (TDC) to bottom dead centre (BDC) or the other way round. The following table explains the spark ignition (SI) and compression ignition (CI) four stroke cycles – for revision purposes! Figure 7.2 shows the SI cycle.

Stroke	Spark ignition	Compression ignition
Induction	The fuel air mixture is forced into the cylinder through the open inlet valve because as the piston moves down it makes a lower pressure. It is acceptable to say the mixture is drawn into the cylinder	Air is forced into the cylinder through the open inlet valve because as the piston moves down it makes a lower pressure. It is acceptable to say the air is drawn into the cylinder
Compression	As the piston moves back up the cylinder the fuel air mixture is compressed to about an eighth of its original volume because the inlet and exhaust valves are closed. This is a	As the piston moves back up the cylinder the fuel air mixture is compressed in some engines to about a sixteenth of its original volume because the inlet and

Figure 7.1 The BMW 2.8 litre six cylinder engine uses variable camshaft control

Engine systems 99

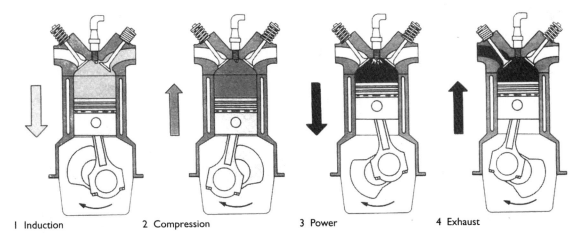

1 Induction 2 Compression 3 Power 4 Exhaust

Figure 7.2 Four stroke cycle

	compression ratio of 8:1, which is typical for many normal engines	exhaust valves are closed. This is a compression ratio of 16:1, which causes a large build up of heat
Power	At a suitable time before top dead centre, a spark at the plug ignites the compressed mixture. The mixture now burns very quickly and the powerful expansion pushes the piston back down the cylinder. Both valves are closed	At a suitable time before top dead centre, very high pressure atomised diesel fuel is injected into the combustion chamber. The mixture burns very quickly and the powerful expansion pushes the piston back down the cylinder. The valves are closed
Exhaust	The final stroke occurs as the piston moves back up the cylinder and pushes the spent gases out of the now open exhaust valve	The final stroke occurs as the piston moves back up the cylinder and pushes the spent gases out of the now open exhaust valve

7.2.2 Cylinder layouts

Great improvements can be made to the performance and balance of an engine by using more than one cylinder. Once this is agreed the actual layout of the cylinders must be considered. The layout can be one of three possibilities as follows.

- **In-line or straight** The cylinders are in a straight line. They can be vertical, inclined or horizontal.
- **Vee** The cylinders are in two rows at a set angle. The actual angle varies but is often 60° or 90°.
- **Opposed** The cylinders are in two rows opposing each other and are usually horizontal.

By far the most common arrangement is the straight four and this is used by all manufacturers in their standard family cars. Larger cars do, however, make use of the 'Vee' configuration. The opposed layout whilst still used is less popular. Engine firing order is important. This means the order in which the power strokes occur. The figures above have the cylinders numbered, and listed by each are some common firing orders. It is important to check in the workshop manual or data book when working on a particular engine.

7.2.3 Camshaft drives

The engine drives the camshaft in one of three ways, gear drive, chain drive or by a drive belt. The last of these is now the most popular as it tends to be simpler and quieter. Note in all cases that the cam is driven at half the engine speed. This is done by the ratio of teeth between the crank and cam cogs which is 1:2, for example 20 crank teeth and 40 cam teeth.

- **Camshaft drive gears** Gears are not used very often on petrol engines but are used on larger diesel engines. They ensure a good positive drive from the crankshaft gear to the camshaft.
- **Camshaft chain drive** Chain drive is still used but was even more popular a few years ago. The problems with it are that a way must be found to tension the chain and also provide lubrication.
- **Camshaft drive belt** Camshaft drive belts have become very popular. The main reasons for this are that they are quieter, do not need lubrication and are less complicated. They do break now and then but this is usually due to lack of servicing. Cam belts should be renewed at set intervals. Figure 7.3 shows an example of the data available relating to camshaft drive

belt fitting. This is one of the many areas where data is essential for diagnostic checks.

7.2.4 Valve mechanisms

A number of methods are used to operate the valves. Three common types are shown as Figure 7.4 and a basic explanation of each follows.

- **Overhead valve with push rods and rockers** The method has been used for many years and although it is not used as much now, many vehicles still on the road are described as overhead valve (OHV). As the cam turns it moves the follower, which in turn pushes the push rod. The push rod moves the rocker, which pivots on the rocker shaft and pushes the valve open. As the cam moves further it allows the spring to close the valve.
- **Overhead cam with followers** Using an overhead cam (OHC) reduces the number of moving parts. In the system shown here the lobe of the cam acts directly on the follower which pivots on its adjuster and pushes the valve open.
- **Overhead cam, direct acting and automatic adjusters** Most new engines now use an OHC with automatic adjustment. This saves on repair and service time and keeps the cost to the customer lower. Systems vary between manufacturers, some use followers and some have the cam acting directly on to the valve. In each case though the adjustment is by oil pressure. A type of plunger, which has a chamber where oil can be pumped under pressure, operates the valve. This expands the plunger and takes up any unwanted clearance.

Valve clearance adjustment is very important. If it is too large the valves will not open fully and will be noisy. If the clearance is too small the valves will not close and no compression will be possible. When an engine is running the valves become very hot and therefore expand. The exhaust valve clearance is usually larger than the inlet, because it gets hotter. Regular servicing is vital for all components but, in particular, the valve operating mechanism needs a good supply of clean oil at all times.

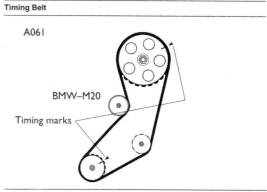

Figure 7.3 Timing belt data is essential

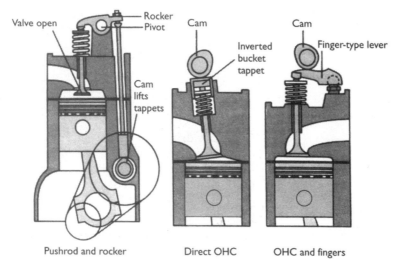

Figure 7.4 Valve operating mechanisms

7.2.5 Valve and ignition timing

Valve timing is important. The diagram of Figure 7.5 shows accurately the degrees of rotation of the crankshaft where the inlet and exhaust valves open and close during the four stroke cycle. The actual position in the cycle of operation when valves open and close depends on many factors and will vary slightly with different designs of engine. Some cars now control valve timing by electronics. The diagram is marked to show what is meant by valve lead, lag and overlap. Ignition timing is marked on the diagram. Note how this changes as engine speed changes.

The valve timing diagram shows that the valves of a four stroke engine open just before and close just after the particular stroke. Looking at the timing diagram, if you start at position IVO, the piston is nearly at the top of the exhaust stroke when the inlet valve opens (IVO). The piston reaches the top and then moves down on the intake stroke. Just after starting the compression stroke the inlet valve closes (IVC). The piston continues upwards and, at a point several degrees before top dead centre, the spark occurs and starts the mixture burning.

The maximum expansion is 'timed' to occur after top dead centre, therefore the piston is pushed down on its power stroke. Before the end of this stroke the exhaust valve opens (EVO). Most of the exhaust gases now leave because of their very high pressure. The piston pushes the rest of the spent gases out as it moves back up the cylinder. The exhaust valve closes (EVC) just after the end of this stroke and the inlet has already opened, ready to start the cycle once again.

The reason for the valves opening and closing like this is that it makes the engine more efficient by giving more time for the mixture to enter and the spent gases to leave. The outgoing exhaust gases in fact help to draw in the fuel air mixture from the inlet. Overall this makes the engine have a better 'volumetric efficiency'.

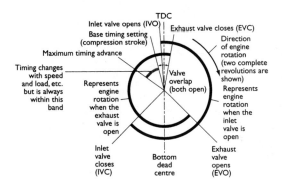

Figure 7.5 Valve and ignition timing

7.3 Diagnostics – engines

7.3.1 Systematic testing example

If the reported fault is excessive use of engine oil proceed as follows.

1. Question the customer to find out how much oil is being used.
2. Examine the vehicle for oil leaks and blue smoke from the exhaust.
3. For example, oil may be leaking from a gasket or seal – if no leaks are found the engine may be burning the oil.
4. A compression test, if the results were acceptable, would indicate a leak to be the most likely fault. Clean down the engine and run for a while, the leak might show up.
5. For example change the gasket or seals.
6. Run a thorough inspection of vehicle systems, particularly those associated with the engine. Double check that the fault has been rectified and that you have not caused any other problems.

7.3.2 Test equipment

Note: You should always refer to the manufacturer's instructions appropriate to the equipment you are using.

Compression tester

With this device the spark plugs are removed and the tester screwed or held in to each spark plug hole in turn. The engine is cranked over by the starter and the gauge will read the compression or pressure of each cylinder.

Cylinder leakage tester

A leakage tester uses compressed air to pressurise each cylinder in turn by a fitting to the spark plug hole. The cylinder under test is set to TDC compression. The percentage of air leaking out and where it is leaking from helps you determine the engine condition. For example if air is leaking through the exhaust pipe then the exhaust valves are not sealing. If air leaks into the cooling system then a leak from the cylinder to the water jacket may be the problem (blown head gasket is possible). Figure 7.6 shows a selection of snap-on diagnostic gauges – vacuum, compression and leakage.

7.3.3 Test results

Some of the information you may have to get from other sources such as data books, or a workshop manual is listed in the following table.

Test carried out	Information required
Compression test	Expected readings for the particular engine under test. For example the pressure reach for each cylinder may be expected to read 800 kPa ± 15%
Cylinder leakage test	The percentage leak that is allowed for the tester you are using; some allow about 15% leakage as the limit

7.3.4 Engine fault diagnosis table 1

Symptom	Possible causes or faults	Suggested action
Oil consumption	Worn piston rings and/or cylinders	Engine overhaul
	Worn valve stems, guides or stem oil seals	Replace valves (guides if possible) and oil seals
Oil on engine or floor	Leaking gaskets or seals	Replace appropriate gasket or seal
	Build up of pressure in the crankcase	Check engine breather system
Mechanical knocking noises	Worn engine bearings (big ends or mains for example)	Replace bearings or overhaul engine. Good idea to also check the oil pressure
	Incorrect valve clearances or defective automatic adjuster	Adjust clearances to correct settings or replace defective adjuster
	Piston slap on side of cylinder	Engine overhaul required now or quite soon
Vibration	Engine mountings loose or worn	Secure or renew
	Misfiring	Check engine ancillary systems such as fuel and ignition

7.3.5 Engine fault diagnosis table 2

Please note that this section covers related engine systems as well as the engine itself.

Symptom	Possible cause
Engine does not rotate when trying to start	Battery connection loose or corroded
	Battery discharged or faulty
	Broken loose or disconnected wiring in the starter circuit
	Defective starter switch or automatic gearbox inhibitor switch

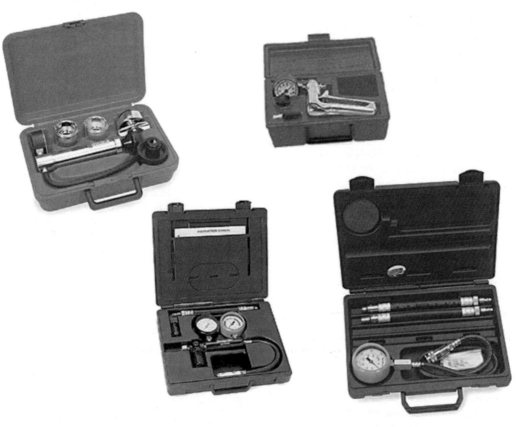

Figure 7.6 Diagnostic gauges

	Starter pinion or flywheel ring gear loose		Incorrect plugs or plug gaps
			HT leads breaking down
	Earth strap broken. Loose or corroded	Engine stalls	Idle speed incorrect
			CO setting incorrect
Engine rotates but does not start	No fuel in the tank!		Fuel filter blocked
	Discharged battery (slow rotation)		Air filter blocked
	Battery terminals loose or corroded		Intake air leak
	Air filter dirty or blocked		Idle control system not working
	Low cylinder compressions	Lack of power	Fuel filter blocked
	Broken timing belt		Air filter blocked
	Damp ignition components		Ignition timing incorrect
	Fuel system fault		Low fuel pump delivery
	Spark plugs worn to excess		Uneven or low cylinder compressions (maybe valves)
	Ignition system open circuit		Fuel injectors blocked
Difficult to start when cold	Discharged battery (slow rotation)		Brakes binding or clutch slipping
	Battery terminals loose or corroded	Backfires	Incorrect ignition timing
	Air filter dirty or blocked		Incorrect valve timing (cam belt not fitted correctly)
	Low cylinder compressions		
	Fuel system fault		Fuel system fault (airflow sensor on some cars)
	Spark plugs worn to excess		
	Enrichment device not working (choke or injection circuit)	Oil pressure gauge low or warning light on	Low engine oil level
			Faulty sensor or switch
Difficult to start when hot	Discharged battery (slow rotation)		Worn engine oil pump and/or engine bearings
	Battery terminals loose or corroded		
	Air filter dirty or blocked		Engine overheating
	Low cylinder compressions		Oil pickup filter blocked
	Fuel system fault		Pressure relief valve not working
Starter noisy	Starter pinion or flywheel ring gear loose	Runs on when switched off	Ignition timing incorrect
			Idle speed too high
	Starter mounting bolts loose		Anti-run on device not working
	Starter worn (bearings, etc.)		Carbon build up in engine
	Discharged battery (starter may jump in and out)		Engine overheating
		Pinking or knocking under load	Ignition timing incorrect
Starter turns engine slowly	Discharged battery (slow rotation)		Ignition system fault
	Battery terminals loose or corroded		Carbon build up in engine
	Earth strap or starter supply loose or disconnected		Knock sensor not working
		Sucking or whistling noises	Leaking exhaust manifold gasket
	Internal starter fault		Leaking inlet manifold gasket
Engine starts but then stops immediately	Ignition wiring connection intermittent		Cylinder head gasket
			Inlet air leak
	Fuel system contamination		Water pump or alternator bearing
	Fuel pump or circuit fault (relay)		
	Intake system air leak	Rattling or tapping	Incorrect valve clearances
	Ballast resistor open circuit (older cars)		Worn valve gear or camshaft
			Loose component
Erratic idle	Air filter blocked	Thumping or knocking noises	Worn main bearings (deep knocking/rumbling noise)
	Incorrect plug gaps		
	Inlet system air leak		Worn big end bearings (heavy knocking noise under load)
	Incorrect CO setting		
	Uneven or low cylinder compressions (maybe valves)		Piston slap (worse when cold)
			Loose component
	Fuel injector fault	Rumbling noises	Bearings on ancillary component
	Incorrect ignition timing		
	Incorrect valve timing		
Misfire at idle speed	Ignition coil or distributor cap tracking		
	Poor cylinder compressions		
	Engine breather blocked		
	Inlet system air leak		
	Faulty plugs		
Misfire through all speeds	Fuel filter blocked		
	Fuel pump delivery low		
	Fuel tank ventilation system blocked		
	Poor cylinder compressions		

7.4 Fuel system

Authors Note: Even though carburettor fuel systems are now very rare, they are still used on some specialist vehicles. For this reason, and because it serves as a good introduction to fuel systems, I decided to include this section.

104 Advanced automotive fault diagnosis

7.4.1 Introduction

All vehicle fuel systems consist of the carburettor or fuel injectors, the fuel tank, the fuel pump, and the fuel filter, together with connecting pipes. An engine works by the massive expansion of an ignited fuel air mixture acting on a piston. The job of the fuel system is to produce this mixture at just the right ratio to run the engine under all operating conditions. There are three main ways this is achieved:

- petrol is mixed with air in a carburettor;
- petrol is injected into the manifold, throttle body cylinder or to mix with the air;
- diesel is injected under very high pressure directly into the air already in the engine combustion chamber.

This section will only examine the carburettor systems; diesel and injection comes under engine management later.

7.4.2 Carburation

Figure 7.7 shows a simple fixed choke carburettor operating under various conditions. The float and needle valve assembly ensures a constant level of petrol in the float chamber. The venturi causes an increase in air speed and hence a drop in pressure in the area of the outlet. The main jet regulates how much fuel can be forced into this intake air stream by the higher pressure now apparent in the float chamber. The basic principle is that as more air is forced into the engine then more fuel will be mixed in to the air stream.

The problem with this system is that the amount of fuel forced into the air stream does not linearly follow the increase in air quantity, unless further compensation fuel and air jets are used. A variable venturi carburettor, which keeps the air pressure in the venturi constant but uses a tapered needle to control the amount of fuel, is another method used to control fuel air ratio.

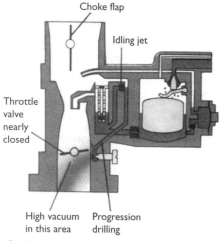

On idling, high vacuum below the throttle valve draws fuel through a separate circuit

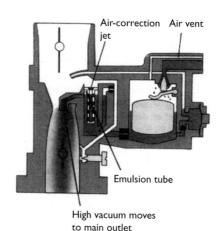

Fuel is drawn from the main outlet as the vacuum around it increases

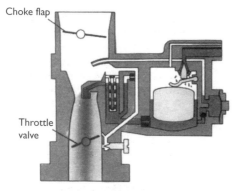

A flap is used to partially block the barrel for cold starts. It increases vacuum around the fuel outlet and draws more fuel to provide a rich mixture

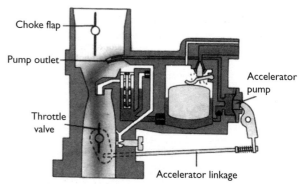

The accelerator pump squirts an enriching shot of fuel down the barrel to provide rapid response when the throttle is opened quickly, when extra power is needed

Figure 7.7 A simple fixed choke carburettor

7.4.3 Stages of carburation

The basic principle of a carburettor is to mix fuel and air together in the correct ratios dependent on engine loads and temperature. Fuel flow is caused by the low air pressure around a spray outlet and atmospheric pressure acting on the fuel in the float chamber.

Whilst the engine's requirement for air fuel mixture is infinitely variable, there are six discernible stages to consider. This can help a lot with diagnosing faults because it helps you 'zoom in' on the area where the problem is likely to be found. Although the examples given relate to a standard fixed choke carburettor, the stages are still relevant to all other types. The six stages along with problems and possible solutions are listed below.

Stage	Query or problem	Possible solution or symptoms
Cold starting when a richer mixture is required to reduce the effects of condensation within the inlet manifold and to aid combustion	What would be the result of the choke not closing fully when starting a cold engine? When starting a cold engine with full choke why must the choke valve partially open when the engine is running?	Difficult starting due to a weak mixture To decrease the depression below the choke valve thus preventing the mixture from becoming too rich
Idle when the minimum amount of fuel should be provided to ensure complete combustion and efficiency	A customer complains of erratic idling and hesitation when moving off from rest? What would be the result of exchanging the idle fuel jet for one of a smaller size?	Idling fuel jet restricted Mixture strength too weak at idle Weak idling air to fuel mixture Erratic idle and flat spots when accelerating from stationary
Progression when a smooth sequence of change is necessary from a range of drillings. This is normally from idle until the main jets come into play	What is the purpose of the progression drilling? What would be the result of the progression drillings becoming restricted?	Ensures an even changeover to the main system Lack of response during initial acceleration with warm engine
Acceleration when a measured increase in the supply of fuel is required to sustain an initial burst of speed	What function does the accelerator pump perform? The customer complains of a flat spot during hard acceleration when engine is hot?	Enriches the mixture for hard acceleration Faulty non-return valve in the pump circuit Accelerator pump diaphragm holed
Cruising where the need is for metered fuel to maintain speed at the most efficient setting	What complaint would the customer have if the main air correction jet was restricted? If mixture was too weak on cruise, what would be the effect on operation?	Poor mid throttle range performance combined with high fuel consumption Possible misfiring and poor performance Flat spots on acceleration
High speeds where a slightly richer mixture is required to maintain efficient combustion and to avoid damage to the engine	What operating symptoms could be caused by blocked full load enrichment tube? What symptoms would be observed if the mixture was too rich at high speed?	Reduced top speed Possible poor fuel consumption and reduced engine performance

7.4.4 Electronic control of carburation

Electronic control of a carburettor is made in the following areas.

Idle speed	Controlled by a stepper motor to prevent stalling but still allow a very low idle speed to improve economy and reduce emissions. Idle speed may also be changed in response to a signal from an automatic gearbox to prevent either the engine from stalling or the car from trying to creep
Fast idle	The same stepper motor as above controls fast idle in response to a signal from the engine temperature sensor during the warm up period
Choke	A rotary choke or some other form of valve or flap operates the choke mechanism depending on engine and ambient temperature conditions
Overrun fuel cut off	A small solenoid operated valve or similar cuts off the fuel under particular conditions. These conditions are often that the engine temperature must be above a set level, engine speed above a set level and the accelerator pedal is in the off position

The air fuel ratio is set by the mechanical design of the carburettor, so it is very difficult to control by electrical means. Some systems have used electronic control of, say, a needle and jet but this has not proved to be very popular.

Figure 7.8 shows the main components of the system used on some vehicles. As with any control system it can be represented as a series of inputs, a form of control and a number of outputs. The inputs to this system are as follows.

Engine speed	From a signal wire to the negative side of the ignition coil as is common with many systems
Engine coolant temperature	This is taken from a thermistor located in the cylinder head waterways. The same sensor is used for the programmed ignition system if fitted
Ambient temperature	A thermistor sensor is placed such as to register the air temperature. A typical position is at the rear of a headlight
Throttle switch	This switch is placed under the actual pedal and only operates when the pedal is fully off, that is when the butterfly valve in the carburettor is closed

The main controlling actuator of this system is the stepper motor. This motor controls by reduction gears a rotary choke valve for cold starting conditions. The same stepper motor controls idle and fast idle with a rod that works on a snail type cam. The system can operate this way because the first part of the movement of the stepper motor does not affect the choke valve; it only affects the idle speed by opening the throttle butterfly slightly. Further rotation then puts on the choke. The extent, to which the choke is on is determined from engine temperature and ambient temperature.

The ECU 'knows' the position of the stepper motor before setting the choke position by a process known as indexing. This involves the stepper motor being driven to say its least setting on switching off the ignition. When the ignition is next turned on the stepper will drive the idle and choke mechanism by a certain number of steps determined as at the end of the last paragraph.

The other main output is the overrun fuel cut off solenoid. This controls the air pressure in the float chamber and when operated causes pressure in the float chamber and pressure in the venturi at the jet outlet to equalise. This prevents any fuel from being 'drawn' into the air stream.

7.5 Diagnostics – fuel system

7.5.1 Systematic testing example

If the reported fault is excessive fuel consumption proceed as follows.

1. Check that the consumption is excessive for the particular vehicle. Test it yourself if necessary.
2. Are there any other problems with the vehicle, misfiring for example or difficult starting?
3. For example if the vehicle is misfiring as well this may indicate that an ignition fault is the cause of the problem.
4. Remove and examine spark plugs, check HT lead resistance and ignition timing. Check CO emissions.
5. Renew plugs and set fuel mixture.
6. Road test vehicle for correct engine operation.

7.5.2 Test equipment

Note: You should always refer to the manufacturer's instructions appropriate to the equipment you are using.

Exhaust gas analyser (Figure 7.9)

This is a sophisticated piece of test equipment used to measure the make up of the vehicle's exhaust gas. The most common requirement is the measuring of carbon monoxide (CO). A sample probe is placed in the exhaust tail pipe or a special position before the catalytic converter if fitted, and the machine reads out the percentage of certain gases produced. A digital readout is most common. The fuel mixture can then be adjusted until the required readings are obtained.

Fuel pressure gauge

The output pressure of the fuel pump can be tested to ensure adequate delivery. The device is a simple pressure gauge but note the added precautions necessary when dealing with petrol.

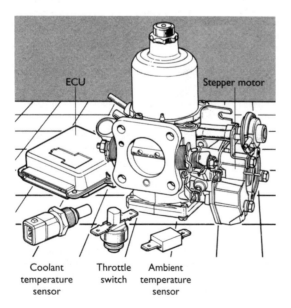

Figure 7.8 HIF variable venturi carburettor with electronic control components

Engine systems **107**

Figure 7.9 Exhaust gas analyser

7.5.3 Test results

Some of the information you may have to get from other sources such as data books or a workshop manual is listed in the following table.

Test carried out	Information required
Exhaust gas analysis	CO setting. Most modern vehicles will have settings of about 1% or less. If a 'cat' is fitted then the readings will be even lower when measured at the tail pipe
Fuel pressure	The expected pressure readings will vary depending on the type of fuel system. Fuel injection pressure will be about 2.5 bar whereas fuel pressure for a carburettor will be about 0.3 bar
Fuel delivery	How much fuel the pump should move in a set time. This will again vary with the type of fuel system. 1 litre in 30 seconds is typical for some injection fuel pumps

7.5.4 Fuel fault diagnosis table 1

Symptom	Possible faults	Suggested action
No fuel at carburettor or injection fuel rail	Empty tank! Blocked filter or line	Fill it! Replace filter, renew/repair line
	Defective fuel pump	Renew/check it is being driven
	No electrical supply to pump	Check fuses/trace fault
Engine will not or is difficult to start	Choke or enrichment device not working	Check linkages or automatic actuator
Engine stalls or will not idle smoothly	Idle speed incorrectly set	Look up correct settings and adjust
	Mixture setting wrong	Look up correct settings and adjust
	Ignition problem	Check ignition system
Poor acceleration	Blockage in carburettor accelerator pump	Strip down and clean out or try a carburettor cleaner first
	Partially blocked filter	Renew
	Injection electrical fault	Refer to specialist information
Excessive fuel consumption	Incorrect mixture settings	Look up correct settings and adjust
	Driving technique!	Explain to the customer – but be diplomatic!
Black smoke from exhaust	Excessively rich mixture	Look up correct settings and adjust
	Flooding	Check and adjust carburettor float settings and operation

7.5.5 Fuel fault diagnosis table 2

Symptom	Possible cause
Excessive consumption	Blocked air filter
	Incorrect CO adjustment
	Fuel injectors leaking
	Ignition timing incorrect
	Temperature sensor fault
	Load sensor fault
	Low tyre pressures
	Driving style!
Fuel leakage	Damaged pipes or unions
	Fuel tank damaged
	Tank breathers blocked
Fuel smell	Fuel leak
	Breather incorrectly fitted
	Fuel cap loose
	Engine flooding
Incorrect emissions	Incorrect adjustments
	Fuel system fault
	Air leak into inlet
	Blocked fuel filter
	Blocked air filter
	Ignition system fault

7.6 Introduction to engine management

Engine management is a general term that describes the control of engine operation. This can range from a simple carburettor to control or manage the fuel, with an ignition distributor with contact breakers to control the ignition to a very sophisticated electronic control system. The fundamental tasks of an engine management system are to manage the ignition and fuelling, as well as other aspects, and to refine the basic control of an engine.

Many of the procedures and explanations in this chapter are generic. In other words the ignition system explained in the next sections may be the same as the system used by a combined ignition and fuel control system.

7.7 Ignition

7.7.1 Basics

The purpose of the ignition system is to supply a spark inside the cylinder, near the end of the compression stroke, to ignite the compressed charge of air fuel vapour. For a spark to jump across an air gap of 0.6 mm under normal atmospheric conditions (1 bar) a voltage of 2 to 3 kV is required. For a spark to jump across a similar gap in an engine cylinder having a compression ratio of 8:1 approximately 8 kV is required. For higher compression ratios and weaker mixtures, a voltage up to 20 kV may be necessary. The ignition system has to transform the normal battery voltage of 12 V to approximately 8 to 20 kV and, in addition, has to deliver this high voltage to the right cylinder, at the right time. Some ignition systems will supply up to 40 kV to the spark plugs.

Conventional ignition is the forerunner of the more advanced systems controlled by electronics. However, the fundamental operation of most ignition systems is very similar; one winding of a coil is switched on and off causing a high voltage to be induced in a second winding. A coil ignition system is composed of various components and sub-assemblies; the actual design and construction of these depend mainly on the engine with which the system is to be used.

7.7.2 Advance angle (timing)

For optimum efficiency the ignition advance angle should be such as to cause the maximum combustion pressure to occur about 10° after TDC. The ideal ignition timing is dependent on two main factors, engine speed and engine load. An increase in engine speed requires the ignition timing to be advanced. The cylinder charge, of air fuel mixture, requires a certain time to burn (normally about 2 ms). At higher engine speeds the time taken for the piston to travel the same distance reduces. Advancing the time of the spark ensures that full burning is achieved.

A change in timing due to engine load is also required as the weaker mixture used on low load conditions burns at a slower rate. In this situation further ignition advance is necessary. Greater load on the engine requires a richer mixture, which burns more rapidly. In this case some retardation of timing is necessary. Overall, under any condition of engine speed and load an ideal advance angle is required to ensure maximum pressure is achieved in the cylinder just after TDC. The ideal advance angle may also be determined by engine temperature and any risk of detonation.

Spark advance is achieved in a number of ways. The simplest of these is the mechanical system comprising a centrifugal advance mechanism and a vacuum (load sensitive) control unit. Manifold depression is almost inversely proportional to the engine load. I prefer to consider manifold pressure, albeit less than atmospheric pressure; the absolute manifold pressure (MAP) is proportional to engine load. Digital ignition systems may adjust the timing in relation to the temperature as well as speed and load. The values of all ignition timing functions are combined either mechanically or electronically in order to determine the ideal ignition point.

The energy storage takes place in the ignition coil. The energy is stored in the form of a magnetic field. To ensure that the coil is charged before the ignition point a dwell period is required. Ignition timing is at the end of the dwell period.

7.7.3 Electronic ignition

Electronic ignition is now fitted to all spark ignition vehicles. This is because the conventional mechanical system has some major disadvantages.

- Mechanical problems with the contact breakers not least of which is the limited lifetime.
- Current flow in the primary circuit is limited to about 4 A or damage will occur to the contacts – or at least the lifetime will be seriously reduced.
- Legislation requires stringent emission limits which means the ignition timing must stay in tune for a long period of time.
- Weaker mixtures require more energy from the spark to ensure successful ignition, even at very high engine speed.

These problems can be overcome by using a power transistor to carry out the switching function and a pulse generator to provide the timing signal. Very early forms of electronic ignition used the existing contact breakers as the signal provider. This was a step in the right direction but did not overcome all the mechanical limitations such as contact bounce and timing slip. All systems nowadays are constant energy ensuring high performance ignition even at high engine speed. Figure 7.10 shows the circuit of a standard electronic ignition system.

The term 'dwell' when applied to ignition is a measure of the time during which the ignition coil is charging, in other words when primary current is flowing. The dwell in conventional systems was simply the time during which the contact breakers were closed. This is now often expressed as a percentage of one charge-discharge cycle. Constant dwell electronic ignition systems have now been replaced almost without exception by constant energy systems discussed in the next section.

Whilst this was a very good system in its time, constant dwell still meant that at very high engine speeds, the time available to charge the coil could only produce a lower power spark. Note that as engine speed increases dwell angle or dwell percentage remains the same but the actual time is reduced.

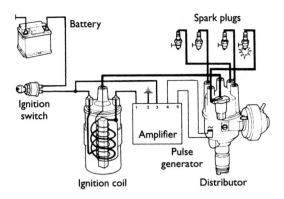

Figure 7.10 Electronic ignition system

In order for a constant energy electronic ignition system to operate the dwell must increase with engine speed. This will only be of benefit, however, if the ignition coil can be charged up to its full capacity, in a very short time (the time available for maximum dwell at the highest expected engine speed). To this end constant energy coils are very low resistance and low inductance. Typical resistance values are less than $1\,\Omega$ (often $0.5\,\Omega$). Constant energy means that, within limits, the energy available to the spark plug remains constant under all operating conditions.

Due to the high energy nature of constant energy ignition coils, the coil cannot be allowed to remain switched on for more than a certain time. This is not a problem when the engine is running, as the variable dwell or current limiting circuit prevents the coil from overheating. Some form of protection must be provided, however, for when the ignition is switched on but the engine is not running. This is known as stationary engine primary current cut off.

7.7.4 Hall effect distributor

The Hall effect distributor has become very popular with many manufacturers. Figure 7.11 shows a typical example. As the central shaft of the distributor rotates, the chopper plate attached under the rotor arm alternately covers and uncovers the Hall chip. The number of vanes corresponds with the number of cylinders. In constant dwell systems the dwell is determined by the width of the vanes. The vanes cause the Hall chip to be alternately in and out of a magnetic field. The result of this is that the device will produce almost a square wave output, which can then easily be used to switch further electronic circuits.

The three terminals on the distributor are marked '+ 0 −'; the terminals + and − are for a voltage supply and terminal '0' is the output

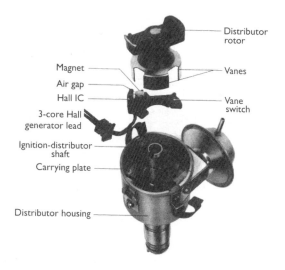

Figure 7.11 Ignition system with Hall generator

signal. Typically the output from a Hall effect sensor will switch between 0 V and about 8 V. The supply voltage is taken from the ignition ECU and on some systems is stabilised at about 10 V to prevent changes to the output of the sensor when the engine is being cranked.

Hall effect distributors are very common due to the accurate signal produced and long-term reliability. They are suitable for use on both constant dwell and constant energy systems. Operation of a Hall effect pulse generator can easily be tested with a DC voltmeter or a logic probe. Note that tests must not be carried out using an ohmmeter as the voltage from the meter can damage the Hall chip.

7.7.5 Inductive distributor

Many forms of inductive type distributors exist and all are based around a coil of wire and a permanent magnet. The example distributor shown in Figure 7.12 has the coil of wire wound on the pick up and as the reluctor rotates the magnetic flux varies due to the peaks on the reluctor. The number of peaks or teeth on the reluctor corresponds to the number of engine cylinders. The gap between the reluctor and pick up can be important and manufacturers have recommended settings.

7.7.6 Current limiting and closed loop dwell

Primary current limiting ensures that no damage can be caused to the system by excessive primary current, but also forms a part of a constant energy system. The primary current is allowed to build up to its pre-set maximum as soon as possible and is then held at this value. The value of this current is calculated and then pre-set during

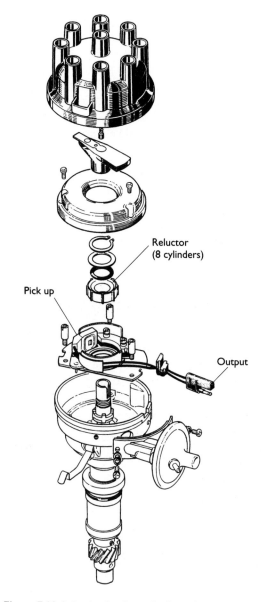

Figure 7.12 Inductive distributor (early type)

construction of the amplifier module. This technique, when combined with dwell angle control, is known as closed loop control as the actual value of the primary current is fed back to the control stages.

A very low resistance, high power precision resistor is used in this circuit. The resistor is connected in series with the power transistor and the ignition coil. A voltage sensing circuit connected across this resistor will be activated at a pre-set voltage (which is proportional to the current), and causes the output stage to hold the current at a constant value. Figure 7.13 shows a block diagram of a closed loop dwell control system.

Stationary current cut off is for when the ignition is on but the engine not running. This is achieved in many cases by a simple timer circuit, which will cut the output stage after about one second.

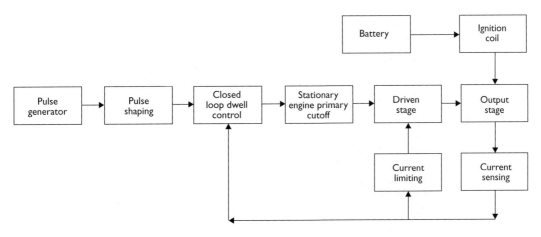

Figure 7.13 Closed loop dwell control system

7.7.7 Programmed ignition

Programmed ignition is the term used by some manufacturers; others call it electronic spark advance (ESA). Constant energy electronic ignition was a major step forwards and is still used on countless applications. However, its limitations lay in still having to rely upon mechanical components for speed and load advance characteristics. In many cases these did not match ideally the requirements of the engine.

Programmed ignition systems have a major difference compared to earlier systems in that they operate digitally. Information about the operating requirements of a particular engine is programmed in to memory inside the ECU. The data for storage in ROM is obtained from rigorous testing on an engine dynamometer and further development work in the vehicle under various operating conditions. Programmed ignition has several advantages.

- The ignition timing can be accurately matched to the individual application under a range of operating conditions.
- Other control input can be utilised such as coolant temperature and ambient air temperature.
- Starting is improved, fuel consumption is reduced as are emissions and idle control is better.
- Other inputs can be taken into account such as engine knock.
- The number of wearing components in the ignition system is considerably reduced.

Programmed ignition or ESA can be a separate system or included as part of the fuel control system. In order for the ECU to calculate suitable timing and dwell outputs, certain input information is required.

The crankshaft sensor consists of a permanent magnet, a winding and a soft iron core. It is mounted in proximity to a reluctor disc. The disc has 34 teeth spaced at 10° intervals around to periphery. It has two teeth missing 180° apart, at a known position BTDC. Many manufacturers use this technique with minor differences. As a tooth from the reluctor disc passes the core of the sensor the reluctance of the magnetic circuit is changed. This induces a voltage in the winding, the frequency of the waveform being proportional to the engine speed. The missing tooth causes a 'missed' output wave and hence engine position can be determined.

Engine load is proportional to manifold pressure in that high load conditions produce high pressure and lower load conditions, such as cruise, produce lower pressure. Load sensors are therefore pressure transducers. They are either mounted in the ECU or as a separate unit and are connected to the inlet manifold with a pipe. The pipe often incorporates a restriction to damp out fluctuations and a vapour trap to prevent petrol fumes reaching the sensor.

Coolant temperature measurement is carried out by a simple thermistor. In many cases the same sensor is used for the operation of the temperature gauge and to provide information to the fuel control system. A separate memory map is used to correct the basic timing settings. Timing may be retarded when the engine is cold to assist in more rapid warm up.

Combustion knock can cause serious damage to an engine if sustained for long periods. This knock or detonation is caused by over advanced ignition timing. At variance with this is that an engine in general will run at its most efficient when the timing is advanced as far as possible. To achieve this the data stored in the basic timing map will be as close to the knock limit of the engine as possible. The knock sensor provides a margin for error. The sensor itself is an accelerometer often of the piezoelectric type. It is fitted in the engine block between cylinders two and three on in-line four cylinder engines. Vee engine's require two sensors, one on each side. The ECU responds to signals from the knock sensor in the engine's knock window for each cylinder; this is often just a few degrees each side of TDC. This prevents clatter from the valve mechanism being interpreted as knock. The signal from the sensor is also filtered in the ECU to remove unwanted noise. If detonation is detected the ignition timing is retarded on the fourth ignition pulse after detection (four cylinder engine), in steps until knock is no longer detected. The steps vary between manufacturers but about 2° is typical. The timing is then advanced slowly in steps of say 1° over a number of engine revolutions, until the advance required by memory is restored. This fine control allows the engine to be run very close to the knock limit without risk of engine damage.

Correction to dwell settings is required if the battery voltage falls, as a lower voltage supply to the coil will require a slightly larger dwell figure. This information is often stored in the form of a dwell correction map.

As the sophistication of systems has increased the information held in the memory chips of the ECU has also increased. The earlier versions of programmed ignition system produced by Rover achieved accuracy in ignition timing of ±1.8° whereas a conventional distributor is ±8°. The information, which is derived from dynamometer tests as well as running tests in the vehicle, is stored in ROM. The basic timing map consists of the correct ignition advance for 16 engine speeds and 16 engine load conditions.

A separate three-dimensional map is used which has eight speed and eight temperature sites. This is used to add corrections for engine coolant temperature to the basic timing settings. This improves driveability and can be used to decrease the warm-up time of the engine. The data is also subjected to an additional load correction below 70°C. Figure 7.14 shows a flow chart representing the logical selection of the optimum ignition setting. Note that the ECU will also make corrections to the dwell angle, both as a function of engine speed to provide constant energy output and due to changes in battery voltage. A lower battery voltage will require a slightly longer dwell and a higher voltage a slightly shorter dwell. A Windows® shareware program that simulates the ignition system (as well as many other systems) is available for download from my web site.

The output of a system such as this programmed ignition is very simple. The output stage, in common with most electronic ignition, consists of a heavy-duty transistor which forms part of, or is driven by, a Darlington pair. This is simply to allow the high ignition primary current to be controlled. The switch off point of the coil will control ignition timing and the switch on point will control the dwell period.

The high tension distribution is similar to a more conventional system. The rotor arm, however, is mounted on the end of the camshaft with the distributor cap positioned over the top. Figure 7.15 shows a programmed ignition system.

7.7.8 Distributorless ignition

Distributorless ignition has all the features of programmed ignition systems but, by using a special type of ignition coil, outputs to the spark plugs without the need for an HT distributor. The system is generally only used on four cylinder engines as the control system becomes too complex for higher numbers. The basic principle is that of the 'lost spark'. The distribution of the spark is achieved by using double ended coils, which are fired alternately by the ECU. The timing is determined from a crankshaft speed and position sensor as well as load and other corrections. When one of the coils is fired a spark is delivered to two engine cylinders, either 1 and 4, or 2 and 3. The spark delivered to the cylinder on the compression stroke will ignite the mixture as normal. The spark produced in the other cylinder will have no effect, as this cylinder will be just completing its exhaust stroke.

Because of the low compression and the exhaust gases in the 'lost spark' cylinder the voltage used for the spark to jump the gap is only about 3 kV. This is similar to the more conventional rotor arm to cap voltage. The spark produced in the compression cylinder is therefore not affected.

An interesting point here is that the spark on one of the cylinders will jump from the earth electrode to the spark plug centre. Many years ago this would not have been acceptable as the spark quality when jumping this way would not have been as good as when it jumps from the centre electrode. However, the energy available

Engine systems 113

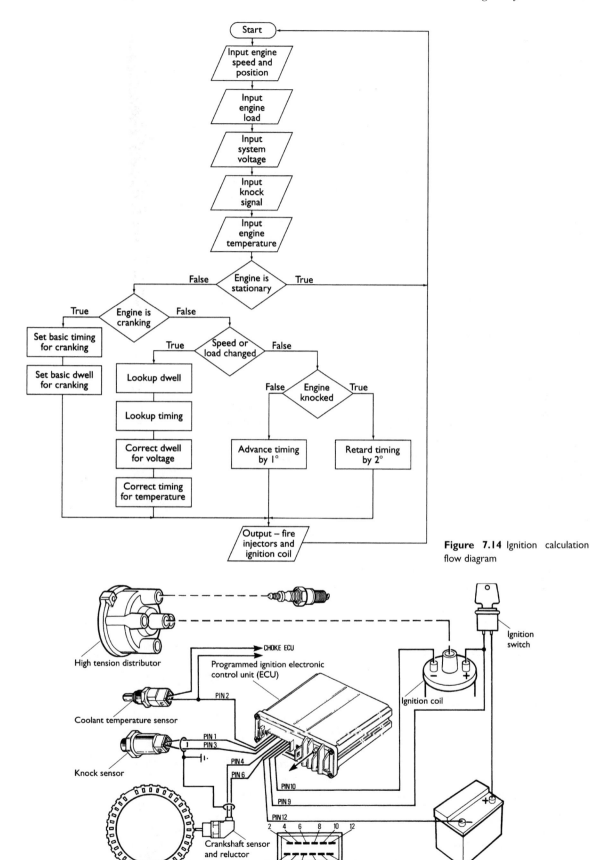

Figure 7.14 Ignition calculation flow diagram

Figure 7.15 Programmed ignition system component layout

from modern constant energy systems will produce a spark of suitable quality in either direction.

The direct ignition system (DIS) consists of three main components, the electronic module, a crankshaft position sensor and the DIS coil. In many systems a MAP sensor is integrated in the module. The module functions in much the same way as has been described for the electronic spark advance system.

The crankshaft position sensor is similar in operation to the one described in the previous section. It is again a reluctance sensor and is positioned against the front of the flywheel or against a reluctor wheel just behind the front crankshaft pulley. The tooth pattern consists of 35 teeth. These are spaced at 10° intervals with a gap where the 36th tooth would be. The missing tooth is positioned at 90° BTDC for numbers 1 and 4 cylinders. This reference position is placed a fixed number of degrees before TDC, in order to allow the timing or ignition point to be calculated as a fixed angle after the reference mark.

The low tension winding is supplied with battery voltage to a centre terminal. The appropriate half of the winding is then switched to earth in the module. The high tension windings are separate and are specific to cylinders 1 and 4, or 2 and 3. Figure 7.16 shows a typical DIS coil for a six cylinder engine.

7.7.9 Direct ignition

Direct ignition is in a way the follow on from distributorless ignition. This system utilises an inductive coil for each cylinder. These coils are mounted directly on the spark plugs. Figure 7.17 shows a cross section of the direct ignition coil. The use of an individual coil for each plug ensures that the rise time for the low inductance primary winding is very fast. This ensures that a very high voltage, high energy spark is produced. This voltage, which can be in excess of 400 kV, provides efficient initiation of the combustion process under cold starting conditions and with weak mixtures. Some direct ignition systems use capacitor discharge ignition.

In order to switch the ignition coils igniter units are used. These can control up to three coils and are simply the power stages of the control unit but in a separate container. This allows less interference to be caused in the main ECU due to heavy current switching and shorter runs of wires carrying higher currents.

Ignition timing and dwell are controlled in a manner similar to the previously described programmed system. The one important addition to

Figure 7.16 DIS coil

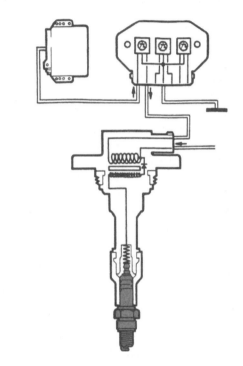

Figure 7.17 Direct ignition system

this on some systems is a camshaft sensor to provide information as to which cylinder is on the compression stroke. A system which does not require a sensor to determine which cylinder is on compression (engine position is known from a crank sensor) determines the information by initially firing all of the coils. The voltage across the plugs allows measurement of the current for each spark and will indicate which cylinder is on its combustion stroke. This works because a burning mixture has a lower resistance. The cylinder with the highest current at this point will be the cylinder on the combustion stroke.

A further feature of some systems is the case when the engine is cranked over for an excessive time making flooding likely. The plugs are all fired with multisparks for a period of time after the ignition is left in the on position for five seconds. This will burn away any excess fuel.

During difficult starting conditions, multisparking is also used by some systems during 70° of crank rotation before TDC. This assists with starting and then once the engine is running, the timing will return to its normal calculated position.

7.7.10 Spark plugs

Figure 7.18 shows a standard spark plug. The centre electrode is connected to the top terminal by a stud. The electrode is constructed of a nickel based alloy. Silver and platinum are also used for some applications. If a copper core is used in the electrode this improves the thermal conduction properties.

The insulating material is ceramic based and of a very high grade. The electrically conductive glass seal between the electrode and terminal stud is also used as a resistor. This resistor has two functions: firstly to prevent burn off of the centre electrode; and secondly to reduce radio interference. In both cases the desired effect is achieved because the resistor damps the current at the instant of ignition.

Flashover or tracking down the outside of the plug insulation is prevented by ribs. These effectively increase the surface distance from the terminal to the metal fixing bolt, which is of course earthed to the engine.

Due to the many and varied constructional features involved in the design of an engine, the range of temperatures a spark plug is exposed to can vary significantly. The operating temperature of the centre electrode of a spark plug is critical. If the temperature becomes too high then pre-ignition may occur as the fuel air mixture may become ignited due to the incandescence of the plug electrode. On the other hand if the electrode temperature is too low then carbon and oil fouling can occur, as deposits are not burnt off. Fouling of the plug nose can cause shunts (a circuit in parallel with the spark gap). It has been shown through experimentation and experience that the ideal operating temperature of the plug electrode is between 400 and 900°C.

The heat range of a spark plug then is a measure of its ability to transfer heat away from the centre electrode. A hot running engine will require plugs with a higher thermal loading ability than a colder running engine. Note that hot and cold running of an engine in this sense refers to the

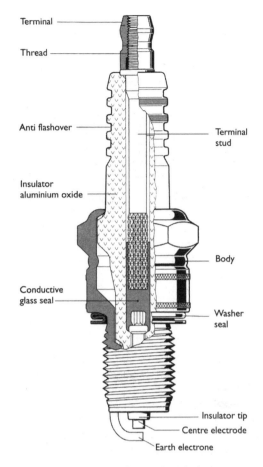

Figure 7.18 Construction of a typical spark plug

combustion temperature and not to the efficiency of the cooling system.

The following factors determine the thermal capacity of a spark plug.

- insulator nose length
- electrode material
- thread contact length
- projection of the electrode

It has been found that a longer projection of the electrode helps to reduce fouling problems due to low power operation, stop go driving and high altitude conditions. In order to use greater projection of the electrode better quality thermal conduction is required to allow suitable heat transfer at higher power outputs. Figure 7.19 shows the heat conducting paths of a spark plug together with changes in design for heat ranges. Also shown is the range of part numbers for NGK plugs.

For normal applications alloys of nickel are used for the electrode material. Chromium, manganese, silicon and magnesium are examples of the alloying constituents. These alloys exhibit

116 Advanced automotive fault diagnosis

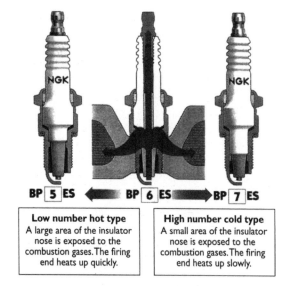

Figure 7.19 'Hot' and 'Cold' spark plugs

excellent properties with respect to corrosion and burn off resistance. To improve on the thermal conductivity compound electrodes are used. This allows a greater nose projection for the same temperature range as discussed in the last section. A common example of this type of plug is the copper core spark plug.

Silver electrodes are used for specialist applications as silver has very good thermal and electrical properties. Again with these plugs nose length can be increased within the same temperature range. Platinum tips are used for some spark plug applications due to the very high burn off resistance of this material. It is also possible because of this to use much smaller diameter electrodes thus increasing mixture accessibility. Platinum also has a catalytic effect further accelerating the combustion process.

Spark plug electrode gaps in general have increased as the power of the ignition systems driving the spark has increased. The simple relationship between plug gap and voltage required is that as the gap increases so must the voltage (leaving aside engine operating conditions). Further, the energy available to form a spark at a fixed engine speed is constant, which means that a larger gap using higher voltage will result in a shorter duration spark. A smaller gap will allow a longer duration spark. For cold starting an engine and for igniting weak mixtures the duration of the spark is critical. Likewise the plug gap must be as large as possible to allow easy access for the mixture to prevent quenching of the flame.

The final choice is therefore a compromise reached through testing and development of a particular application. Plug gaps in the region of 0.6 to 1.2 mm seem to be the norm at present.

7.8 Diagnostic – ignition systems

7.8.1 Testing procedure

> **Warning**
> Caution/Achtung/Attention – high voltages can seriously damage your health!

The following procedure is generic and with a little adaptation can be applied to any ignition system. Refer to manufacturer's recommendations if in any doubt.

Ignition systems diagnostic chart

Start
↓
Hand and eye checks (loose wires, loose switches and other obvious faults) – all connections clean and tight
↓
Check battery – must be 70% charged
↓
Check supply to ignition coil (within 0.5 V of battery)
↓
Spark from coil via known good HT lead (jumps about 10 mm, but do not try a greater distance)?
  - No → If no spark or it will only jump a short distance continue with this procedure (colour of spark is not relevant)
    ↓
    Check continuity of coil windings (primary 0.5 to 3 ohms, secondary several k ohms)
    ↓
    Supply and earth to 'module' (12 V minimum supply, earth drop 0.5 V maximum)
    →
    Supply to pulse generator if appropriate (5 or 10 to 12 V)
    ↓
    Output of pulse generator (inductive about 1 V AC when cranking, Hall type switches 0 V to 8 V DC)?
      - No → Replace pulse generator
      - Yes ↓
    Continuity of LT wires (0 to 0.1 ohm)
    ↓
    Replace 'module' but only if all tests above are satisfactory
  - Yes → If good spark then check HT system for tracking and open circuits
    ↓
    Check plug condition (leads should be a maximum resistance of about 30 k/m)
    ↓
    Check/replace plugs
↓
End

7.8.2 Ignition fault diagnosis table

Symptom	Possible fault
Engine rotates but does not start	Damp ignition components Spark plugs worn to excess Ignition system open circuit
Difficult to start when cold	Spark plugs worn to excess High resistance in ignition circuit
Engine starts but then stops immediately	Ignition wiring connection intermittent
Erratic idle	Incorrect plug gaps Incorrect ignition timing
Misfire at idle speed	Ignition coil or distributor cap tracking Spark plugs worn to excess
Misfire through all speeds	Incorrect plugs or plug gaps HT leads breaking down Timing incorrect
Lack of power	Ignition timing incorrect HT components tracking
Backfires	Incorrect ignition timing Tracking
Runs on when switched off	Ignition timing incorrect Carbon buildup in engine Idle speed too high Anti-run on device inoperative
Pinking or knocking under load	Ignition timing incorrect Ignition system electronic fault Knock sensor not working

Figure 7.20 shows a typical ignition timing light, essential to ensure correct settings where these are adjustable, or to check programmed advance systems for correct operation.

7.8.3 Ignition components and testing

Component	Description	Test method
Spark plug	Seals electrodes for the spark to jump across in the cylinder. Must withstand very high voltages, pressures and temperatures	Compare nose condition to a manufacturer's chart Inspect ignition secondary waveform when the engine is under load
Ignition coil	Stores energy in the form of magnetism and delivers it to the distributor via the HT lead Consists of primary and secondary windings	Resistance checks of the primary and secondary windings Primary: 1.5 Ω (ballasted) to 3 Ω Secondary: 5 to 10 kΩ
Ignition switch	Provides driver control of the ignition system and is usually also used to cause the starter to crank	Voltage drop across the contacts
HT distributor	Directs the spark from the coil to each cylinder in a pre-set sequence	Visual inspection tracking (conducting lines) and contamination
Engine speed advance	Changes the ignition timing with engine speed As speed increases the timing is advanced	Measure the timing at certain speeds using an 'advance' timing light. Refer to data
Engine load (Vacuum) advance	Changes timing depending on engine load	Apply a known vacuum and note timing changes

7.8.4 DIS diagnostics

The DIS system is very reliable due to the lack of any moving parts. Some problems can be experienced when trying to examine HT oscilloscope patterns due to the lack of a king lead. This can often be overcome with a special adapter but it is still necessary to move the sensing clip to each lead in turn.

The DIS coil can be tested with an ohmmeter. The resistance of each primary winding should be 0.5 Ω and the secondary windings between 11 and 16 kΩ. The coil will produce in excess of 37 kV in an open circuit condition. The plug leads have integral retaining clips to prevent water ingress and vibration problems. The maximum resistance for the HT leads is 30 kΩ per lead.

No service adjustments are possible with this system with the exception of octane adjustment on some models. This involves connecting two pins together on the module for normal operation, or earthing one pin or the other to change to a different fuel. The actual procedure must be checked with the manufacturer for each particular model.

7.8.5 Spark plug diagnostics

Examination of the spark plugs is a good way of assessing engine and associated systems condition. Figure 7.21 a) shows a new plug and b) to f) show various conditions with diagnostic notes added.

Figure 7.20 Ignition timing light

a) New spark plug

Use this image to compare with used spark plugs. Not in particular, on this standard design, how the end of the nose is flat and that the earth/ground electrode is a consistent size and shape.

b) Carbon fouled (standard plug)

This plug has black deposits over the centre electrode and insulator in particular. It is likely that this engine was running too rich – or on an older vehicle the choke was used excessively. However, carbon fouling may also be due to:

- Poor quality spark due to ignition fault
- Incorrect plug gap
- Over retarded timing
- Loss of cylinder compression
- Prolonged low speed driving
- Incorrect (too cold) spark plug fitted.

c) Deposits

The deposits on this plug are most likely to be caused by oil leaking into the cylinder. Alternatively, poor quality fuel mixture supply or very short, cold engine operation could result in a similar condition.

Figure 7.21(a,b,c)

Engine systems **119**

d) Damaged insulation

A plug that is damaged in this way is because of either overheating or impact damage. Impact is most likely in this case. The damage can of course be caused as the plug is being fitted! However, in this case a possible cause would be that the reach was too long for the engine and the piston hit the earth/ground electrode, closing up the gap and breaking the insulation.

e) Carbon fouled (platinum plug)

The carbon build upon this plug would suggest an incorrect mixture. However, before diagnosing a fault based on spark plug condition, make sure the engine has been run up to temperature – ideally by a good road test. The engine from which this plug was removed is in good condition – it had just been started from cold and only run for a few minutes.

f) Overheating

When a plug overheats the insulator tip becomes glossy and/or they are blistered or melted away. The electrodes also wear quickly. Excessive overheating can result in the electrodes melting and serious piston damage is likely to occur. Causes of overheating are:

- Over advanced ignition
- Mixture too lean
- Cooling system fault
- Incorrect plug (too hot)
- Incorrect fuel (octane low).

Figure 7.21 (d,e,f) Spark plug diagnostics

7.9 Emissions

7.9.1 Introduction

The following table lists the four main exhaust emissions which are hazardous to health, together with a short description of each.

Substance	Description
Carbon monoxide (CO)	This gas is very dangerous even in low concentrations. It has no smell or taste and is colourless. When inhaled it combines in the body with the red blood cells preventing them from carrying oxygen. If absorbed by the body it can be fatal in a very short time
Nitrogen oxides (NO_x)	Oxides of nitrogen are colourless and odourless when they leave the engine but as soon as they reach the atmosphere and mix with more oxygen, nitrogen oxides are formed. They are reddish brown and have an acrid and pungent smell. These gasses damage the body's respiratory system when inhaled. When combined with water vapour nitric acid can be formed which is very damaging to the windpipe and lungs. Nitrogen oxides are also a contributing factor to acid rain
Hydrocarbons (HC)	A number of different hydrocarbons are emitted from an engine and are part or unburnt fuel. When they mix with the atmosphere they can help to form smog. It is also believed that hydrocarbons may be carcinogenic
Particulate matter (PM)	This heading in the main covers lead and carbon. Lead was traditionally added to petrol to slow its burning rate to reduce detonation. It is detrimental to health and is thought to cause brain damage especially in children. Lead will eventually be phased out as all new engines now run on unleaded fuel. Particles of soot or carbon are more of a problem on diesel fuelled vehicles and these now have limits set by legislation

The following table describes two further sources of emissions from a vehicle.

Source	Comments
Fuel evaporation from the tank and system	Fuel evaporation causes hydrocarbons to be produced. The effect is greater as temperature increases. A charcoal canister is the preferred method for reducing this problem. The fuel tank is usually run at a pressure just under atmospheric by a connection to the intake manifold drawing the vapour through the charcoal canister. This must be controlled by the management system, however, as even a 1% concentration of fuel vapour would shift the lambda value by 20%. This is done by using a 'purge valve', which under some conditions is closed (full load and idle for example) and can be progressively opened under other conditions. The system monitors the effect by use of the lambda sensor signal
Crankcase fumes (blow by)	Hydrocarbons become concentrated in the crankcase mostly due to pressure blowing past the piston rings. These gases must be conducted back into the combustion process. This is usually via the air intake system. This is described as positive crankcase ventilation

7.9.2 Exhaust gas recirculation (EGR)

This technique is used primarily to reduce peak combustion temperatures and hence the production of nitrogen oxides (NO_x). EGR can be either internal due to valve overlap, or external via a simple arrangement of pipes and a valve (Figure 7.22 shows an example) connecting the exhaust manifold back to the inlet manifold. A proportion of exhaust gas is simply returned to the inlet side of the engine.

This process is controlled electronically as determined by a ROM in the ECU. This ensures that driveability is not effected and also that the rate of EGR is controlled. If the rate is too high, then the production of hydrocarbons increases.

One drawback of EGR systems is that they can become restricted by exhaust residue over a period of time thus changing the actual percentage of recirculation. However, valves are now available that reduce this particular problem.

Figure 7.22 EGR valve

7.9.3 Catalytic converters

Stringent regulations in most parts of the world have made the use of a catalytic converter almost indispensable. The three-way catalyst (TWC) is used to great effect by most manufacturers. It is a very simple device and looks similar to a standard exhaust box. Note that in order to operate correctly, however, the engine must be run at or very near to stoichiometry. This is to ensure that the right 'ingredients' are available for the catalyst to perform its function.

Figure 7.23 shows some new metallic substrates for use inside a catalytic converter. There are many types of hydrocarbons but the example illustrates the main reaction. Note that the reactions rely on some CO being produced by the engine in order to reduce the NO_x. This is one of the reasons that manufacturers have been forced to run the engine at stoichiometry. The legislation has tended to stifle the development of lean burn techniques. The fine details of the emission regulations can in fact have a very marked effect on the type of reduction techniques used. The main reactions in the 'cat' are as follows:

- $2CO + O_2 \rightarrow 2CO_2$
- $2C_2H_6 + 2CO \rightarrow 4CO_2 + 6H_2O$
- $2NO + 2CO \rightarrow N_2 + 2CO_2$

Noble metals are used for the catalysts; platinum promotes the oxidation of HC and CO, and rhodium helps the reduction of NO_x. The whole three-way catalytic converter only contains about three to four grams of the precious metals.

The ideal operating temperature range is from about 400 to 800°C. A serious problem to counter is the delay in the catalyst reaching this temperature. This is known as catalyst light off time. Various methods have been used to reduce this time as significant emissions are produced before light off occurs. Electrical heating is one solution, as is a form of burner which involves lighting fuel inside the converter. Another possibility is positioning the converter as part of the exhaust manifold and down pipe assembly. This greatly reduces light off time but gas flow problems, vibration and excessive temperature variations can be problems that reduce the potential life of the unit.

Catalytic converters can be damaged in two ways. The first is by the use of leaded fuel which causes lead compounds to be deposited on the active surfaces thus reducing effective area. The second is engine misfire which can cause the catalytic converter to overheat due to burning inside the unit. BMW, for example, use a system on some vehicles where a sensor monitors output of the ignition HT system and will not allow fuel to be injected if the spark is not present.

For a catalytic converter to operate at its optimum conversion rate to oxidise CO and HC whilst reducing NOx, a narrow band within 0.5% of lambda value 1 is essential. Lambda sensors in use at present tend to operate within about 3% of the lambda mean value. When a catalytic converter is in prime condition this is not a problem due to storage capacity within the converter for CO and O_2. Damaged converters, however, cannot store sufficient quantity of these gases and hence become less efficient. The damage as suggested earlier in this section can be due to overheating or by 'poisoning' due to lead or even silicon. If the control can be kept within 0.5% of lambda

Figure 7.23 Catalytic converter metallic substrates

the converter will continue to be effective even if damaged to some extent. Sensors are becoming available which can work to this tolerance. A second sensor fitted after the converter can be used to ensure ideal operation.

7.10 Diagnostics – emissions

7.10.1 Testing procedure

If the reported fault is incorrect exhaust emissions the following procedure should be utilised.

Figure 7.24 is typical of the emissions data available.

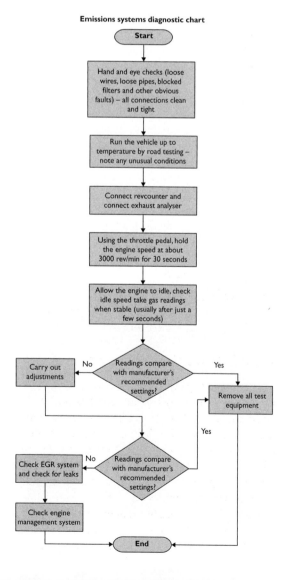

7.10.2 Emissions fault diagnosis table

Symptom	Possible cause
EGR valve sticking	Buildup of carbon
	Electrical fault
High CO and high HC	Rich mixture
	Blocked air filter
	Damaged catalytic converter
	Engine management system fault
Low CO and high HC	Misfire
	Fouled plug(s)
	Weak mixture
Low CO and low or normal HC	Exhaust leak
	Fouled injector

7.11 Fuel injection

7.11.1 Introduction

The ideal air fuel ratio is about 14.7:1. This is the theoretical amount of air required to completely burn the fuel. It is given a 'lambda (λ)' value of 1.

$$\lambda = \text{actual air quantity} \div \text{theoretical air quantity}$$

Air fuel ratio is altered during the following operating conditions of an engine to improve its performance, driveability, consumption and emissions:

- **cold starting** richer mixture is needed to compensate for fuel condensation and improve driveability;
- **load or acceleration** richer to improve performance;
- **cruise or light loads** weaker for economy;
- **overrun** very weak (if any) fuel, to improve emissions and economy.

The more accurately the air fuel ratio is controlled to cater for external conditions, then the better the overall operation of the engine.

The major advantage, therefore, of a fuel injection system is accurate control of the fuel quantity injected into the engine. The basic principle of fuel injection is that if petrol is supplied to an injector (electrically controlled valve), at a constant differential pressure, then the amount of fuel injected will be directly proportional to the injector open time.

Most systems are now electronically controlled even if containing some mechanical metering components. This allows the operation of the injection system to be very closely matched to the requirements of the engine. This matching process is carried out during development on test beds and

Mazda 626 2.0i GX 16 valve   **1996–1997**

Adjustment Data (Fuel)

DESCRIPTION	SETTING	UNITS
Carburettor/injection make	Mazda	
Carburettor/injection type	MPi	
Fuel pump pressure	5.10 ± 0.70	bar
Injection pressure	1.50	bar
Idle speed	700 ± 50	rev/min
Raised idle speed	Not applicable	rev/min
CO at idle speed	0.50 maximum	%
carbon dioxide at idle speed	14.50/16.00	%
HC at idle speed	100	ppm
Oxygen at idle speed	0.10/0.50	%

Figure 7.24 Typical emissions data

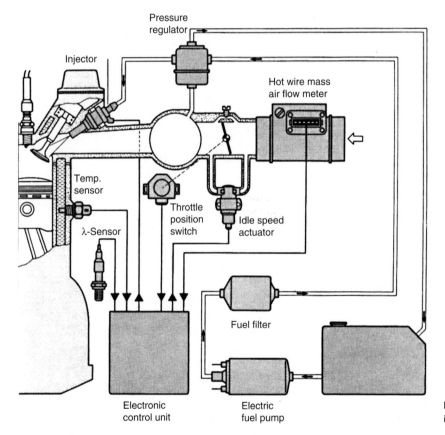

Figure 7.25 LH-Jetronic fuel injection

dynamometers, as well as development in the car. The ideal operating data for a large number of engine operating conditions is stored in a ROM in the ECU. Close control of fuel quantity injected allows the optimum setting for mixture strength when all operating factors are taken into account.

Further advantages of electronic fuel injection control are that overrun cut off can easily be implemented, fuel can be cut at the engines rev limit and information on fuel used can be supplied to a trip computer. Figure 7.25 shows a typical fuel injection system layout for LH-Jetronic.

7.11.2 Injection systems

Fuel injection systems can be classified into two main categories, single point injection and multipoint injection. Figure 7.26 shows these techniques together with a third known as continuous injection (K/KE Jetronic). Depending on the sophistication of the system, idle speed and idle mixture adjustment can be either mechanically or electronically controlled.

Figure 7.27 shows a block diagram of inputs and outputs common to most fuel injection systems. Note that the two most important input sensors to

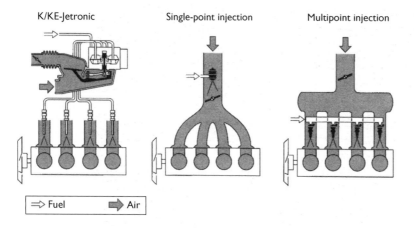

Figure 7.26 Petrol injection systems

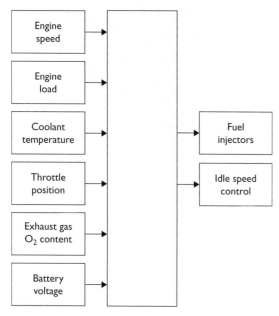

Figure 7.27 Fuel injection system block diagram

Figure 7.28 Engine management ECU

7.11.3 Fuel injection components

Many of the sensors and actuators associated with fuel injection are covered in Chapter 6. Figure 7.29 shows those associated with the LH-Jetronic injection system. Starting at the top left and working clockwise the components are as follows.

Air flow meter

The type shown is a hot-wire meter. This allows direct measurement of air mass as temperature compensation is built in. The air quantity helps to determine the fuel required.

Electronic control unit (ECU)

This is also referred to as the electronic control module (ECM). The circuitry to react to the sensor signals by controlling the actuators is in the ECU. The data is stored in ROM.

Fuel pump

Pressurised fuel is supplied to the injectors. Most pumps work on the centrifugal roller principle. The pump ensures a constant supply of fuel to the fuel rail. The volume in the rail acts as a swamp to prevent pressure fluctuations as the injectors

the system are speed and load. The basic fuelling requirement is determined from these inputs in a similar way to the determination of ignition timing.

An engine's fuelling requirements are stored as part of a ROM chip in the ECU. When the ECU has determined the 'look up value' of the fuel required (injector open time), corrections to this figure can be added for battery voltage, temperature, throttle change or position and fuel cut off. Figure 7.28 shows an injection system ECU.

Idle speed and fast idle are also generally controlled by the ECU and a suitable actuator. It is also possible to have a form of closed loop control with electronic fuel injection. This involves a lambda sensor to monitor exhaust gas oxygen content. This allows very accurate control of the mixture strength, as the oxygen content of the exhaust is proportional to the air fuel ratio. The signal from the lambda sensor is used to adjust the injector open time.

Figure 7.29 Injection components in position

operate. The pump must be able to maintain a pressure of about 3 bar.

Fuel filter

The fuel supplied to the injectors must be free from any contamination or else the injector nozzle will be damaged or blocked.

Lambda sensor

The quantity of oxygen in the exhaust, when accurately measured, ensures that the fuel air mixture is kept within the lambda window (0.97 to 1.03).

Temperature sensor

A simple thermistor is used to determine the engine coolant temperature.

Fuel injectors

These are simple solenoid operated valves designed to operate very quickly and produce a finely atomised spray pattern.

Idle or fast idle control actuator

The rotary actuator is shown here. It is used to provide extra air for cold fast idle conditions as well as controlling idle speed. It is supplied with a variable duty cycle square wave.

Fuel pressure regulator

This device is to ensure a constant differential pressure across the injectors. It is a mechanical device and has a connection to the inlet manifold.

Throttle position switch

This is used to supply information as to whether the throttle is at idle, full load or somewhere in between.

7.11.4 Fuel mixture calculation

The quantity of fuel to be injected is determined primarily by the quantity of air drawn into the engine. This is dependent on two factors:

- engine speed (rev/min);
- engine load (inlet manifold pressure).

This speed load characteristic is held in the ECU memory in ROM lookup tables.

A sensor connected to the manifold by a pipe senses manifold absolute pressure. The sensor is fed with a stabilised 5 V supply and transmits an output voltage according to the pressure. The sensor is fitted away from the manifold and hence a pipe is required to connect it. The output signal varies between about 0.25 V at 0.17 bar to about 4.75 V at 1.05 bar. The density of air varies with temperature such that the information from the

MAP sensor on air quantity will be incorrect over wide temperature variations. An air temperature sensor is used to inform the ECU of the inlet air temperature such that the quantity of fuel injected may be corrected. As the temperature of air decreases its density increases and hence the quantity of fuel injected must also be increased. The other method of sensing engine load is direct measurement of air intake quantity using a hot-wire meter or a flap type air flow meter.

In order to operate the injectors the ECU needs to know, in addition to air pressure, the engine speed to determine the injection quantity. The same flywheel sensor used by the ignition system provides this information. All four injectors operate simultaneously once per engine revolution, injecting half of the required fuel. This helps to ensure balanced combustion. The start of injection varies according to ignition timing.

A basic open period for the injectors is determined by using the ROM information relating to manifold pressure and engine speed. Two corrections are then made, one relative to air temperature and another depending on whether the engine is idling, at full or partial load.

The ECU then carries out another group of corrections, if applicable:

- after start enrichment;
- operational enrichment;
- acceleration enrichment;
- weakening on deceleration;
- cut off on overrun;
- reinstatement of injection after cut off;
- correction for battery voltage variation.

Under starting conditions the injection period is calculated differently. This is determined mostly from a set figure varied as a function of temperature.

The coolant temperature sensor is a thermistor and is used to provide a signal to the ECU relating to engine coolant temperature. The ECU can then calculate any corrections to fuel injection and ignition timing. The operation of this sensor is the same as the air temperature sensor.

The throttle potentiometer is fixed on the throttle butterfly spindle and informs the ECU of throttle position and rate of change of throttle position. The sensor provides information on acceleration, deceleration and whether the throttle is in the full load or idle position. It comprises a variable resistance and a fixed resistance. As is common with many sensors a fixed supply of 5 V is provided and the return signal will vary approximately between 0 and 5 V. The voltage increases as the throttle is opened.

7.12 Diagnostics – fuel injection systems

7.12.1 Testing procedure

Warning
Caution/Achtung/Attention – burning fuel can seriously damage your health!

The following procedure is generic and with a little adaptation can be applied to any fuel injection system. Refer to manufacturer's recommendations if in any doubt. It is assumed that the ignition system is operating correctly. Most tests are carried out while cranking the engine.

7.12.2 Fuel injection fault diagnosis table

Symptom	Possible fault
Engine rotates but does not start	No fuel in the tank! Air filter dirty or blocked Fuel pump not running No fuel being injected
Difficult to start when cold	Air filter dirty or blocked Fuel system wiring fault Enrichment device not working Coolant temperature sensor short circuit
Difficult to start when hot	Air filter dirty or blocked Fuel system wiring fault Coolant temperature sensor open circuit
Engine starts but then stops immediately	Fuel system contamination Fuel pump or circuit fault (relay) Intake system air leak
Erratic idle	Air filter blocked Inlet system air leak Incorrect CO setting Idle air control valve not operating Fuel injectors not spraying correctly
Misfire through all speeds	Fuel filter blocked Fuel pump delivery low Fuel tank ventilation system blocked
Engine stalls	Idle speed incorrect CO setting incorrect Fuel filter blocked Air filter blocked Intake air leak Idle control system not working
Lack of power	Fuel filter blocked Air filter blocked Low fuel pump delivery Fuel injectors blocked
Backfires	Fuel system fault (air flow sensor on some cars) Ignition timing

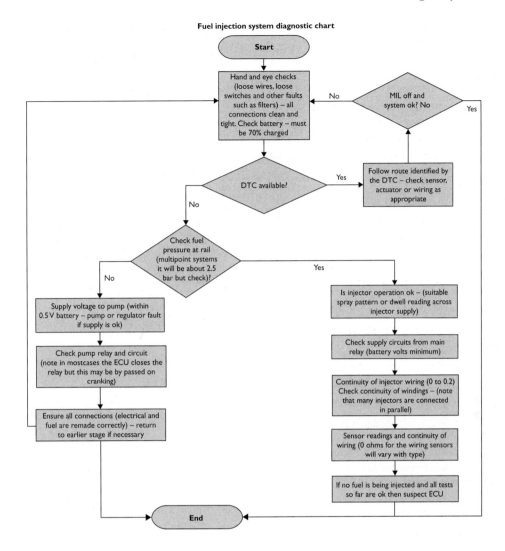

Fuel injection system diagnostic chart

7.13 Diesel injection

7.13.1 Introduction

The basic principle of the four stroke diesel engine is very similar to the petrol system. The main difference is that the mixture formation takes place in the cylinder combustion chamber as the fuel is injected under very high pressure. The timing and quantity of the fuel injected is important from the usual issues of performance, economy and emissions.

Fuel is metered into the combustion chamber by way of a high pressure pump connected to injectors via heavy duty pipes. When the fuel is injected it mixes with the air in the cylinder and will self ignite at about 800°C. The mixture formation in the cylinder is influenced by the following factors.

The timing of a diesel fuel injection pump to an engine is usually done using start of delivery as the reference mark. The actual start of injection, in other words when fuel starts to leave the injector, is slightly later than start of delivery, as this is influenced by the compression ratio of the engine, the compressibility of the fuel and the length of the delivery pipes. This timing has a great effect on the production of carbon particles (soot), if too early, and increases the hydrocarbon emissions, if too late.

The duration of the injection is expressed in degrees of crankshaft rotation in milliseconds. This clearly influences fuel quantity but the rate of discharge is also important. This rate is not constant due to the mechanical characteristics of the injection pump.

Pressure of injection will affect the quantity of fuel but the most important issue here is the effect

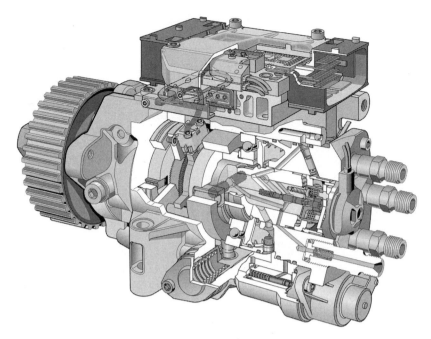

Figure 7.30 Bosch VR Pump (*Source*: Bosch Press)

on atomisation. At higher pressures the fuel will atomise into smaller droplets with a corresponding improvement in the burn quality. Indirect injection systems use pressures up to about 350 bar and direct injection systems can be up to about 1000 bar. Emissions of soot are greatly reduced by higher pressure injection.

The direction of injection must match very closely the swirl and combustion chamber design. Deviations of only 2° from the ideal can greatly increase particulate emissions.

Diesel engines do not in general use a throttle butterfly as the throttle acts directly on the injection pump to control fuel quantity. At low speeds in particular the very high excess air factor ensures complete burning and very low emissions. Diesel engines operate where possible with an excess air factor even at high speeds.

Figure 7.30 shows a typical diesel fuel injection pump. Detailed operation of the components is beyond the scope of this book. The principles and problems are the issues under consideration, in particular the way electronics can be employed to solve some of the problems.

7.13.2 Electronic control of diesel injection

The advent of electronic control over the diesel injection pump has allowed many advances over the purely mechanical system. The production of high pressure and injection is, however, still mechanical with all current systems. The following advantages are apparent over the non-electronic control system:

- more precise control of fuel quantity injected;
- better control of start of injection;
- idle speed control;
- control of exhaust gas recirculation;
- drive by wire system (potentiometer on throttle pedal);
- an anti-surge function;
- output to data acquisition systems, etc.;
- temperature compensation;
- cruise control.

Because fuel must be injected at high pressure the hydraulic head, pressure pump and drive elements are still used. An electromagnetic moving iron actuator adjusts the position of the control collar, which in turn controls the delivery stroke and therefore the injected quantity of fuel. Fuel pressure is applied to a roller ring and this controls the start of injection. A solenoid operated valve controls the supply to the roller ring. These actuators together allow control of start of injection and injection quantity.

Ideal values for fuel quantity and timing are stored in memory maps in the ECU. The injected fuel quantity is calculated from the accelerator position and the engine speed. Start of injection is determined from the following:

- fuel quantity;
- engine speed;

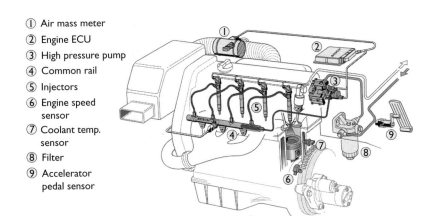

① Air mass meter
② Engine ECU
③ High pressure pump
④ Common rail
⑤ Injectors
⑥ Engine speed sensor
⑦ Coolant temp. sensor
⑧ Filter
⑨ Accelerator pedal sensor

Figure 7.31 Common rail diesel system components (*Source*: Bosch Press)

- engine temperature;
- air pressure.

The ECU is able to compare start of injection with actual delivery from a signal produced by the needle motion sensor in the injector.

Control of exhaust gas recirculation is a simple solenoid valve. This is controlled as a function of engine speed, temperature and injected quantity. The ECU is also in control of the stop solenoid and glow plugs via a suitable relay.

7.13.3 Common rail (CR) diesel systems

The development of diesel fuel systems is continuing, with many new electronic changes to the control and injection processes. One of the latest developments is the 'common rail' system, operating at very high injection pressures. It also has piloted and phased injection to reduce noise and vibration.

The common rail system (Figure 7.31) has made it possible, on small high speed diesel engines, to have direct injection when previously they would have been of indirect injection design. These developments are showing improvements in fuel consumption and performance of up to 20% over the earlier indirect injection engines of a similar capacity. The common rail injection system can be used on the full range of diesel engine capacities.

The combustion process, with common rail injection, is improved by a pilot injection of a very small quantity of fuel, at between 40° and 90° btdc. This pilot fuel ignites in the compressing air charge so that the cylinder temperature and pressure are higher than in a conventional diesel injection engine at the start of injection. The higher temperature and pressure reduces ignition lag to a minimum, so that the controlled combustion phase during the main injection period, is softer and more efficient.

Fuel injection pressures are varied – throughout the engine speed and load range – to suit the instantaneous conditions of driver demand and engine speed and load conditions. Data input, from other vehicle system ECUs, is used to further adapt the engine output, to suit changing conditions elsewhere on the vehicle. Examples are traction control, cruise control and automatic transmission gearshifts.

The electronic diesel control (EDC) module carries out calculations to determine the quantity of fuel delivered. It also determines the injection timing based on engine speed and load conditions.

The actuation of the injectors, at a specific crankshaft angle, and for a specific duration is made by signal currents from the EDC module. A further function of the EDC module is to control the accumulator (rail) pressure.

In summary, common rail diesel fuel injection systems consist of four main component areas.

1. Low pressure delivery.
2. High pressure delivery with a high pressure pump and accumulator (the rail).
3. Electronically controlled injectors (Figure 7.32).
4. Electronic control unit and associated sensors and switches.

The main sensors for calculation of the fuel quantity and injection advance requirements are the accelerator pedal sensor, crankshaft speed and position sensor, air-mass meter and the engine coolant temperature sensor.

7.13.4 Diesel exhaust emissions

Exhaust emissions from diesel engines have been reduced considerably by changes in the design of combustion chambers and injection techniques.

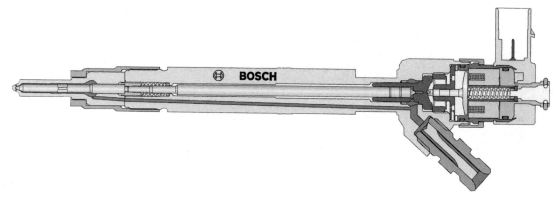

Figure 7.32 Sectioned details of the CR injector (*Source*: Bosch Press)

More accurate control of start of injection and spill timing has allowed further improvements to be made. Electronic control has also made a significant contribution. A number of further techniques can be employed to control emissions.

Overall the emissions from diesel combustion are far lower than emission from petrol combustion. The CO, HC and NO_x emissions are lower mainly due to the higher compression ratio and excess air factor. The higher compression ratio improves the thermal efficiency and thus lowers the fuel consumption. The excess air factor ensures more complete burning of the fuel.

The main problem area is that of particulate emissions. These particle chains of carbon molecules can also contain hydrocarbons, aldehydes mostly. The dirt effect of this emission is a pollution problem but the possible carcinogenic effect of this soot gives cause for concern. The diameter of these particles is only a few ten thousandths of a millimetre. This means they float in the air and can be inhaled.

In much the same way as with petrol engines EGR is employed primarily to reduce NO_x emissions by reducing the reaction temperature in the combustion chamber. However if the percentage of EGR is too high, increased hydrocarbons and soot are produced. This is appropriate to turbocharged engines such that if the air is passed through an intercooler, as well as improvements in volumetric efficiency, lower temperature will again reduce the production of NO_x. The intercooler is fitted in the same way as the cooling system radiator.

7.13.5 Catalytic converter – diesel

On a diesel engine a catalyst can be used to reduce the emission of hydrocarbons but will have less effect on nitrogen oxides. This is because diesels are always run with excess air to ensure better and more efficient burning of the fuel. A normal catalyst therefore will not strip the oxygen off the NO_x to oxidise the hydrocarbons because the excess oxygen will be used. Special NO_x converters are becoming available.

7.13.6 Filters

To reduce the emission of particulate matter (soot), filters can be used. These can vary from a fine grid design made from a ceramic material to centrifugal filters and water trap techniques. The problem to overcome is that the filters can get blocked, which adversely affects the overall performance. Several techniques are employed including centrifugal filters.

7.14 Diagnostics – diesel injection systems

7.14.1 Test equipment

Note: You should always refer to the manufacturer's instructions appropriate to the equipment you are using.

Smoke meter (Figure 7.33)

This is an essential device in the UK and other countries where the level of smoke in the exhaust forms part of the annual test. Most devices use infrared light to 'count' the number of soot particles in the exhaust sample. This particulate matter (PM) is highly suspected of being carcinogenic.

Injector tester (Figure 7.34)

The pressure required to 'crack' (lift the nozzle) on an injector can be tested.

Engine systems **131**

7.14.3 Diesel engine smoke

Diesel fuel is a hydrocarbon fuel. When it is burned in the cylinder it will produce carbon dioxide and water. There are, however, many circumstances under which the fuel may not be completely burned and one of the results is smoke. Despite the fact that diesel engines are designed to run under all conditions with an excess of air, problems still occur. Very often these smoke problems are easily avoided by proper maintenance and servicing of the engine and its fuel system. The emission of smoke is usually due to a shortage of air (oxygen). If insufficient air is available for complete combustion then unburnt fuel will be exhausted as tiny particles of fuel (smoke).

The identification of the colour of diesel smoke and under what conditions it occurs can be helpful in diagnosing what caused it in the first place. Poor quality fuel reduces engine performance, increases smoke and reduces engine life. There are three colours of smoke; white, blue and black. All smoke diagnosis tests must be carried out with the engine at normal operating temperature.

Figure 7.33 Smoke meter

White or grey smoke

White smoke is vaporised unburnt fuel and is caused by there being sufficient heat in the cylinder to vaporise it but not enough remaining heat to burn it. All diesel engines generate white smoke when starting from cold and it is not detrimental to the engine in any way – it is a diesel characteristic. Possible causes of white smoke are listed below.

Figure 7.34 Diesel injector tester

- **Faulty cold starting equipment** cold engines suffer from a delay in the combustion process. A cold start unit is fitted to advance the injection timing to counteract this delay. This means that white smoke could be a cold start unit problem.
- **Restrictions in the air supply** a partially blocked air cleaner will restrict the air supply; an easy cause to rectify but often overlooked. Incidentally, a blocked air cleaner element at light load in the workshop becomes a black smoke problem when the engine is under load. In both cases there will not be sufficient air entering the cylinder for the piston to compress and generate full heat for combustion.
- **Cold running** check the cooling system thermostat to see if the correct rated thermostat is fitted.
- **Incorrect fuel injection pump timing** if fuel is injected late (retarded timing) it may be vaporised but not burned.

7.14.2 Diesel injection fault diagnosis table

Symptom	Possible fault
Engine rotates but does not start	No fuel in the tank! Cam belt broken Fuel pump drive broken Open circuit supply to stop solenoid Fuel filter blocked
Excessive smoke	Refer to the next section
Lack of power	Timing incorrect Governor set too low Injector nozzles worn Injector operating pressure incorrect
Difficult to start	Timing incorrect Glow plugs not working
Fuel smell in the car	Fuel lines leaking Leak off pipes broken
Diesel knock (particularly when cold)	Timing incorrect Glow plug hold on for idle circuit not working
Engine oil contaminated with fuel	Piston broke (like me after a good holiday!) Worn piston rings Excessive fuel injected

- **Poor compressions** poor compressions may lead to leakage during the compression stroke and inevitably less heat would be generated.
- **Leaking cylinder head gasket** if coolant were leaking into the combustion area, the result would be less temperature in the cylinder causing white smoke. Steam may also be generated if the leak is sufficient. All internal combustion engines have water as a by-product from burning fuel – you will have noticed your own car exhaust, especially on a cold morning.

Blue smoke

Blue smoke is almost certainly a lubricating oil burning problem. Possible causes of blue smoke are listed:

- incorrect grade of lubricating oil;
- worn or damaged valve stem oil seals, valve guides or stems where lubricating oil is getting into the combustion chamber;
- worn or sticking piston rings;
- worn cylinder bores.

Black smoke

Black smoke is partly burned fuel. Possible causes are listed below.

- **Restriction in air intake system** a blocked air cleaner element will not let enough air in to burn all the fuel.
- **Incorrect valve clearances** excessive valve clearances will cause the valves not to fully open and to close sooner. This is another form of insufficient air supply.
- **Poor compressions** air required for combustion may leak from the cylinder.
- **Defective or incorrect injectors** check the injector to see if the spray is fully atomised and solid fuel is not being injected.
- **Incorrect fuel injection pump timing** this is less likely because the timing would need to be advanced to the point where additional engine noise would be evident.
- **Low boost pressure** if a turbocharger is fitted and is not supplying enough air for the fuel injected this is another form of air starvation.

7.14.4 Glow plug circuit

Figure 7.35 shows a typical glow plug circuit controlled by an ECU. Most timer circuits put the glow plugs on for a few seconds prior to cranking. A warning light may be used to indicate the 'ready' condition to the driver. Take care to note the type

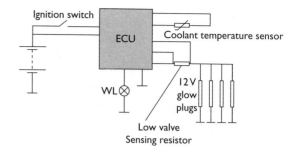

Figure 7.35 Glow plug circuit

of glow plugs used; most are 12 V and connected in parallel but some are connected in series (4 × 3 V plugs). To check the operation of most 12 V glow plug circuits, use the following steps.

1. Hand and eye checks.
2. Battery condition – at least 70%.
3. Engine must be cold – it may be possible to simulate this by disconnecting the temperature sensor.
4. Voltage supplied to plugs when ignition is switched on (spring loaded position in some cases) – 10 to 12 V.
5. Warning light operation – should go out after a few seconds.
6. Voltage supplied to plugs while cranking – 9 to 11 V.
7. Voltage supplied to plugs after engine has started – 0 V or if silent idle system is used 5 to 6 V for several minutes.
8. Same tests with engine at running temperature – glow plugs may not be energised or only for the starting phase.

7.14.5 Diesel systems

It is recommended that when the injection pump or the injectors are diagnosed as being at fault, reconditioned units should be fitted. Other than basic settings of timing, idle speed and governor speed, major overhaul is often required. Figure 7.36 shows a general diagnosis pattern for diesel systems produced by 'Autologic'.

7.15 Engine management

7.15.1 Introduction

As the requirement for lower and lower emissions continues together with the need for better performance, other areas of engine control are constantly being investigated. This is becoming even more important as the possibility of carbon

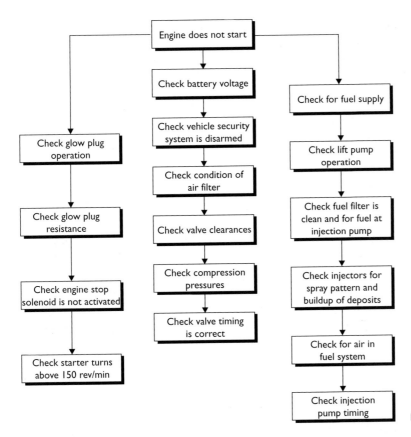

Figure 7.36 General diagnosis pattern

dioxide emissions being included in the regulations increases. Some of the current and potential areas for further control of engine operation are included in this section. Most of the common areas of 'control' have been covered under either Ignition (Section 7.7) or Fuel injection (Section 7.11). The main areas of control are as follows:

- ignition timing;
- dwell angle;
- fuel quantity;
- idle speed.

Further areas of engine control may include:

- EGR;
- canister purge;
- valve timing;
- inlet manifold length;
- closed loop lambda control.

It is not possible for an engine to operate at its best volumetric efficiency with fixed manifolds. This is because the length of the inlet tract determines the velocity of the intake air and in particular the propagation of the pressure waves set up by the pumping action of the cylinders. These standing waves can be used to improve the ram effect of the charge as it enters the cylinder but only if they coincide with the opening of the inlet valves. The length of the inlet tract has an effect on the frequency of these waves.

With the widespread use of twin cam engines, one cam for the inlet valves and one for the exhaust valves, it is possible to vary the valve overlap while the engine is running. Honda has a system that improves the power and torque range by opening both of the inlet valves only at higher speed.

A BMW system uses oil pressure controlled by valves to turn the cam with respect to its drive gear. This alters the cam phasing or relative position. The position of the cams is determined from a suitable map held in ROM in the control unit.

7.15.2 Closed loop lambda control

Current regulations have almost made closed loop control of air fuel mixture in conjunction with a three way catalytic converter mandatory. It is under discussion that a lambda value of one may become compulsory for all operating conditions, but this is yet to be agreed.

Lambda control is a closed loop feedback system in that the signal from a lambda sensor in the exhaust can directly affect the fuel quantity injected. The lambda sensor is described in more detail in Chapter 6.

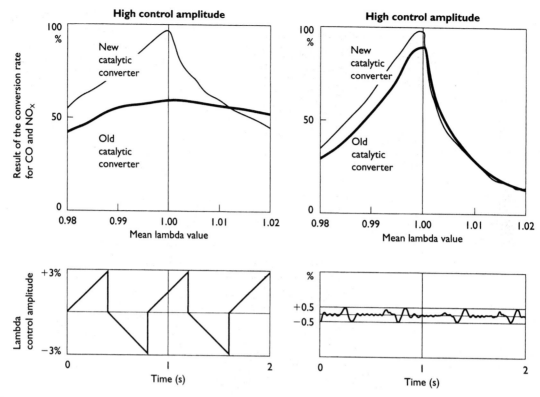

Figure 7.37 Catalytic convertor conversion rate – effect of lambda control

A graph to show the effect of lambda control in conjunction with a catalytic converter is shown in Figure 7.37. The principle of operation is as follows; the lambda sensor produces a voltage which is proportional to the oxygen content of the exhaust which is in turn proportional to the air fuel ratio. At the ideal setting this voltage is about 450 mV. If the voltage received by the ECU is below this value (weak mixture) the quantity of fuel injected is increased slightly. If the signal voltage is above the threshold (rich mixture) the fuel quantity is reduced. This alteration in air fuel ratio must not be too sudden as it could cause the engine to buck. To prevent this the ECU contains an integrator, which changes the mixture over a period of time.

A delay also exists between the mixture formation in the manifold and the measurement of the exhaust gas oxygen. This is due to the engine's working cycle and the speed of the inlet mixture, the time for the exhaust to reach the sensor and the sensor's response time. This is sometimes known as dead time and can be as much as one second at idle speed but only a few hundred milliseconds at higher engine speeds.

Due to the dead time the mixture cannot be controlled to an exact value of lambda equals one. If the integrator is adjusted to allow for engine speed then it is possible to keep the mixture in the lambda window (0.97 to 1.03), which is the region in which the TWC is at its most efficient.

7.15.3 Engine management operation

The combination of ignition and injection control has several advantages. The information received from various sensors is used for computing both fuelling and ignition requirements. Perhaps more importantly ignition and injection are closely linked. The influence they have on each other can be easily taken into account to ensure that the engine is working at its optimum, under all operating conditions. Figure 7.38 shows the layout and components of a common system.

Overall this type of system is less complicated than separate fuel and ignition systems and in many cases the ECU is able to work in an emergency mode by substituting missing information from sensors with pre-programmed values. This will allow limited but continued operation in the event of certain system failures.

The ignition system is integrated and is operated without a high tension (HT) distributor. The ignition process is controlled digitally by the ECU. The data for the ideal characteristics are stored in ROM from information gathered during

Engine systems 135

Figure 7.38 Engine management – Motronic

both prototyping and development of the engine. The main parameters for ignition advance are engine speed and load but greater accuracy can be achieved by taking further parameters into account such as engine temperature. This provides both optimum output and close control of anti-pollution levels. Performance and pollution level control means that the actual ignition point must be in many cases a trade off between the two.

The injection system is multipoint and, as is the case for all fuel systems, the amount of fuel delivered is primarily determined by the amount of air 'drawn' into the engine. The method for measuring this data is indirect in the case of this system as a pressure sensor is used to determine the air quantity.

Electromagnetic injectors control fuel supply into the engine. The injector open period is determined by the ECU. This will obtain very accurate control of the air fuel mixture under all operating conditions of the engine. The data for this is stored in ROM in the same way as for the ignition.

The main source of reference for the ignition system is from the crankshaft position sensor. This is a magnetic inductive pick up sensor positioned next to a flywheel ring containing 58 teeth. Each tooth takes up a 6° angle of the flywheel with one 12° gap positioned 114° before top dead centre (BTDC) for number 1 cylinder. The signal produced by the flywheel sensor is essentially a sine wave with a cycle missing corresponding to the gap in the teeth of the reluctor plate. The information provided to the ECU is engine speed from the frequency of the signal and engine position from the number of pulses before or after the missed pulses.

The basic ignition advance angle is obtained from a memorised cartographic map. This is held in a ROM chip within the ECU. The parameters for this are:

- **engine rev/min** given by the flywheel sensor;
- **inlet air pressure** given by the MAP sensor.

The above two parameters (speed and load) give the basic setting but to ensure optimum advance angle the timing is corrected by:

- coolant temperature
- air temperature
- throttle butterfly position

The ignition is set to a predetermined advance during the starting phase. Figure 7.39 shows a typical advance, fuelling and other maps used by the Motronic system. This data is held in ROM. For full ignition control, the ECU has to first

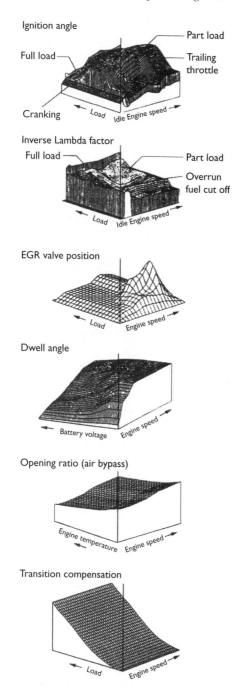

Figure 7.39 Motronic control maps (samples)

determine the basic timing for three different conditions.

- Under idling conditions ignition timing is often moved very quickly by the ECU in order to control idle speed. When timing is advanced engine speed will increase within certain limits.
- Full load conditions require careful control of ignition timing to prevent combustion knock. When a full load signal is sensed by the ECU (high manifold pressure) the ignition advance angle is reduced.
- Partial throttle is the main area of control and, as already stated, the basic timing is set initially by a programme as a function of engine speed and manifold pressure.

Corrections are the added according to:

- operational strategy
- knock protection
- phase correction

The ECU will also control ignition timing variation during overrun fuel cut off and reinstatement and also to ensure anti-jerk control. When starting the ignition timing plan is replaced by a specific starting strategy. Phase correction is when the ECU adjusts the timing to take into account the time taken for the HT pulse to reach the spark plugs. To ensure good driveability the ECU can limit the variations between the two ignition systems to a maximum value, which varies according to engine speed and the basic injection period.

An anti jerk function operates when the basic injection period is less than 2.5 ms and the engine speed is between 720 and 3200 rev/min. This function operates to correct the programmed ignition timing in relation to the instantaneous engine speed and a set filtered speed. This is done to stabilise the engine rotational characteristics as much as possible.

In order to maintain constant energy HT the dwell period must increase in line with engine speed. So that the ignition primary current reaches its maximum at the point of ignition the ECU controls the dwell by use of another memory map, which takes battery voltage into account.

Fuel is collected from the tank by a pump either immersed in it or outside, but near the tank. The immersed type is quieter in operation and has better cooling and no internal leaks. The fuel is directed forwards to the fuel rail or manifold, via a paper filter.

Fuel pressure is maintained at about 2.5 bar above manifold pressure by a regulator mounted on the fuel rail. Excess fuel is returned to the tank. The fuel is usually picked up via a swirl pot in the tank to prevent aeration of the fuel. Each of the four inlet manifold tracts has its own injector.

The fuel pump is a high pressure type and is a two stage device: a low pressure stage created by a turbine drawing fuel from the tank; and a high pressure stage created by a gear pump delivering fuel to the filter. It is powered by a 12 V supply from the fuel pump relay, which is controlled by the ECU as a safety measure.

The rotation of the turbine draws fuel in via the inlet. The fuel passes through the turbine and enters the pump housing where it is pressurised

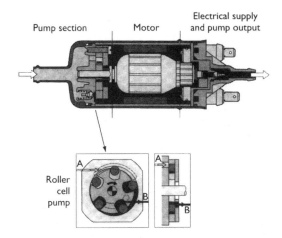

Figure 7.40 Roller cell fuel pump

by rotation of the pump and the reduction of the volume in the gear chambers. This pressure opens a residual valve and fuel passes to the filter. When the pump stops, pressure is maintained by this valve, which prevents the fuel returning. If, due to a faulty regulator or a blockage in the line, fuel pressure rises above 7 bar an over pressure valve will open releasing fuel back to the tank. Figure 7.40 shows this type of pump.

The fuel filter is placed between the fuel pump and the fuel rail. It is fitted one way only to ensure that the outlet screen traps any paper particles from the filter element. The filter will stop contamination down to between 8 and 10 microns. Replacement varies between manufacturers but 80 000 km (50 000 miles) is often recommended.

The fuel rail, in addition to providing a uniform supply to the injectors, acts as an accumulator. Depending on the size of the fuel rail some systems also use an extra accumulator. The volume of the fuel rail is large enough to act as a pressure fluctuation damper ensuring that all injectors are supplied with fuel at a constant pressure.

Multipoint systems use one injector for each cylinder although very high performance vehicles may use two. The injectors are connected to the fuel rail by a rubber seal. The injector is an electrically operated valve manufactured to a very high precision. The injector comprises a body and needle attached to a magnetic core. When the winding in the injector housing is energised the core or armature is attracted and the valve opens compressing a return spring. The fuel is delivered in a fine spray to wait behind the closed inlet valve until the induction stroke begins. Providing the pressure across the injector remains constant the quantity of fuel admitted is related to the open period, which in turn is determined by the time the electromagnetic circuit is energised.

The purpose of the fuel pressure regulator is to maintain differential pressure across the injectors at a predetermined constant. This means the regulator must adjust the fuel pressure in response to changes in manifold pressure. It is made of two compressed cases containing a diaphragm, a spring and a valve.

The calibration of the regulator valve is determined by the spring tension. Changes in manifold pressure vary the basic setting. When the fuel pressure is sufficient to move the diaphragm, the valve opens and allows fuel to return to the tank. The decrease in pressure in the manifold, also acting on the diaphragm at say idle speed, will allow the valve to open more easily, hence maintaining a constant differential pressure between the fuel rail and the inlet manifold. This is a constant across the injectors and hence the quantity of fuel injected is determined only by the open time of the injectors. The differential pressure is maintained at about 2.5 bar.

The air supply circuit will vary considerably between manufacturers but an individual manifold from a collector housing, into which the air is fed via a simple butterfly, essentially supplies each cylinder. The air is supplied from a suitable filter. A supplementary air circuit is utilised during the warm-up period after a cold start and to control idle speed.

7.15.4 Gasoline direct injection (GDI)

High-pressure injection systems for petrol/gasoline engines is based on a pressure reservoir and a fuel rail, which a high-pressure pump charges to a regulated pressure of up to 120 bar. The fuel can therefore be injected directly into the combustion chamber via electromagnetic injectors.

The air mass drawn in can be adjusted through the electronically controlled throttle valve and is measured with the help of an air-mass meter. For mixture control, a wide-band oxygen sensor is used in the exhaust, before the catalytic converters. This sensor can measure a range between a lambda value of 0.8 and infinity. The engine electronic control unit regulates the operating modes of the engine with gasoline direct injection in three ways.

1. Stratified charge operation – with lambda values greater than 1.
2. Homogeneous operation – at lambda = 1.
3. Rich homogeneous operation – with lambda = 0.8.

Compared to the traditional manifold injection system, the Bosch DI-Motronic must inject the

138 *Advanced automotive fault diagnosis*

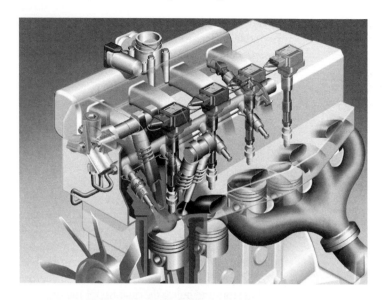

Figure 7.41 GDI system showing the main components (*Source*: Bosch Press)

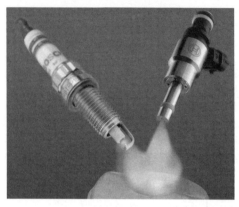

Figure 7.42 Atomization in a GDI engine (*Source*: Bosch Press)

entire fuel amount in full-load operation in a quarter of the time. The available time is significantly shorter during stratified charge operation in part-load. Especially at idle, injection times of less than 0.5 ms are required due to the lower fuel consumption (Figure 7.41). This is only one-fifth of the available time for manifold injection.

The fuel must be atomised very finely in order to create an optimal mixture in the brief moment between injection and ignition (Figure 7.42). The fuel droplets for direct injection are on average smaller than 20 μm. This is only one-fifth of the droplet size reached with the traditional manifold injection and one-third of the diameter of a single human hair. This improves efficiency considerably.

Direct injection engines operate according to the stratified charge concept in the part-load range and function with high excess air. In return, very low fuel consumption is achieved.

The engine operates with an almost completely opened throttle valve, which avoids additional alternating charge losses. With stratified charge operation, the lambda value in the combustion chamber is between about 1.5 and 3. In the part-load range, gasoline direct injection achieves the greatest fuel savings with up to 40% at idle compared to conventional petrol injection processes. With increasing engine load, and therefore increasing injection quantities, the stratified charge cloud becomes even richer and emission characteristics become worse.

Because soot may form under these conditions, the DI-Motronic engine control converts to a homogeneous cylinder charge at a pre-defined engine load. The system injects very early during the intake process in order to achieve a good mixture of fuel and air at a ratio of lambda = 1. As is the case for conventional manifold injection systems, the amount of air drawn in for all operating modes, is adjusted through the throttle valve according to the desired torque specified by the driver.

Diagnosing faults with a GDI system is little different from the manifold injection types. Extra care is needed because of the higher fuel pressures of course. Injector waveforms can be checked as can those associated with the other sensors and actuators.

7.16 Diagnostics – combined injection and fuel control systems

7.16.1 Testing procedure

Warning
Caution/Achtung/Attention – burning fuel can seriously damage your health!

Engine systems

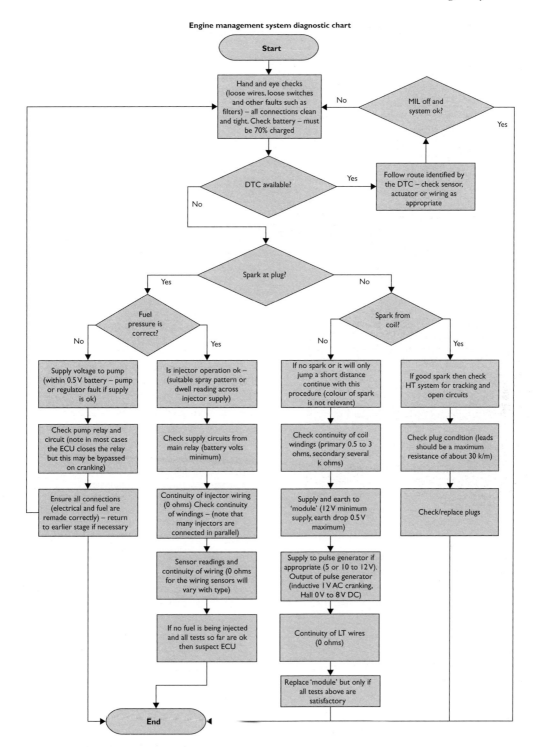

Engine management system diagnostic chart

Warning

Caution/Achtung/Attention – high voltages can seriously damage your health!

The following procedure is very generic but with a little adaptation can be applied to any system. Refer to manufacturer's recommendations if in any doubt.

7.16.2 Combined injection and fuel control fault diagnosis table

Symptom	Possible fault
Engine will not start	Engine and battery earth connections Fuel filter and fuel pump Air intake system for leaks Fuses/fuel pump/system relays

Symptom	Possible faults
Engine difficult to start when cold	Fuel injection system wiring and connections Coolant temperature sensor Auxiliary air valve/idle speed control valve Fuel pressure regulator and delivery rate ECU and connector Limp home function – if fitted Engine and battery earth connections Fuel injection system wiring and connections Fuses/fuel pump/system relays Fuel filter and fuel pump Air intake system for leaks Coolant temperature sensor Auxiliary air valve/idle speed control valve Fuel pressure regulator and delivery rate ECU and connector Limp home function – if fitted
Engine difficult to start when warm	Engine and battery earth connections Fuses/fuel pump/system relays Fuel filter and fuel pump Air intake system for leaks Coolant temperature sensor Fuel injection system wiring and connections Air-mass meter Fuel pressure regulator and delivery rate Air sensor filter ECU and connector Knock control – if fitted
Engine starts then stops	Engine and battery earth connections Fuel filter and fuel pump Air intake system for leaks Fuses/fuel pump/system relays Idle speed and CO content Throttle potentiometer Coolant temperature sensor Fuel injection system wiring and connections ECU and connector Limp home function – if fitted
Erratic idling speed	Engine and battery earth connections Air intake system for leaks Auxiliary air valve/idle speed control valve Idle speed and CO content Fuel injection system wiring and connections Coolant temperature sensor Knock control – if fitted Air-mass meter Fuel pressure regulator and delivery rate ECU and connector Limp home function – if fitted
Incorrect idle speed	Air intake system for leaks Vacuum hoses for leaks Auxiliary air valve/idle speed control valve Idle speed and CO content Coolant temperature sensor
Misfire at idle speed	Engine and battery earth connections Air intake system for leaks Fuel injection system wiring and connections Coolant temperature sensor Fuel pressure regulator and delivery rate Air-mass meter Fuses/fuel pump/system relays
Misfire at constant speed	Air flow sensor
Hesitation when accelerating	Engine and battery earth connections Air intake system for leaks Fuel injection system wiring and connections Vacuum hoses for leaks Coolant temperature sensor Fuel pressure regulator and delivery rate Air-mass meter ECU and connector Limp home function – if fitted
Hesitation at constant speed	Engine and battery earth connections Throttle linkage Vacuum hoses for leaks Auxiliary air valve/idle speed control valve Fuel lines for blockage, fuel filter and fuel pump Injector valves ECU and connector Limp home function – if fitted
Hesitation on overrun	Air intake system for leaks Fuel injection system wiring and connections Coolant temperature sensor Throttle potentiometer Fuses/fuel pump/system relays Air sensor filter Injector valves Air-mass meter
Knock during acceleration	Knock control – if fitted Fuel injection system wiring and connections Air-mass meter ECU and connector
Poor engine response	Engine and battery earth connections Air intake system for leaks Fuel injection system wiring and connections Throttle linkage Coolant temperature sensor Fuel pressure regulator and delivery rate Air-mass meter ECU and connector Limp home function – if fitted
Excessive fuel consumption	Engine and battery earth connections Idle speed and CO content Throttle potentiometer Throttle valve/housing/sticking/initial position Fuel pressure regulator and delivery rate Coolant temperature sensor Air-mass meter Limp home function – if fitted
CO level too high	Limp home function – if fitted ECU and connector Emission control and EGR valve – if fitted Fuel injection system wiring and connections Air intake system for leaks Coolant temperature sensor Fuel pressure regulator and delivery rate
CO level too low	Engine and battery earth connections Air intake system for leaks Idle speed and CO content Coolant temperature sensor Fuel injection system wiring and connections Injector valves ECU and connector Limp home function – if fitted Air-mass meter Fuel pressure regulator and delivery rate
Poor performance	Engine and battery earth connections Air intake system for leaks Throttle valve/housing/sticking/initial position Fuel injection system wiring and connections Coolant temperature sensor Fuel pressure regulator/fuel pressure and delivery rate Air-mass meter ECU and connector Limp home function – if fitted

Figure 7.43 Air flow meter under test

7.16.3 On-board diagnostics (OBD)

Figure 7.44 shows the Bosch Motronic M5 with the OBD2 system. On-board diagnostics are becoming essential for the longer term operation of a system for it to produce a clean exhaust. Many countries now require a very comprehensive diagnosis of all components in the system which affect the exhaust. Any fault detected will be indicated to the driver by a warning light. OBD2 is intended to standardise the many varying methods used by different manufacturers. EOBD is the European version; however, it is also thought that an extension to total vehicle diagnostics through a common interface is possible in the near future.

Digital electronics allow both sensors and actuators to be monitored. Allocating values to all operating states of the sensors and actuators does this. If a deviation from these figures is detected this is stored in memory and can be output in the workshop to assist with faultfinding.

Monitoring of the ignition system is very important as misfiring not only produces more emissions of hydrocarbons but the unburned fuel enters the catalytic converter and burns there. This can cause higher than normal temperatures and may damage the catalytic converter.

An accurate crankshaft speed sensor is used to monitor ignition and combustion in the cylinders. Misfiring alters the torque of the crankshaft for an instant, which causes irregular rotation. This allows a misfire to be recognised instantly.

A number of further sensors are required for the OBD2 functions. Another lambda sensor after the catalytic converter monitors its operation. An intake pressure sensor and a valve are needed to control the activated charcoal filter to reduce and monitor evaporative emissions from the fuel tank. A differential pressure sensor also monitors the fuel tank permeability. Besides the driver's fault lamp a considerable increase in the electronics is required in the control unit in order to operate this system. For more information on OBD, please refer to Chapter 6.

7.16.4 Fuel pump testing

Typical high-pressure fuel pump characteristics are:

- **delivery** 120 litres per hour (1 litre in 30 s) at 3 bar;
- **resistance** 0.8 Ω (static);
- **voltage** 12 V;
- **current** 10.5 A.

An ideal test for a fuel pump is its delivery. Using a suitable measuring receptacle, bypass the pump relay and check the quantity of fuel delivered in a set time (refer to manufacturer's specifications). A reduced amount would indicate either a fuel blockage, a reduced electrical supply to the pump or an inefficient pump.

7.16.5 Injector testing

Injectors typically have the characteristics as listed:

- **supply voltage** 12 V;
- **resistance** 16 Ω;
- **static output** 150 cc per minute at 3 bar.

142 Advanced automotive fault diagnosis

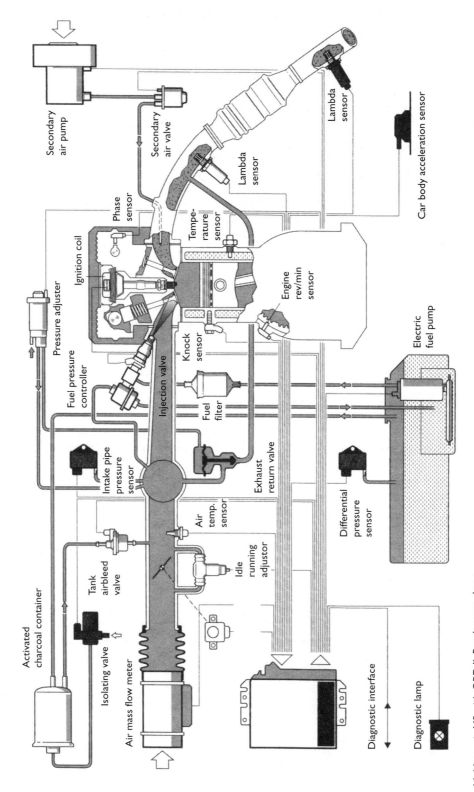

Figure 7.44 Motronic M5 with OBD II. Bosch press photo

Engine systems **143**

Resistance checks (with the supply disconnected) are an ideal start to testing injectors. Further tests with the fuel pressurised by the pump and each injector in turn held in a suitable receptacle, include:

- **spray pattern** usually a nice cone shape with good atomisation;
- **delivery** set quantity over a set time;
- **leakage** any more than two drops a minute is considered excessive (zero is desirable).

7.17 Engine management and faultfinding information

7.17.1 Diagnostic charts

'Autodata' supply diagnostic charts specific to particular management systems. Figure 7.45 is a typical example. Note that some boxes refer you to a further publication.

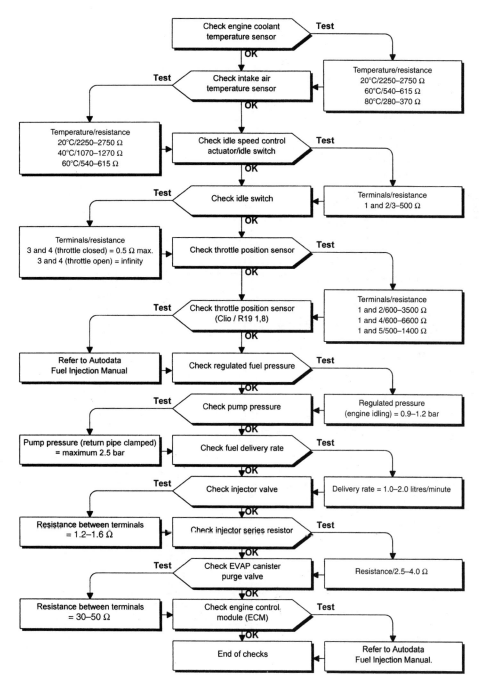

Figure 7.45 Bosch Mono-Jetronic (single point injection) (*Source*: Autodata)

7.17.2 Circuit diagrams

Circuit diagrams can be printed out from some 'Workshop manual CDs'. Figure 7.46 is an example of this.

7.17.3 Component testing data

Figure 7.47 is a printout from a CD, showing what data is available for testing a specific component.

7.18 Air supply and exhaust systems

7.18.1 Exhaust system

A vehicle exhaust system directs combustion products away from the passenger compartment, reduces combustion noise and, on most modern vehicles, reduces harmful pollutants in the exhaust stream. The main parts of the system are the exhaust manifold, the silencer or muffler,

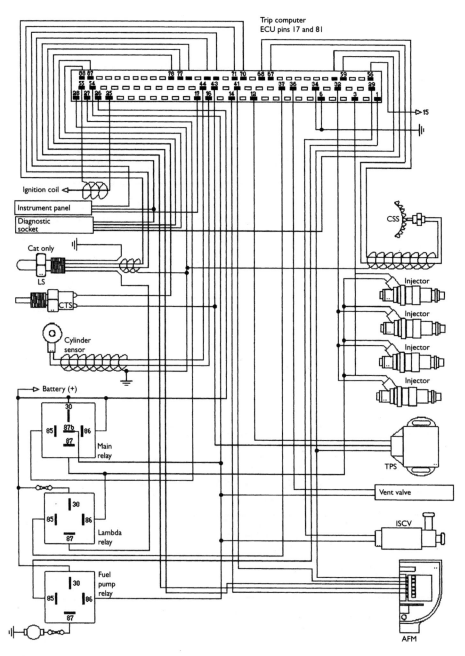

Figure 7.46 Engine management circuit diagram

the pipes connecting them, and a catalytic-converter.

Most exhaust systems are made from mild steel, but some are made from stainless steel which lasts much longer. The system is suspended under the vehicle on rubber mountings. These allow movement because the engine is also rubber mounted, and they also reduce vibration noise.

An exhaust manifold links the engine exhaust ports to the down pipe and main system. It also reduces combustion noise and transfers heat downstream to allow the continued burning of hydrocarbons and carbon monoxide. The manifold is connected to the down pipe, which in turn can be connected to the catalytic converter. Most exhaust manifolds are made from cast iron, as this has the necessary strength and heat transfer properties.

The silencer's main function is to reduce engine noise to an acceptable level. Engine noise is a mixed up collection of its firing frequencies (the number of times per second each cylinder fires). These range from about 100 to 400 Hz (cycles/sec). A silencer reduces noise in two main ways:

- interior chambers using baffles, which are tuned to set up cancelling vibrations;
- absorptive surfaces function like sound-deadening wall and ceiling panels to absorb noise.

When the exhaust gases finally leave the exhaust system, their temperature, pressure and noise have been reduced considerably. The overall length of an exhaust system including the silencers can affect the smooth flow of gases. For this reason do not alter the length or change the layout of an exhaust system. Figure 7.48 shows a silencer/muffler box.

Mazda 626 2.0i GX 16 valve   1993–1998

Fuel and Ignition Diagnostics

Air Mass Meter
position   in the intake tubing
condition
ignition on / meter plug disconnected
connection 5 / green / red v+
          from   4.8   to   5.2   volts
connection 3 / black / orange v−
condition
ignition on / meter plug disconnected
connection 2 / brown / blue v+
          from   4   to   6   volts
connection 3 / black / orange v−
condition
ignition on / meter plug connected / engine off
connection 1 / white / blue v+
          from   0   to   0.3   volts
connection 3 / black / orange v−
condition
meter plug connected / engine running at idle
connection 1 / white / blue v+
          from   0.4   to   1   volts
connection 3 / black / orange v−

Figure 7.47 Testing data

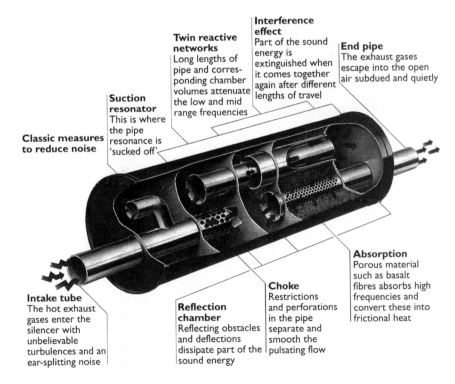

Figure 7.48 Exhaust noise reduction methods

7.18.2 Catalytic converters

Stringent regulations in most parts of the world have made the use of a catalytic converter necessary. The three-way catalyst (TWC) is used to great effect by most manufacturers. It is in effect a very simple device; it looks similar to a standard exhaust silencer box. Note that in order for the 'cat' to operate correctly, the engine must be always well tuned. This is to ensure that the right 'ingredients' are available for the catalyst to perform its function. A catalytic converter works by converting the dangerous exhaust gases into gases which are non-toxic.

The core has traditionally been made from ceramic of magnesium aluminium silicate. Due to the several thousand very small channels, this provides a large surface area. It is coated with a wash coat of aluminium oxide, which again increases its effective surface area by about several thousand times. 'Noble' metals are used for the catalysts. Platinum helps to burn off the hydrocarbons (HC) and carbon monoxide (CO), and rhodium helps in the reduction of nitrogen oxides (NO_x). The whole three-way catalytic converter only contains about three to four grams of these precious metals. Some converters now use metal cores (substrates). Figure 7.49 shows the

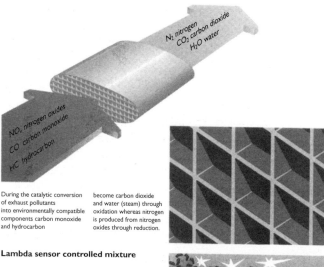

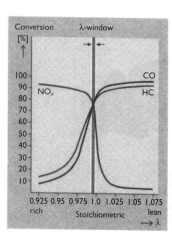

Figure 7.49 Catalytic converter operation

operation of a catalytic converter and how the lambda window is important.

The engine can damage a catalytic converter in one of two ways:

- firstly by the use of leaded fuel which can cause lead compounds to be deposited on the active surfaces;
- secondly by engine misfire which can cause the catalytic converter to overheat due to burning fuel inside the unit.

Some manufacturers use a system on some vehicles where a sensor checks the output of the ignition HT system and, if the spark is not present, will not allow fuel to be injected. Misfire detection is also part of current on-board diagnostic (OBD) legislation in some countries and future legislation in others.

7.18.3 Air supply system

There are three purposes of the complete air supply system:

- clean the air;
- control air temperature;
- reduce noise.

A filter does the air cleaning and drawing air from around the exhaust manifold helps to control air temperature. When large quantities of air are drawn into the engine it causes the air to vibrate and this makes it noisy. In the same way as with the exhaust system, baffles are used to stop resonance. Resonance means that when vibrations reach a natural level they tend to increase and keep going. A good example of how much noise is reduced by the air intake system is to compare the noise when an engine is run with the air filter removed.

Two types of air filter are in use, the first of these being by far the most popular:

- paper element;
- oil bath and mesh.

The paper element is made of resin-impregnated paper. Air filters using this type of replaceable element are used both for car and commercial vehicles. They provide a very high filtering efficiency and reasonable service life. They can be mounted in any position available under the bonnet. Service intervals vary so check recommendations.

The oil bath and mesh type of air cleaner was widely used on non-turbo charged commercial vehicles. However, it is not very practical for modern low styled bonnets. Because it can be cleaned and fresh oil added, an oil bath air cleaner might still be used for vehicles operating in dusty conditions.

Air temperature control is used to help the vehicle conform to emission control regulations and for good driveability when the engine is cold. Good vaporisation of the fuel is the key. An automatic control is often fitted to make sure that the air intake temperature is always correct. The air cleaner has two intake pipes, one for cold air and the other for hot air from the exhaust manifold or hot box. The proportion of hot and cold air is controlled by a flap, which is moved by a diaphragm acted on by low pressure from the inlet manifold. The flap rests in the hot air pick up position.

A thermo-valve in the air stream senses the temperature of the air going into the engine. When a temperature of about 25°C is reached, the valve opens. This removes the connection to the manifold, which in turn increases the pressure acting on the diaphragm. The flap is now caused to move and the pick up is now from the cool air position. The flap is constantly moving ensuring that the temperature of air entering the engine remains constant. Picking up hot air when the engine is still cold can also help to prevent icing in a carburettor.

7.19 Diagnostics – exhaust and air supply

7.19.1 Systematic testing

If the reported fault is a noisy exhaust proceed as follows.

1. Check if the noise is due to the exhaust knocking or blowing.
2. Examine the vehicle on the lift.
3. Are further tests required or is it obvious?
4. Cover the end of the exhaust pipe with a rag for a second or two to highlight where the exhaust may be blowing.
5. Renew the exhaust section or complete system as appropriate.
6. Run and test for leaks and knocking.

7.19.2 Test results

Some of the information you may have to get from other sources such as data books or a workshop manual is listed in the following table.

Test carried out	Information required
Air filter condition	Clearly a physical examination but note the required service intervals
Exhaust noise	An idea of the normal noise level – note that 'big bore' exhausts will make more noise than the 'correct' type

7.19.3 Exhaust and air supply fault diagnosis table 1

Symptom	Possible faults	Suggested action
Exhaust noise	Hole in pipe, box or at joints	Renew as appropriate
Knocking noise	Exhaust incorrectly positioned	Reposition
	Broken mountings	Renew
Rich mixture/smoke	Blocked air filter	Replace
Noisy air intake	Intake trunking or filter box leaking or loose	Repair or secure as required
Poor cold driveability	Hot air pick up not operating	Check pipe connections to inlet manifold for leaks. Renew temperature valve or actuator

7.19.4 Exhaust fault diagnosis table 2

Symptom	Possible cause
Excessive noise	Leaking exhaust system or manifold joints Hole in exhaust system
Excessive fumes in car	Leaking exhaust system or manifold joints
Rattling noise	Incorrect fitting of exhaust system Broken exhaust mountings Engine mountings worn

7.20 Cooling

7.20.1 Air-cooled system

Air-cooled engines with multi-cylinders, especially under a bonnet, must have some form of fan cooling and ducting. This is to make sure all cylinders are cooled evenly. The cylinders and cylinder heads are finned. Hotter areas, such as near the exhaust ports on the cylinders, have bigger fins.

Fan-blown air is directed by a metal cowling so it stays close to the finned areas. A thermostatically controlled flap will control airflow. When the engine is warming up, the flap will be closed to restrict the movement of air. When the engine reaches its operating temperature the flap opens and allows the air to flow over the engine. The cooling fan is a large device and is driven from the engine by a belt. This belt must not be allowed to slip or break, because serious damage will occur.

Car heating is not easy to arrange with an air-cooled engine. Some vehicles use a heat exchanger around the exhaust pipe. Air is passed through this device where it is warmed. It can then be used for demisting and heating with the aid of an electric motor and fan.

7.20.2 Water cooled system

The main parts of a water cooled system are as follows:

- water jacket;
- water pump;
- thermostat;
- radiator;
- cooling fan.

Water-cooled engines work on the principle of surrounding the hot areas inside the engine with a water jacket. The water takes on heat from the engine and, as it circulates through the radiator, gives it off to atmosphere. The heat concentration around the top of the engine means a water pump is needed to ensure proper circulation.

The water pump circulates water through the radiator and around the engine when the thermostat is open. Water circulates only round the engine when the thermostat is closed and not through the radiator. Forcing water around the engine prevents vapour pockets forming in very hot areas. This circulation is assisted by the thermo-siphon action. The thermo-siphon action causes the water to circulate because as the water is heated it rises and moves to the top of the radiator. This pushes down on the colder water underneath which moves into the engine. This water is heated, rises and so on.

Coolant from the engine water jacket passes through a hose to the radiator at the top. It then passes through thin pipes called the radiator matrix to the lower tank and then back to the lower part of the engine.

Many water passages between the top and bottom tanks of the radiator are used, to increase the surface area. Fins further increase the surface area to make the radiator even more efficient. A cooling fan assists air flow. The heat from the coolant passes to the pipes and fins and then to the air as it is blown by a fan over the fins.

Many modern radiators are made from aluminium pipes and fins with plastic tanks top and bottom (down flow), or at each end (cross flow). The cross flow radiators with tanks at each end are becoming the most popular. The more traditional method was to use copper and brass.

A thermostat is a temperature controlled valve. Its purpose is to allow coolant to heat up more quickly and then be kept at a constant temperature. The total coolant volume in an engine takes time to heat up. Modern engines run more efficiently when at the correct operating temperature. The action of the thermostat is such as to prevent water circulation from the engine to the radiator, until a set temperature is reached. When the valve opens there is a full circuit for the coolant and a good cooling action occurs because of full flow through the radiator. The constant action of the thermostat ensures that the engine temperature remains at a constant level. The thermostat used by almost all modern engine is a wax capsule type. If the thermostat is faulty ensure that the correct type for the engine is fitted as some work at different temperatures.

The water pump is driven by a V-belt or multi V-belt from the crankshaft pulley or by the cam belt. The pump is a simple impeller type and is usually fitted at the front of the engine (where the pulleys are). It assists with the thermo-siphon action of the cooling system, forcing water around the engine block and radiator.

The engine fan, which maintains the flow of air through the radiator, is mounted on the water-pump pulley on older systems. Most cooling fans now are electric. These are more efficient because they only work when needed. The forward motion of the car also helps the air movement through the radiator.

7.20.3 Sealed and semi-sealed systems

Cooling systems on most vehicles today are sealed or semi-sealed. This allows them to operate at pressures as much as $100\,N/m^2$ (100 Pascal) over atmospheric pressure, raising the boiling point of the coolant to as much as 126.6°C (remember water boils at 100°C at atmospheric pressure). The system can therefore operate at a higher temperature and with greater efficiency.

The pressure buildup is made possible by the radiator pressure cap. The cap contains a pressure valve which opens at a set pressure, and a vacuum valve which opens at a set vacuum. On a semi-sealed system, air is pushed out to atmosphere through the pressure valve as the coolant expands. Air is then drawn back into the radiator through the vacuum valve as the coolant cools and contracts. A sealed system has an expansion tank into which coolant is forced as it expands, and when the engine cools, coolant can flow from the tank back into the cooling system. Figure 7.50 shows a semi-sealed type cooling system.

Correct levels in the expansion tank or in an unsealed radiator are very important. If too much coolant is used it will be expelled on to the floor when the engine gets hot. If not enough is used then the level could become low and overheating could take place.

> **Warning**
> If a pressure cap is removed from a hot system, hot water under pressure will boil the instant pressure is released. This can be very dangerous.

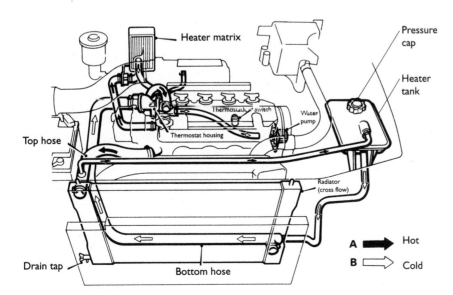

Figure 7.50 Semi-sealed cooling system

Heat from the engine can be used to increase the temperature of the car interior. This is achieved by use of a heat exchanger, often called the heater matrix. Due to the action of the thermostat in the engine cooling system the water temperature remains nearly constant. The air being passed over the heater matrix is therefore heated to a set level.

A source of hot air is now available for heating the vehicle interior. Some form of control is required over how much heat is required. The method used on most modern vehicles is blending. This is a control flap that determines how much of the air being passed into the vehicle is directed over the heater matrix. Some systems use a valve to control the hot coolant flowing to the heater matrix.

By a suitable arrangement of flaps it is possible to direct air of the chosen temperature to selected areas of the vehicle interior. In general, basic systems allow the warm air to be adjusted between the inside of the windscreen and the driver and passenger footwells. Fresh cool air outlets with directional nozzles are also fitted.

One final facility, which is available on many vehicles, is the choice between fresh or recirculated air. The primary reason for this is to decrease the time taken to demist or defrost the vehicle windows and simply to heat the car interior more quickly, and to a higher temperature. The other reason is that for example, in heavy congested traffic, the outside air may not be very clean.

7.21 Diagnostics – cooling

7.21.1 Systematic testing

If the reported fault is loss of coolant proceed as follows.

1. Check coolant level and discuss with customer how much is being lost.
2. Run the engine to see if it is overheating.
3. If the engine is not overheating a leak would seem to be most likely.
4. Pressure test the cooling system and check for leaks from hoses, gaskets and the radiator.
5. Renew a gasket or the radiator, clips or hoses as required. Top up the coolant and check antifreeze content.
6. Road test the vehicle to confirm the fault is cured and that no other problems have occurred.

7.21.2 Test equipment

Note: You should always refer to the manufacturer's instructions appropriate to the equipment you are using.

Cooling system pressure tester (Figure 7.51)

This is a pump with a pressure gauge built in, together with suitable adapters for fitting to the header tank or radiator filler. The system can then be pressurized to check for leaks. The pressure can be looked up or it is often stamped on the filler cap. A good way of doing this test is to pressurise the system when cold and then start the engine and allow it to warm up. You can be looking for leaks but beware of rotating components.

Antifreeze tester

This piece of equipment is a hydrometer used to measure the relative density of the coolant. The relative density of coolant varies with the amount of antifreeze. A table can be used to determine how much more antifreeze should be added to give the required protection.

Temperature meter/thermometer

Sometimes the dashboard temperature gauge reading too high can create the symptoms of an overheating problem. A suitable meter or thermometer can be used to check the temperature. Note though

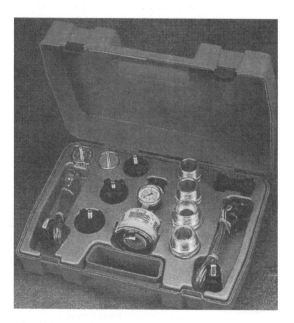

Figure 7.51 Cooling system pressure tester

that normal operating temperature is often well above 90°C (hot enough to burn badly).

7.21.3 Test results

Some of the information you may have to get from other sources such as data books or a workshop manual is listed in the following table.

Test carried out	Information required
Leakage test	System pressure. Printed on the cap or from data books. About 1 bar is normal
Antifreeze content	Cooling system capacity and required percentage of antifreeze. If the system holds six litres then for a 50% antifreeze content you will need to add three litres of antifreeze. Don't forget you will need to drain out three litres of water to make room for the antifreeze!
Operating temperature	This is about the same as the thermostat opening temperature – 88 to 92°C is a typical range

7.21.4 Cooling fault diagnosis table 1

Symptom	Possible faults	Suggested action
Overheating	Lack of coolant	Top up but then check for leaks
	Thermostat stuck closed	Renew
	Electric cooling fan not operating	Check operation of thermal switch
	Blocked radiator	Renew
	Water pump/fan belt slipping	Check, adjust/renew
Loss of coolant	Leaks	Pressure test when cold and hot, look for leaks and repair as required
Engine does not reach normal temperature or it takes a long time	Thermostat stuck in the open position	Renew

7.21.5 Cooling fault diagnosis table 2

Symptom	Possible cause
Overheating	Low coolant level (maybe due to a leak)
	Thermostat stuck closed
	Radiator core blocked
	Cooling fan not operating
	Temperature gauge inaccurate
	Airlock in system (some systems have a complex bleeding procedure)
	Pressure cap faulty
Overcooling	Thermostat stuck open
	Temperature gauge inaccurate
	Cooling fan operating when not needed
External coolant leak	Loose or damaged hose
	Radiator leak
	Pressure cap seal faulty
	Water pump leak from seal or bearing
	Boiling due to overheating or faulty pressure cap
	Core plug leaking
Internal coolant leak	Cylinder head gasket leaking
	Cylinder head cracked
Corrosion	Incorrect coolant (antifreeze, etc.)
	Infrequent flushing
Freezing	Lack of antifreeze
	Incorrect antifreeze

7.22 Lubrication

7.22.1 Lubrication system

From the sump reservoir under the crankshaft oil is drawn through a strainer into the pump. Oil pumps have an output of tens of litres per minute and operating pressures of over 5 bar at high speeds. A pressure relief valve limits the pressure of the lubrication system to between 2.5 and bar^2. This control is needed because the pump would produce excessive pressure at high speeds. After leaving the pump, oil passes into a filter and then into a main oil gallery in the engine block or crankcase.

Drillings connect the gallery to the crankshaft bearing housings and, when the engine is running, oil is forced under pressure between the rotating crank journals and the main bearings. The crankshaft is drilled so that the oil supply from the main bearings is also to the big end bearing bases of the connecting rods.

The con-rods are often drilled near the base so that a jet of oil sprays the cylinder walls and the underside of the pistons. In some cases the con-rod may be drilled along its entire length so that oil from the big end bearing is taken directly to the gudgeon pin (small end). The surplus then splashes out to cool the underside of the piston and cylinder.

The camshaft operates at half crankshaft speed, but it still needs good lubrication because as the high pressure loads on the cams. It is usual to supply pressurised oil to the camshaft bearings and splash or spray oil on the cam lobes. On overhead camshaft engines, two systems are used. In the simplest system the rotating cam lobes dip into a trough of oil. Another method is to spray the cam lobes with oil. This is usually done by an oil pipe with small holes in it alongside the camshaft. The small holes in the side of the pipe aim a jet of oil at each rotating cam lobe. The surplus splashes over the valve assembly and then falls back into the sump.

On cars where a chain drives the cam, a small tapping from the main oil gallery sprays oil on the chain as it moves past or the chain may simply dip in the sump oil. Figure 7.52 shows a typical lubrication system.

7.22.2 Oil filters

Even new engines can contain very small particles of metal left over from the manufacturing process or grains of sand which have not been removed from the crankcase after casting. Old engines continually deposit tiny bits of metal worn from highly loaded components such as the piston rings. To prevent any of these lodging in bearings or blocking oil ways, the oil is filtered.

The primary filter is a wire mesh strainer that stops particles of dirt or swarf from entering the oil pump. This is normally on the end of the oil pick-up pipe. An extra filter is also used that stops very fine particles. The most common type has a folded, resin impregnated paper element. Pumping oil through it removes all but smallest solids from the oil.

Most engines use a full-flow system to filter all of the oil after it leaves the pump. The most popular method is to pump the oil into a canister containing a cylindrical filter. From the inner walls of the canister, the oil flows through the filter and out from the centre to the main oil gallery. Full-flow filtration works well provided the filter is renewed at regular intervals. If it is left in service too long it may become blocked. When this happens the buildup of pressure inside the filter forces open a spring-loaded relief valve in the housing and the oil bypasses the filter. This valve prevents engine failure, but the engine will be lubricated with dirty oil until the filter is renewed. This is better than no oil!

A bypass filtration system was used on older vehicles. This system only filters a proportion of the oil pump output. The remainder is fed directly to the oil gallery. At first view this seems a strange

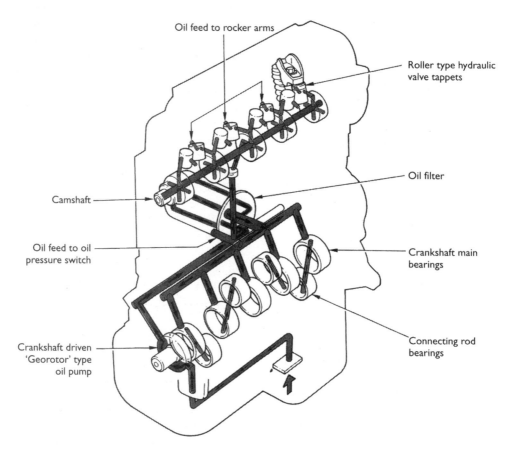

Figure 7.52 Lubrication system

idea but all of the oil does eventually get filtered. The smaller amount through the filter allows a higher degree of filtration.

7.22.3 Oil pumps

In its simplest form an oil pump consists of two gear wheels meshed together in a tight space so that oil cannot escape past the sides. The engine drives one wheel. As the gears rotate in opposite directions, the gap between each tooth in each wheel traps a small quantity of oil from an inlet port. The trapped oil is carried round by each wheel towards an outlet port on the opposite side where it is forced out by the meshing teeth.

The principle of squeezing oil from an ever-decreasing space is also used in the rotor type pump. An inner and outer rotor are mounted on different axes in the same cylinder. The inner rotor, which commonly has four lobes, is driven by the engine. It meshes with an outer rotor, which has five lobes. As they rotate, the spaces between them change size. The inlet port is at a point where the space between the rotor lobes is increasing. This draws the oil in to the pump. The oil is then carried round the pump. As rotation continues, the space between the lobes gets smaller. This compresses the oil out of the outlet port.

Oil pumps can produce more pressure than is required. A valve is used to limit this pressure to a set value. The pressure relief valve is a simple device, which in most cases works on the ball and spring principle. This means that when the pressure on the ball is greater than the spring, the ball moves. The pressure relief valve is placed in the main gallery so that excess pressure is prevented. When the ball moves oil is simply returned to the sump.

7.22.4 Crankcase ventilation – engine breather systems

Breathing is very important; without being able to breathe, we would die! It is almost as important for an engine breathing system to work correctly. There are two main reasons for engine breathers.

- Prevent pressure buildup inside the engine crankcase due to combustion gases blowing past the pistons. The buildup of pressure will blow gaskets and seals but also there is a high risk of explosion.
- Prevent toxic emissions from the engine. Emission limits are now very strict, for good reason – our health!

Crankase breathing or ventilation of the engine was first done by what is known as an open system, but this has now been completely replaced by the closed system. The gases escaping from an engine with open crankcase ventilation as described above are very toxic. Legislation now demands a positive closed system of ventilation. This makes the pollution from cylinder blow-by gases negligible. Positive crankcase ventilation is the solution to this problem.

In early types of closed system crankcase ventilation, the lower pressure at the carburettor air cleaner was used to cause an airflow through the inside of the engine. The breather outlet was simply connected by a pipe to the air cleaner. This caused the crankcase gases to be circulated and then burned in the engine cylinders. A flame trap was included in the system, to prevent a crankcase explosion if the engine backfired.

In modern closed systems the much lower pressure within the inlet manifold is used to extract crankcase gases. This has to be controlled in most cases by a variable regulator valve or pressure conscious valve (PCV). The valve is fitted between the breather outlet and the inlet manifold. It consists of a spring-loaded plunger, which opens as the inlet manifold pressure reduces. When the engine is stationary the valve is closed. Under normal running conditions the valve opens to allow crankcase gases to enter the inlet manifold with minimum restriction. At low manifold pressures during idling and overrun (pressure is less than atmospheric), further travel of the valve plunger against its spring closes it in the opposite direction. This reduces gas flow to the inlet manifold. This feature makes sure that the fuel control process is not interfered with under these conditions. The valve also acts as a safety device in case of a backfire. Any high pressure created in the inlet manifold will close the valve completely. This will isolate the crankcase and prevent the risk of explosion.

7.23 Diagnostics – lubrication

7.23.1 Systematic testing

If the reported fault is that the oil pressure light comes on at low speed proceed as follows.

1. Run the engine and see when the light goes off or comes on.
2. Is the problem worse when the engine is hot? Check the oil level! When was it last serviced?

3. If oil level is correct then you must investigate further.
4. Carry out an oil pressure test to measure the actual pressure.
5. If pressure is correct then renew the oil pressure switch. If not engine strip down is likely.
6. Run and test for leaks.

7.23.2 Test equipment

Note: You should always refer to the manufacturer's instructions appropriate to the equipment you are using.

Oil pressure test gauge (Figure 7.53)

This is a simple pressure gauge that can be fitted with suitable adapters into the oil pressure switch hole. The engine is then run and the pressure readings compared to data.

Vacuum gauge

A simple 'U' tube full of water is often used. This is connected to the oil dipstick tube and the engine is run. The gauge should show a pressure less than atmospheric (a partial vacuum). This checks the operation of the crankcase ventilation system.

Figure 7.53 Oil pressure gauge

7.23.3 Test results

Some of the information you may have to get from other sources such as data books, or a workshop manual is listed in the following table.

Test carried out	Information required
Oil pressure	Oil pressure is measured in bars. A typical reading would be about 3 bar
Crankcase pressure	By tradition pressures less than atmosphere are given in strange ways, such as inches of mercury or inches of water! This is why I like to stick to absolute pressure and the bar! Zero bar is no pressure, 1 bar is atmospheric pressure and so on. Two to 3 bar is more than atmospheric pressure like in a tyre. The trouble is standards vary so make sure you compare like with like! Back to crankcase pressure – it should be less than atmospheric, check data
Oil condition	Recommended type of lubricant

7.23.4 Lubrication fault diagnosis table 1

Symptom	Possible faults	Suggested action
Low oil pressure	Lack of oil	Top up
	Blocked filter	Renew oil and filter
	Defective oil pump	Renew after further tests
	Defective oil pressure relief valve	Adjust if possible or renew
High crankcase pressure	Blocked crankcase breather	Clean or replace
	Blocked hose	Clean or renew hose
	Pressure blowing by pistons	Engine overhaul may be required
Loss of oil	Worn piston rings	Engine overhaul may be required
	Leaks	Renew seals or gaskets

7.23.5 Lubrication fault diagnosis table 2

Symptom	Possible cause
Oil leaks	Worn oil seal (check breather system)
	Gasket blown
	Cam or rocker cover loose
	Oil filter seal
Blue smoke	Piston rings
	Valve stem seals
	Head gasket

7.24 Batteries

7.24.1 Safety

The following points must be observed when working with batteries:

- good ventilation
- protective clothing
- supply of water available (running water preferable)
- first aid equipment available, including eyewash
- no smoking or naked lights permitted.

7.24.2 Lead-acid batteries

Incremental changes over the years have made the sealed and maintenance-free battery, now in common use, very reliable and long lasting. This may not always appear to be the case to some end users, but note that quality is often related to the price the customer pays. Many bottom of the range cheap batteries with a 12 month guarantee will last for 13 months!

The basic construction of a nominal 12 V lead-acid battery consists of six cells connected in series. Each cell producing about 2 V is housed in an individual compartment within a polypropylene or similar case. Figure 7.54 shows a cut-away battery with its main component parts. The active material is held in grids or baskets to form the positive and negative plates. Separators made from a microporous plastic insulate these plates from each other.

The grids, connecting strips and the battery posts are made from a lead alloy. For many years this was lead antimony (PbSb) but this has now been largely replaced by lead calcium (PbCa). The newer materials cause less gassing of the electrolyte when the battery is fully charged. This has been one of the main reasons why sealed batteries became feasible as water loss is considerably reduced.

Modern batteries described as sealed do still have a small vent to stop the pressure buildup due to the very small amount of gassing. A further requirement of sealed batteries is accurate control of charging voltage.

7.24.3 Battery rating

In simple terms the characteristics or rating of a particular battery are determined by how much current it can produce and how long it can sustain this current. The rate at which a battery can produce current is determined by the speed of the chemical reaction. This in turn is determined by a number of factors:

- surface area of the plates;
- temperature;
- electrolyte strength;
- current demanded.

The actual current supplied therefore determines the overall capacity of a battery. The rating

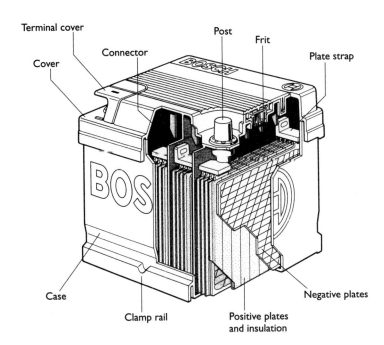

Figure 7.54 Lead-acid battery

of a battery has to specify the current output and the time.

Ampere hour capacity	This is now seldom used but describes how much current the battery is able to supply for either 10 or 20 hours. The 20-hour figure is the most common. For example, a battery quoted as being 44 Ah (ampere-hour) will be able, if fully charged, to supply 2.2 A for 20 hours before being completely discharged (cell voltage above 1.75 V)
Reserve capacity	A system used now on all new batteries is reserve capacity. This is quoted as a time in minutes for which the battery will supply 25 A at 25°C to a final voltage of 1.75 V per cell. This is used to give an indication of how long the battery could run the car if the charging system was not working. Typically a 44 Ah battery will have a reserve capacity of about 60 minutes.
Cold cranking amps	Batteries are given a rating to indicate performance at high current output and at low temperature. A typical value of 170 A means that the battery will supply this current for one minute at a temperature of −18°C at which point the cell voltage will fall to 1.4 V (BS).

These cold cranking amps (CCA) capacity rating methods do vary to some extent: British standards, DIN standards and SAE standards are the three main examples!

- BS 60 seconds
- DIN 30 seconds
- SAE 30 seconds

In summary, the capacity of a battery is the amount of electrical energy that can be obtained from it. It is usually given in ampere-hours, reserve capacity (RC) and cold cranking amps.

- A 40 Ah battery means it should give 2 A for 20 hours.
- The reserve capacity indicates the time in minutes for which the battery will supply 25 A at 25°C.
- Cold cranking current indicates the maximum battery current at −18°C (0°F) for a set time (standards vary).

A battery for normal light vehicle use may be rated as follows: 44 Ah, 60 RC and 170 A CCA (BS). A 'heavy duty' battery will have the same Ah rating as its 'standard duty' counterpart, but it will have a higher CCA and RC.

7.25 Diagnosing battery faults

7.25.1 Servicing batteries

In use a battery requires very little attention other than the following when necessary.

- Corrosion should be cleaned from terminals using hot water.
- Terminals should be smeared with petroleum jelly or vaseline *not* ordinary grease.
- Battery tops should be clean and dry.
- If not sealed, cells should be topped up with distilled water 3 mm above the plates.
- Battery should be securely clamped in position.

7.25.2 Maintenance-free

By far the majority of batteries now available are classed as 'maintenance-free'. This implies that little attention is required during the life of the battery. Earlier batteries and some heavier types do, however, still require the electrolyte level to be checked and topped up periodically. Battery posts are still a little prone to corrosion and hence the usual service of cleaning with hot water if appropriate and the application of petroleum jelly or proprietary terminal grease is still recommended. Ensuring that the battery case and in particular the top remains clean will help to reduce the rate of self-discharge.

The state of charge of a battery is still very important and in general it is not advisable to allow the state of charge to fall below 70% for long periods as the sulphate on the plates can harden, making recharging difficult. If a battery is to be stored for a long period (more than a few weeks), then it must be recharged every so often to prevent it from becoming sulphated. Recommendations vary but a recharge every six weeks is a reasonable suggestion.

7.25.3 Charging

The recharging recommendations of battery manufacturers vary slightly. The following methods, however, are reasonably compatible and should not cause any problems. The efficiency of a battery is not 100%. Therefore the recharging process must 'put back' the same Ah capacity as was used on discharge plus a bit more to allow for losses. It is therefore clear that the main question about charging is not how much, but at what rate.

The old recommendation was that the battery should be charged at a tenth of its Ah capacity for about 10 hours or less. This is assuming that the Ah capacity is quoted at the twenty hour rate, as a tenth of this figure will make allowance for the charge factor. This figure is still valid but as Ah capacity is not always used nowadays, a different method of deciding the rate is necessary. One way is to set a rate at a sixteenth of the reserve capacity, again for up to 10 hours. The final suggestion is to set a charge rate at one fortieth of the cold start performance figure, also for up to 10 hours. Clearly if a battery is already half charged, half the time is required to recharge to full capacity.

In summary the ideal charge rate is determined from:

- 1/10 of the Ah capacity;
- 1/16 of the RC;
- 1/40 of the CCA.

The above suggested charge rates are to be recommended as the best way to prolong battery life. They do all, however, imply a constant current charging source. A constant voltage charging system is often the best way to charge a battery. This implies that the charger, an alternator on a car for example, is held at a constant level and the state of charge in the battery will determine how much current will flow. This is often the fastest way to recharge a flat battery. If a constant voltage of less than 14.4 V is used then it is not possible to cause excessive gassing and this method is particularly appropriate for sealed batteries.

Boost charging is a popular technique often applied in many workshops. It is not recommended as the best method but, if correctly administered and not repeated too often, it is suitable for most batteries. The key to fast or boost charging is that the battery temperature should not exceed 43°C. With sealed batteries it is particularly important not to let the battery gas excessively in order to prevent the buildup of pressure. A rate of about five times the 'normal' charge setting will bring the battery to 70–80% of its full capacity within approximately one hour. The table below summarises the charging techniques for a lead-acid battery. Figures 7.55 and 7.56 show two typical battery chargers.

Charging method	Notes
Constant voltage	Will recharge any battery in seven hours or less without any risk of overcharging (14.4 V maximum)
Constant current	Ideal charge rate can be estimated as: 1/10 of Ah capacity, 1/16 of RC or 1/40 of cold start current (charge time of 10 to 12 hours or pro rata original state)
Boost charging	At no more than five times the ideal rate, a battery can be brought up to about 70% of charge in about one hour

7.25.4 Battery faults

Any electrical device can suffer from two main faults; open circuit or short circuit. A battery is no exception but it can also suffer from other problems such as low charge or low capacity. Often a problem which seems to be with a vehicle battery can be traced to another part of the vehicle such as the charging system. The following table lists all of the common problems encountered with lead-acid batteries, together with typical causes.

Symptom or fault	Likely causes
Low state of charge	Charging system fault Unwanted drain on battery Electrolyte diluted Incorrect battery for application
Low capacity	Low state of charge Corroded terminals Impurities in the electrolyte Sulphated Old age – active material fallen from the plates

Figure 7.55 Battery charger

Excessive gassing and temperatures	Overcharging Positioned too near exhaust component
Short circuit cell	Damaged plates and insulators Buildup of active material in sediment trap
Open circuit cell	Broken connecting strap Excessive sulphation Very low electrolyte
Service life shorter than expected	Excessive temperature Battery has too low a capacity Vibration excessive Contaminated electrolyte Long periods of not being used Overcharging

Repairing modern batteries is not possible. Most of the problems listed will require the battery to be replaced. In the case of sulphation it is sometimes possible to bring the battery back to life with a very long low current charge. A fortieth of the Ah capacity or about a two hundredth of the cold start performance for about 50 hours is an appropriate rate.

7.25.5 Testing batteries

For testing the state of charge of a non-sealed type of battery, it was traditional to use a hydrometer. The Hydrometer is a syringe which draws electrolyte from a cell and a float which will float at a particular depth in the electrolyte according to its density. The relative density or specific gravity is then read from the graduated scale on the float. A fully charged cell should show 1.280, when half charged 1.200 and if discharged 1.120.

Most vehicles are now fitted with maintenance free batteries and a hydrometer cannot be used to find the state of charge.

This can, however, be determined from the voltage of the battery, as given in the following table. An accurate voltmeter is required for this test (Figure 7.57) – note the misleading surface charge shown here.

Battery Volts at 20°C	State of Charge
12.0 V	Discharged (20% or less)
12.3 V	Half charged (50%)
12.7 V	Charged (100%)

Figure 7.56 Battery charger and engine starter

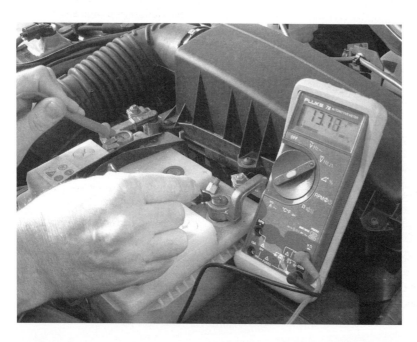

Figure 7.57 Battery voltage testing

Figure 7.58 'MicroVAT', charging system, starter and battery tester (*Source*: Snap-on).

To test a battery more thoroughly, it is now preferred to use a volt, amp tester (VAT). There are many variations on the market; however, this section will outline just one type. Snap-on produce a compact and very useful tester called the MicroVAT (Figure 7.58). This equipment will carry out a range of diagnostic tests.

The device, as with many similar types, will do not only battery condition tests, but also tests on the charging and starting system.

This VAT takes advantage of new impedance/current test technology to detect the full range of battery failure modes including bad cells, sulphation, internal short circuits, and other chemical and physical failures. Testing takes less than 5 seconds and will even work on batteries discharged down to as low as one volt.

Some of the key features of this tester are:

- Automated system test of battery, alternator and starter in under a minute.
- Detailed test data: alternator ripple, internal resistance, starter draw, state of charge, charging amps, and volts.
- Tests discharged batteries down to one volt.
- Impedance/current (IC) test technology.
- Wireless printer option.
- Integrated high and low amp probe options.

MicroVAT uses a fan cooled 50 A load and integrated amp probe to test the quantity and quality of alternator output with an alternator ripple test. Many late model computer-controlled charging systems virtually shut down under no load conditions.

Diagnostic tests that can be carried out with this tester, when an amps probe is also used, are as follows.

Starting test data
- Average cranking current
- Maximum cranking current
- Pre-set voltage
- Pre-set load voltage
- Average cranking voltage
- Minimum cranking voltage.

Battery test data
- Diagnosis
- Actual CCA
- Percentage capacity
- Open circuit voltage
- Impedance (often described as internal resistance).

Alternator test data
- Diagnosis
- Failure mode
- Charging at idle
- Charging volts under load
- Average current at idle
- Peak current
- Peak to peak ripple at idle
- Peak to peak ripple under load.

7.26 Starting

7.26.1 Starter circuit

In comparison with most other circuits on the modern vehicle the starter circuit is very simple. The problem to overcome, however, is that of volt drop in the main supply wires. A spring-loaded key switch usually operates the starter; the same switch also controls the ignition and accessories. The supply from the key switch, via a relay in many cases, causes the starter solenoid to operate and this in turn, by a set of contacts, controls the heavy current. In some cases an extra terminal on the starter solenoid provides an output when cranking, usually used to bypass a dropping resistor on the ignition or fuel pump circuits. The basic circuit for the starting system is shown in Figure 7.59. The problem of volt drop in the main supply circuit is due to the high current required by the starter particularly under adverse starting conditions such as very low temperatures.

A typical cranking current for a light vehicle engine is in the order of 150 A but this may peak in excess of 500 A to provide the initial stalled torque. It is generally accepted that a maximum volt drop of only 0.5 V should be allowed between the battery and starter when operating. An Ohm's law calculation indicates that the

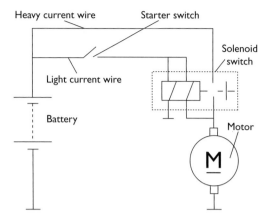

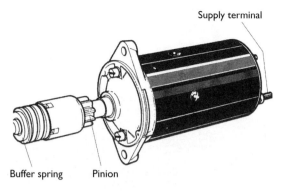

Figure 7.59 Starter circuit

Figure 7.60 Inertia type starter

maximum allowed circuit resistance is 2.5 mΩ, when using a 12 V supply. This is a worst case situation and lower resistance values are used in most applications. The choice of suitable conductors is therefore very important.

7.26.2 Inertia starters

In all standard motor vehicle applications it is necessary to connect the starter to the engine ring gear only during the starting phase. If the connection remained permanent, the excessive speed at which the starter would be driven by the engine would destroy the motor almost immediately.

The inertia type of starter motor has been the technique used for over 85 years, but it is now becoming redundant.

The starter engages with the flywheel ring gear by means of a small pinion. The toothed pinion and a sleeve splined on to the armature shaft are threaded such that when the starter is operated via a remote relay, the armature will cause the sleeve to rotate inside the pinion. The pinion remains still due to its inertia and, because of the screwed sleeve rotating inside it, the pinion is moved into mesh with the ring gear.

When the engine fires and runs under its own power the pinion is driven faster than the armature shaft. This causes the pinion to be screwed back along the sleeve and out of engagement with the flywheel. The main spring acts as a buffer when the pinion first takes up the driving torque and also acts as a buffer when the engine throws the pinion back out of mesh.

7.26.3 Pre-engaged starters

Pre-engaged starters are fitted to the majority of vehicles in use today. They provide a positive engagement with the ring gear, as full power is not applied until the pinion is fully in mesh. They prevent premature ejection as the pinion is held into mesh by the action of a solenoid. A one-way clutch is incorporated into the pinion to prevent the starter motor being driven by the engine. An example of a pre-engaged starter in common use is shown in Figure 7.61.

Figure 7.62 shows the circuit associated with operating this type of pre-engaged starter. The basic operation of the pre-engaged starter is as follows. When the key switch is operated a supply is made to terminal 50 on the solenoid. This causes two windings to be energised, the hold-on winding and the pull-in winding. Note that the pull-in winding is of very low resistance and hence a high current flows. This winding is connected in series with the motor circuit and the current flowing will allow the motor to rotate slowly to facilitate engagement. At the same time the magnetism created in the solenoid attracts the plunger and via an operating lever pushes the pinion into mesh with the flywheel ring gear. When the pinion is fully in mesh the plunger at the end of its travel causes a heavy-duty set of copper contacts to close. These contacts now supply full battery power to the main circuit of the starter motor. When the main contacts are closed the pull-in winding is effectively switched off due to equal voltage supply on both ends. The hold-on winding holds the plunger in position as long as the solenoid is supplied from the key switch.

When the engine starts and the key is released, the main supply is removed and the plunger and pinion return to their rest positions under spring tension. A lost motion spring located on the plunger ensures that the main contacts open before the pinion is retracted from mesh.

During engagement if the teeth of the pinion hit the teeth of the flywheel (tooth to tooth abutment), the main contacts are allowed to close due to the engagement spring being compressed. This allows the motor to rotate under power and the pinion will slip into mesh.

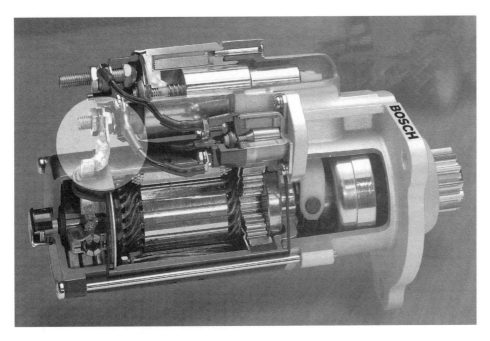

Figure 7.61 Supply link wire from solenoid to starter motor

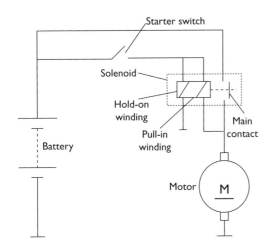

Figure 7.62 Starter circuit (Pre-engaged)

The torque developed by the starter is passed through a one-way clutch to the ring gear. The purpose of this free wheeling device is to prevent the starter being driven at excessively high speed if the pinion is held in mesh after the engine has started. The clutch consists of a driving and driven member with several rollers in between the two. The rollers are spring loaded and either wedge-lock the two members together by being compressed against the springs, or free wheel in the opposite direction.

Many variations of pre-engaged starter are in common use but all work on similar lines to the above description. The wound field type of motor has now largely been replaced by the permanent magnet version.

7.26.4 Permanent magnet (PM) starters

Permanent magnet starters began to appear on production vehicles in the late eighties. The two advantages of these motors, compared to conventional types, are less weight and smaller size. This makes the PM starter a popular choice by vehicle manufacturers as, due to the lower lines of today's cars, less space is now available for engine electrical systems. The reduction in weight provides a contribution towards reducing fuel consumption.

The principle of operation is similar in most respects to the conventional pre-engaged starter motor, the main difference being the replacement of field windings and pole shoes with high quality permanent magnets. The reduction in weight is in the region of 15% and the diameter of the yoke can be reduced by a similar factor.

Permanent magnets provide constant excitation and it would be reasonable to expect the speed and torque characteristic to be constant. However, due to the fall in battery voltage under load and the low resistance of the armature windings, the characteristic is comparable to series wound motors.

Development by some manufacturers has also taken place in the construction of the brushes. A copper and graphite mix is used but the brushes are made in two parts allowing a higher copper content in the power zone and a higher graphite content in the commutation zone. This results in increased service life and a reduction in volt drop giving improved starter power. Figure 7.63 shows three modern PM starters.

162 Advanced automotive fault diagnosis

Figure 7.63 Modern permanent magnet starters

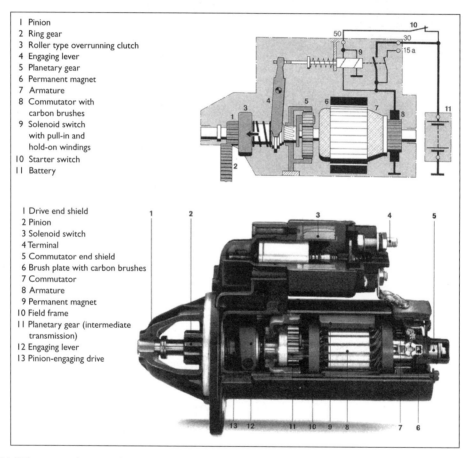

1 Pinion
2 Ring gear
3 Roller type overrunning clutch
4 Engaging lever
5 Planetary gear
6 Permanent magnet
7 Armature
8 Commutator with carbon brushes
9 Solenoid switch with pull-in and hold-on windings
10 Starter switch
11 Battery

1 Drive end shield
2 Pinion
3 Solenoid switch
4 Terminal
5 Commutator end shield
6 Brush plate with carbon brushes
7 Commutator
8 Armature
9 Permanent magnet
10 Field frame
11 Planetary gear (intermediate transmission)
12 Engaging lever
13 Pinion-engaging drive

Figure 7.64 PM starter with intermediate transmission (*Source*: Bosch)

For applications with a higher power requirement PM motors with intermediate transmission have been developed. This allows the armature to rotate at a higher and more efficient speed whilst still providing the torque, due to the gear reduction. Permanent magnet starters with intermediate transmission are available with power outputs of about 1.7 kW, suitable for spark ignition engines up to about five litres or compression ignition engines up to about 1.6 litres. This form of PM motor can give a weight saving of up to 40%. The principle of operation is again similar to the conventional pre-engaged starter.

The sun gear is on the armature shaft and the planet carrier drives the pinion. The ring gear or annulus remains stationary and also acts as an intermediate bearing. This arrangement of gears gives a reduction ratio of about 5:1. Figure 7.64

shows a PM starter with intermediate transmission, together with its circuit and operating mechanism.

7.27 Diagnosing starting system faults

7.27.1 Circuit testing procedure

The process of checking a 12 V starting system operation is as follows.

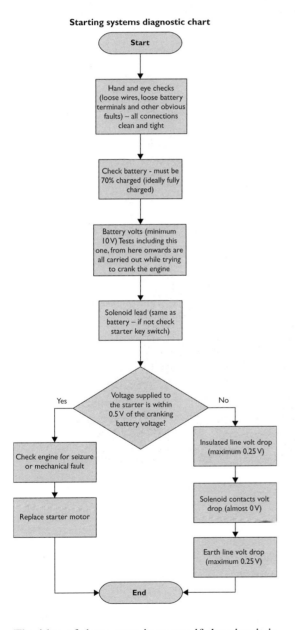

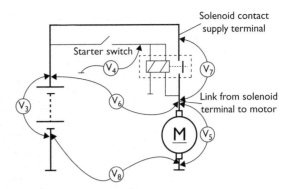

Figure 7.65 Starter circuit volt drop testing

a circuit with voltmeters connected. The numbered voltmeters relate to the number of the test in the above list.

Note that connections to the starter should be made to the link between the solenoid contacts and the motor *not* to the main supply terminal. Figure 7.66 shows an example of this link.

7.27.2 Starting fault diagnosis table

Symptom	Possible fault
Engine does not rotate when trying to start	Battery connection loose or corroded
	Battery discharged or faulty
	Broken loose or disconnected wiring in the starter circuit
	Defective starter switch or automatic gearbox inhibitor switch
	Starter pinion or flywheel ring gear loose
	Earth strap broken. Loose or corroded
Starter noisy	Starter pinion or flywheel ring gear loose
	Starter mounting bolts loose
	Starter worn (bearings, etc.)
	Discharged battery (starter may jump in and out)
Starter turns engine slowly	Discharged battery (slow rotation)
	Battery terminals loose or corroded
	Earth strap or starter supply loose or disconnected
	High resistance in supply or earth circuit
	Internal starter fault

7.28 Charging

7.28.1 Introduction

The 'current' demands made by modern vehicles are considerable. The charging system must be able to meet these demands under all operating conditions and still fast charge the battery.

The idea of these tests is to see if the circuit is supplying all the available voltage at the battery to the starter. If it is, then the starter is at fault, if not, then the circuit is at fault. Figure 7.65 shows

Figure 7.66 Supply link/wire from solenoid terminal to starter motor

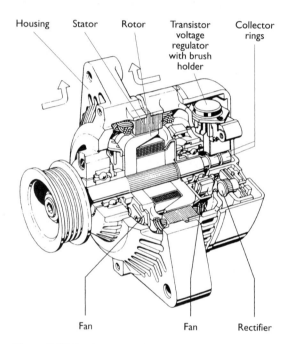

Figure 7.67 Bosch compact alternator

The main component of the charging system is the alternator and on most modern vehicles, with the exception of its associated wiring, it is the only component in the charging system. Figure 7.67 shows an alternator in common use. The alternator generates AC but must produce DC at its output terminal as only DC can be used to charge the battery and run electronic circuits. The output of the alternator must be a constant voltage regardless of engine speed and current load.

The charging system must meet the following criteria (when the engine is running):

- supply the current demands made by some or all loads;
- supply whatever charge current the battery demands;
- operate at idle speed;
- constant voltage under all conditions;
- efficient power to weight ratio;
- reliable, quiet, resistance to contamination;
- low maintenance;
- provide indication of correct operation.

7.28.2 Basic principles

When the alternator voltage is less than the battery (engine slow or not running for example), the direction of current flow is from the battery to the vehicle loads. The alternator diodes prevent current flowing into the alternator. When the alternator output is greater than the battery voltage, current will flow from the alternator to the vehicle loads and the battery.

It is clear therefore, that the alternator output voltage must be above battery voltage at all times when the engine is running. The actual voltage used is critical and depends on a number of factors.

The main consideration for charging voltage is the battery terminal voltage when fully charged. If the charging system voltage is set to this value then there can be no risk of overcharging the battery. This is known as the constant voltage charging technique. The figure of $14.2 \pm 0.2\,\text{V}$ is the accepted charging voltage for a 12 V system. Commercial vehicles generally employ two batteries in series at a nominal voltage of 24 V; therefore the accepted charge voltage would be doubled. These voltages are used as the standard input for all vehicle loads. For the purpose of clarity the text will just consider a 12 V system.

The other areas for consideration when determining charging voltage are any expected volt drops in the charging circuit wiring and the operating temperature of the system and battery. The voltage drops must be kept to a minimum but it is important to note that the terminal voltage of the alternator may be slightly above that supplied to the battery. Figure 7.68 shows the basic principle of an alternator.

7.28.3 Rectification of AC to DC

In order to full wave rectify the output of a three phase machine six diodes are needed. These are connected in the form of a bridge, which consists of three positive diodes and three negative diodes. The output produced by this configuration is shown compared to the three phase signals.

A further three positive diodes are often included in a rectifier pack. These are usually smaller than the main diodes and are only used to supply a small current back to the field windings in the rotor. The extra diodes are known as the auxiliary, field or excitation diodes.

When a star wound stator is used the addition of the voltages at the neutral point of the star is in theory 0 V. In practice, however, due to slight inaccuracies in the construction of the stator and rotor a potential develops at this point. By employing two extra diodes, one positive and one negative connected to the star point the energy can be collected. This can increase the power output of an alternator by up to 15%.

Figure 7.69 shows the full circuit of an alternator using an eight diode main rectifier and three field diodes. The voltage regulator, which forms the starting point for the next section, is also shown in this diagram. The warning light in an alternator circuit, in addition to its function in warning of charging faults, also acts to supply the initial excitation to the field windings. An alternator will not always self-excite as the residual magnetism in the fields is not usually enough to produce a voltage which will overcome the 0.6 or 0.7 V needed to forward bias the rectifier diodes. A typical wattage for the warning light bulb is 2 W. Many manufacturers also connect a resistor in parallel with the bulb to assist in excitation and allow operation if the bulb blows. The charge warning light bulb is extinguished when the alternator produces an output from the field diodes as this causes both sides of the bulb to take on the same voltage (a potential difference across the bulb of 0 V).

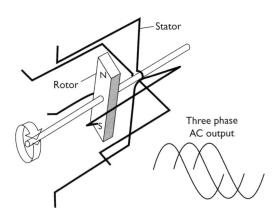

Figure 7.68 Basic alternator principle

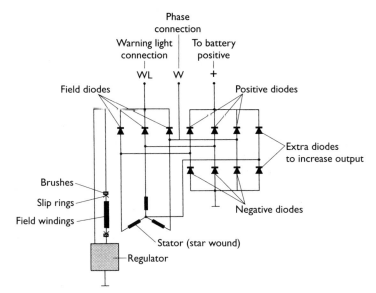

Figure 7.69 Complete internal alternator circuit

7.28.4 Regulation of output voltage

To prevent the vehicle battery from being overcharged the regulated system voltage should be kept below the gassing voltage of the lead-acid battery. A figure of $14.2 \pm 0.2\,V$ is used for all 12 V charging systems. Accurate voltage control is vital with the ever-increasing use of electronic systems. It has also enabled the wider use of sealed batteries, as the possibility of overcharging is minimal.

Voltage regulation is a difficult task on a vehicle alternator because of the constantly changing engine speed and loads on the alternator. The output of an alternator without regulation would rise linearly in proportion with engine speed. Alternator output is also proportional to magnetic field strength and this in turn is proportional to the field current. It is the task of the regulator to control this field current in response to alternator output voltage. The abrupt switching of the field current does not cause abrupt changes in output voltage due to the very high inductance of the field (rotor), windings. The whole switching process also only takes a few milliseconds.

Regulators can be mechanical or electronic, the latter now almost universal on modern cars. The mechanical type uses a winding connected across the output of the alternator. The magnetism produced in this winding is proportional to output voltage. A set of normally closed contacts is attached to an armature, which is held in position by a spring. The supply to the field windings is via these contacts. When the output voltage rises beyond a pre-set level, say 14 V, the magnetism in the regulator winding will overcome spring tension and open the contacts. This switches off the field correct and causes alternator output to fall. As output falls below a pre-set level the spring will close the regulator contacts again and so the process continues.

The problem with mechanical regulators is the wear on the contacts and other moving parts. This has been overcome with the use of electronic regulators which, due to more accurate tolerances and much faster switching, are far superior, producing a more stable output. Due to the compactness and vibration resistance of electronic regulators they are now fitted almost universally on the alternator reducing the number of connecting cables required.

The key to electronic voltage regulation is the zener diode. This diode can be constructed to breakdown and conduct in the reverse direction at a precise level. This is used as the sensing element in an electronic regulator. Figure 7.70 shows two common electronic voltage regulators.

Figure 7.70 Voltage regulator

Electronic regulators can be made to sense either the battery voltage or the machine voltage (alternator) or a combination of the two. Most systems in use at present tend to be machine sensed as this offers some protection against over voltage in the event of the alternator being driven with the battery disconnected.

Over voltage protection is required in some applications to prevent damage to electronic components. When an alternator is connected to a vehicle battery system voltage, even in the event of regulator failure, will not often exceed about 20 V due to the low resistance and swamping effect of the battery. If an alternator is run with the battery disconnected (which is not recommended), a heavy duty Zener diode connected across the output will offer some protection as, if the system voltage exceeds its breakdown figure, it will conduct and cause the system voltage to be kept within reasonable limits. This device is often referred to as a surge protection diode.

7.28.5 Charging circuits

On many applications the charging circuit is one of the simplest on the vehicle. The main output is connected to the battery via suitable size cable (or in some cases two cables to increase reliability and flexibility). The warning light is connected to an ignition supply on one side and to the alternator terminal at the other. A wire may also be connected to the phase terminal if it is utilised. Figure 7.71 shows two typical wiring circuits. Note the output of the alternator is often connected to the starter main supply simply for convenience of wiring. If the wires are kept as short as possible this will reduce voltage drop in

Engine systems **167**

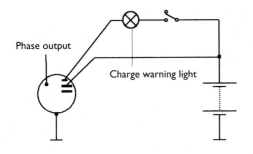

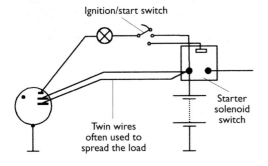

Figure 7.71 Example charging circuits

the circuit. The volt drop across the main supply wire when the alternator is producing full output current should be less than 0.5 V.

Some systems have an extra wire from the alternator to 'sense' battery voltage directly. An ignition feed may also be found and this is often used to ensure instant excitement of the field windings. A number of vehicles link a wire from the engine management ECU to the alternator. This is used to send a signal to increase engine idle speed if the battery is low on charge.

7.29 Diagnosing charging system faults

7.29.1 Testing procedure

After connecting a voltmeter across the battery and an ammeter in series with the alternator output wire(s), as shown in Figure 7.72, the process of checking the charging system operation is as follows.

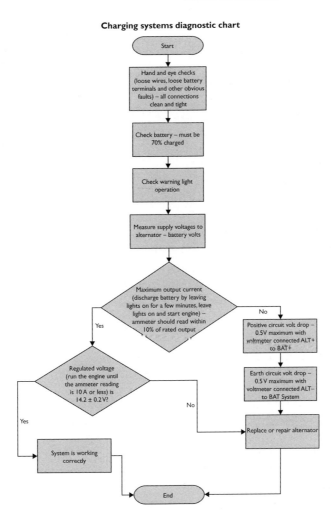

168 Advanced automotive fault diagnosis

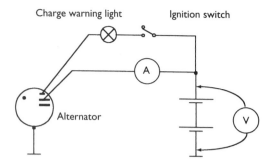

Figure 7.72 Alternator circuit testing

If the alternator is found to be defective then a quality replacement unit is the normal recommendation. Repairs are possible but only if the general state of the alternator is good.

	Alternator internal fault (diode open circuit, brushes worn or regulator fault, etc.)
	Open circuit in alternator wiring, either main supply, ignition or sensing wires if fitted
	Short circuit component causing battery drain even when all switches are off
	High resistance in the main charging circuit
Charge warning light stays on when engine is running	Slipping or broken alternator drive belt
	Alternator internal fault (diode open circuit, brushes worn or regulator fault, etc.)
	Loose or broken wiring/connections
Charge warning light does not come on at any time	Alternator internal fault (brushes worn, open circuit or regulator fault, etc.)
	Blown warning light bulb
	Open circuit in warning light circuit

7.29.2 Charging fault diagnosis table

Symptom	Possible fault
Battery loses charge	Defective battery
	Slipping alternator drive belt
	Battery terminals loose or corroded

Knowledge check questions

To use these questions, you should first try to answer them without help but if necessary, refer back to the content of the chapter. Use notes, lists and sketches to answer them. It is not necessary to write pages and pages of text!

1. Describe how a VAT is used to check battery condition.
2. List in a logical sequence, a series of tests that would determine why an engine, which is cranking over correctly, will not start.
3. Describe how the colour of smoke from a diesel engine can be used as an aid to fault diagnosis.
4. Make a block diagram to show the main components of an engine management system and how it can be considered as a series of inputs and outputs.
5. Describe how a cylinder leakage tester is used to check the condition of an engine.

8
Chassis systems

8.1 Brakes

8.1.1 Introduction

The main braking system of a car works by hydraulics. This means that when the driver presses the brake pedal, liquid pressure forces pistons to apply brakes on each wheel. A handbrake system, usually operated by a lever and cables, is used for parking. Most hand brakes operate on the rear wheels.

Two types of light vehicle brakes are used. Disc brakes are used on the front wheels of some cars and on all wheels of sports and performance cars. Braking pressure forces brake pads against both sides of a steel disc. Drum brakes are fitted on the rear wheels of some cars and on all wheels of older vehicles. Braking pressure forces brake shoes to expand outwards into contact with a drum. The important part of brake pads and shoes is a friction lining that grips well and withstands wear.

8.1.2 Principle of hydraulic braking

A complete system includes a master cylinder operating several wheel cylinders. The system is designed to give the power amplification needed for braking the particular vehicle. On any vehicle, when braking, a lot of the weight is transferred to the front wheels. Most braking effort is therefore designed to work on the front brakes. Some cars have special hydraulic valves to limit rear wheel braking. This reduces the chance of the rear wheels locking and skidding.

The main merits of hydraulic brakes are as follows:

- almost immediate reaction to pedal pressure (no free play as with mechanical linkages);
- automatic even pressure distribution (fluid pressure effectively remains the same in all parts of the system);
- increase in force (liquid lever).

Caution and regular servicing is required to ensure the following:

- no air must be allowed in the hydraulic circuits (air compresses and would not transfer the force);
- correct adjustment must be maintained between shoe linings to drums and pads to discs (otherwise the pedal movement would be too large);
- lining materials must be free from contamination (such as oil, grease or brake fluid).

A separate mechanical system is a good safety feature. Most vehicles have the mechanical hand brake working on the rear wheels but a few have it working on the front – take care.

Note the importance of flexible connections to allow for suspension and steering movement. These flexible pipes are made of high quality rubber and are covered in layers of strong mesh to prevent expansion when under pressure.

Extra safety is built into braking systems by using a double acting master cylinder (Figure 8.1). This is often described as tandem and can be thought of as two cylinders in one housing. The pressure from the pedal acts on both cylinders but fluid cannot pass from one to the other. Each cylinder is then connected to a complete circuit. This can be by a number of methods:

- diagonal split;
- separate front and rear;
- duplicated front.

8.1.3 Disc and drum brake systems

Figure 8.2 shows a typical disc brake, calliper pads and disc. The type shown is known as single acting sliding calliper. This is because only one cylinder is used but pads are still pressed equally on both sides of the disc by the sliding action. Disc brakes keep cooler because they are in the air stream and only part of the disc is heated as the brakes are applied. They also throw off water better than drum brakes. In most cases servicing is

170 Advanced automotive fault diagnosis

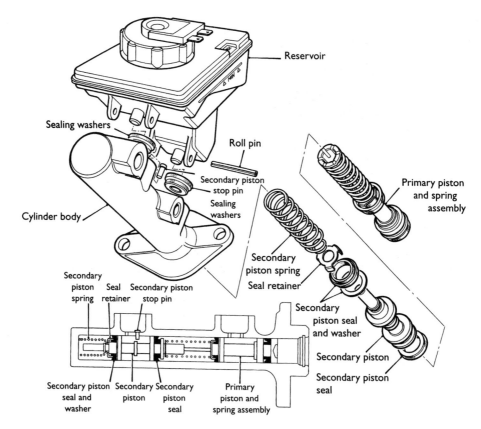

Figure 8.1 Double acting master cylinder

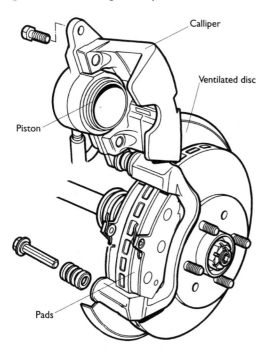

Figure 8.2 Disc brake calliper and brake pads

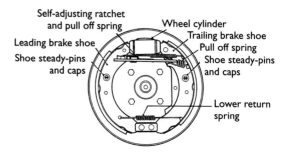

Figure 8.3 Drum brake adjusters – ratchet type

Disc brakes provide for good braking and are less prone to brake fade than drum brakes. This is because they are more exposed and can get rid of heat more easily. Brake fade occurs when the brakes become so hot they cannot transfer energy any more, and stop working! This type of problem can happen say after keeping the car brakes on for a long time when travelling down a long steep hill. This is why a lower gear should be used to employ the engine as a brake. It is clearly important to use good quality pads and linings because inferior materials can fail if overheated.

Drum brakes operate by shoes being forced on to the inside of the drum. A common type with a ratchet for automatic adjustment is shown as Figure 8.3. Shoes can be moved by double or

minimal. Disc brakes are self adjusting and replacing pads is usually a simple task. In the type shown just one bolt has to be removed to hinge the calliper upwards.

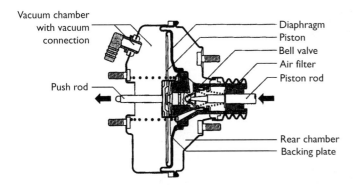

Figure 8.4 Vacuum operated brake servo (booster)

single acting cylinders. The most common layout is to use one double acting cylinder and brake shoes on each rear wheel of the vehicle, and disc brakes on the front wheels. A double acting cylinder simply means that as fluid pressure acts through a centre inlet, pistons are forced out of both ends.

Drum brakes more affected by wet and heat than disc brakes because both water and heat are trapped inside the drum. However, they are easier to fit with a mechanical hand brake linkage.

8.1.4 Brake adjustments

Brakes must be adjusted so that the minimum movement of the pedal starts to apply the brakes. The adjustment in question is the gap between the pads and disc and/or the shoes and drum.

Disc brakes are self-adjusting because as pressure is released it moves the pads just away from the disc. Drum brakes are different because the shoes are moved away from the drum to a set position by a pull off spring. The set position is adjustable and this can be done in a number of ways.

- Self-adjusting drum brakes are almost universal now. On light vehicles a common type uses an offset ratchet which clicks to a wider position if the shoes move beyond a certain amount when operated. Modern cars frequently have a self-adjusting hand brake.
- Screwdriver adjustment through a hole in the wheel and drum is also used. This is often a type of nut on a threaded bar which pushes the shoes out as it is screwed along the thread. This method can also have an automatic adjuster fitted.
- An adjustment screw on the back plate is now quite an old method in which a screw or square head protruding from the back plate moves the shoes by a snail cam.

The adjustment procedure stated by the manufacturer must be followed. As a guide, though, most recommend tightening the adjuster until the wheels lock and then moving it back until the wheel is just released. You must ensure that the brakes are not rubbing as this would build up heat and wear the friction material very quickly.

As an aid to fault diagnosis, the effects of incorrect adjustment are as follows:

- reduced braking efficiency;
- unbalanced braking;
- excessive pedal travel.

8.1.5 Servo-assisted braking

Servo systems are designed to give little assistance for light braking but increase the assistance as pedal pressure is increased. A common servo system uses low pressure (vacuum) from the manifold on one side, and the higher atmospheric pressure on the other side of a diaphragm. The low pressure is taken via a non-return safety valve from the engine inlet manifold. This pressure difference causes a force, which is made to act on the master cylinder. Figure 8.4 shows a vacuum servo.

Hydraulic power brakes use the pressure from an engine driven pump. This pump will often be the same one used to supply the power assisted steering. Pressure from the pump is made to act on a plunger in line with the normal master cylinder. As the driver applies force to the pedal, a servo valve opens in proportion to the force applied by the driver. The hydraulic assisting force is therefore also proportional. This maintains the important 'driver feel'.

A hydraulic accumulator (a reservoir for fluid under pressure) is incorporated into many systems. This is because the pressure supplied by the pump varies with engine speed. The pressure

Figure 8.5 Checking a brake disk with a dial gauge

in the accumulator is kept between set pressures in the region of 70 bar.

> **Warning**
> If you have to disconnect any components from the braking system on a vehicle fitted with an accumulator, you must follow the manufacturer's recommendations on releasing the pressure first.

8.1.6 Brake fluid

Always use new and approved brake fluid when topping up or renewing the system. The manufacturer's recommendations must always be followed. Brake fluid is hygroscopic which means that over a period of time it absorbs water. This increases the risk of the fluid boiling due to the heat from the brakes. Pockets of steam in the system would not allow full braking pressure to be applied. Many manufacturers recommend that the fluid should be changed at regular intervals – in some cases once per year or every 30 000 km. Make sure the correct grade of fluid is used.

8.2 Diagnostics – brakes

8.2.1 Systematic testing

If the reported fault is the hand brake not holding proceed as follows.

1. Confirm the fault by trying to pull away with the hand brake on.
2. Check the foot brake operation. If correct this suggests the brake shoes and drums (or pads and discs) are likely to be in good order.
3. Consider this: do you need to remove the wheels and drums or could it be a cable fault?

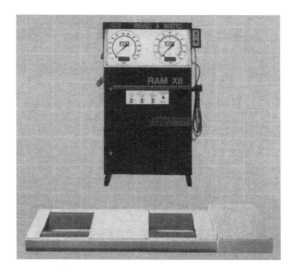

Figure 8.6 Rolling road brake tester display

4. Check cable operation by using an assistant in the car while you observe.
5. Renew the cable if seized.
6. Check hand brake operation and all associated systems.

8.2.2 Test equipment

> *Note:* You should always refer to the manufacturer's instructions appropriate to the equipment you are using.

Brake fluid tester

Because brake fluid can absorb a small amount of water it must be renewed or tested regularly. It becomes dangerous if the water turns into steam inside the cylinders or pipes, causing the brakes to become ineffective. The tester measures the moisture content of the fluid.

Brake roller test (Figure 8.6)

This is the type of test carried out as part of the annual safety test. The front or rear wheels are driven into a pair of rollers. The rollers drive each wheel of the car and as the brakes are applied the braking force affects the rotation. A measure of braking efficiency can then be worked out.

8.2.3 Dial gauge (Figure 8.5)

A dial gauge, sometimes called a clock gauge or a dial test indicator (DTI), is used to check the brake disk for run out. The symptoms of this would often be vibration or pulsation when braking. Manufacturers recommend maximum run out figures. In some cases the disk can be re-ground but often it is safer and more cost effective to fit new disks. This would also be done in pairs.

8.2.4 Test results

Some of the information you may have to get from other sources such as data books or a workshop manual is listed in the following table.

Test carried out	Information required
Brake roller test	Required braking efficiency: 50% for first line brakes, 25% for second line brakes and 16% for the parking brake. On modern vehicles half of the main system is the second line (dual line brakes). Old vehicles had to use the parking brake as the second line, therefore it had to work at 25%
Brake fluid condition	Manufacturers specify maximum moisture content

8.2.5 Brakes fault diagnosis Table 1

Symptom	Possible faults	Suggested action
Excessive pedal travel	Incorrect adjustment	Adjust it! But check condition as well
Poor performance when stopping	Pad and/or shoe linings worn	Renew
	Seized calliper or wheel cylinders	Renew or free off if possible and safe
	Contaminated linings	Renew (both sides)
Car pulls to one side when braking	Seized calliper or wheel cylinder on one side	Overhaul or renew if piston or cylinder is worn
	Contaminated linings on one side	Renew (both sides)
Spongy pedal	Air in the hydraulic system	Bleed system and then check for leaks
	Master cylinder seals failing	Overhaul or renew
Pedal travels to the floor when pressed	Fluid reservoir empty	Refill, bleed system and check for leaks
	Failed seals in master cylinder	Overhaul or renew
	Leak from a pipe or union	Replace or repair as required
Brakes overheating	Shoe return springs broken	Renew (both sides)
	Callipers or wheel cylinders sticking	Free off or renew if in any doubt
Brake judder	Linings worn	Renew
	Drums out of round	Renew
	Discs have excessive run-out	Renew
Squeaking	Badly worn linings	Renew
	Dirt in brake drums	Clean out with proper cleaner
	Anti-squeal shims missing at rear of pads	Replace and smear with copper grease

8.2.6 Brakes fault diagnosis Table 2

Symptom	Possible cause
Brake fade	Incorrect linings Badly lined shoes Distorted shoes Overloaded vehicle Excessive braking
Spongy pedal	Air in system Badly lined shoes Shoes distorted or incorrectly set Faulty drums Weak master cylinder mounting
Long pedal	Discs running out pushing pads back Distorted damping shims Misplaced dust covers Drum brakes need adjustment Fluid leak Fluid contamination Worn or swollen seals in master cylinder Blocked filler cap vent
Brakes binding	Brakes or handbrake maladjusted No clearance at master cylinder push rod Seals swollen Seized pistons Shoe springs weak or broken Servo faulty
Hard pedal – poor braking	Incorrect linings Glazed linings Linings wet, greasy or not bedded correctly Servo unit inoperative Seized calliper pistons Worn dampers causing wheel bounce
Brakes pulling	Seized pistons Variation in linings Unsuitable tyres or pressures Loose brakes Greasy linings Faulty drums, suspension or steering
Fall in fluid level	Worn disc pads External leak Leak in servo unit

Disc brake squeal – pad rattle	Worn retaining pins Worn discs No pad damping shims or springs
Uneven or excessive pad wear	Disc corroded or badly scored Incorrect friction material
Brake judder	Excessive disc or drum run-out Calliper mounting bolts loose Worn suspension or steering components

8.2.7 Brake hydraulic faults

Brake hose clamps will assist in diagnosing hydraulic faults and enable a fault to be located quickly. Proceed as follows.

1. Clamp all hydraulic flexible hoses and check the pedal.
2. Remove the clamps one at a time and check the pedal again (each time).
3. The location of air in the system or the faulty part of the system will now be apparent.

8.3 Anti-lock brakes

8.3.1 Introduction

The reason for the development of anti-lock brakes (ABS) is very simple. Under braking conditions if one or more of the vehicle wheels locks (begins to skid) then this has a number of consequences:

- braking distance increases;
- steering control is lost;
- tyre wear is abnormal.

The obvious consequence is that an accident is far more likely to occur. The maximum deceleration of a vehicle is achieved when maximum energy conversion is taking place in the brake system. This is the conversion of kinetic energy to heat energy at the discs and brake drums. The potential for this conversion process between a tyre skidding, even on a dry road, is far less. A good driver can pump the brakes on and off to prevent locking but electronic control can achieve even better results.

ABS is becoming more common on lower price vehicles, which should be a contribution to safety. It is important to remember, however, that for normal use, the system is not intended to allow faster driving and shorter braking distances. It should be viewed as operating in an emergency only. Figure 8.7 shows how ABS can help to maintain steering control even under very heavy braking conditions.

8.3.2 Requirements of ABS

A good way of considering the operation of a complicated system is to ask: 'What must the system be able to do?' In other words, 'what are the requirements?' These can be considered for ABS under the following headings:

Fail safe system	In the event of the ABS system failing then conventional brakes must still operate to their full potential. In addition a warning must be given to the driver. This is normally in the form of a simple warning light
Manoeuvrability must be maintained	Good steering and road holding must continue when the ABS system is operating. This is arguably the key issue as being able to swerve round a hazard whilst still braking hard is often the best course of action
Immediate response must be available	Even over a short distance the system must react such as to make use of the best grip on the road. The response

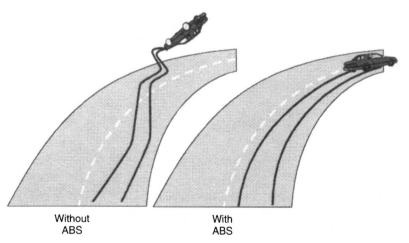

Figure 8.7 Advantages of ABS

	must be appropriate whether the driver applies the brakes gently or slams them on hard
Operational influences	Normal driving and manoeuvring should produce no reaction on the brake pedal. The stability and steering must be retained under all road conditions. The system must also adapt to braking hysteresis when the brakes are applied, released and then re-applied. Even if the wheels on one side are on dry tarmac and the other side on ice, the yaw (rotation about the vertical axis of the vehicle) of the vehicle must be kept to a minimum and only increase slowly to allow the driver to compensate
Controlled wheels	In its basic form at least one wheel on each side of the vehicle should be controlled on a separate circuit. It is now general for all four wheels to be controlled on passenger vehicles
Speed range of operation	The system must operate under all speed conditions down to walking pace. At this very slow speed even when the wheels lock the vehicle will come to rest very quickly. If the wheels did not lock then in theory the vehicle would never stop!
Other operating conditions	The system must be able to recognise aquaplaning and react accordingly. It must also still operate on an uneven road surface. The one area still not perfected is braking from slow speed on snow. The ABS will actually increase stopping distance in snow but steering will be maintained. This is considered to be a suitable trade off

A number of different types of anti-lock brake systems are in use, but all try to achieve the requirements as set out above.

8.3.3 General system description

As with other systems ABS can be considered as a central control unit with a series of inputs and outputs. An ABS system is represented by the closed loop system block diagram shown in Figure 8.8. The most important of the inputs are the wheel speed sensors and the main output is some form of brake system pressure control.

The task of the control unit is to compare signals from each wheel sensor to measure the acceleration or deceleration of an individual wheel. From this data and pre-programmed look up tables, brake pressure to one or more of the wheels can be regulated. Brake pressure can be reduced, held

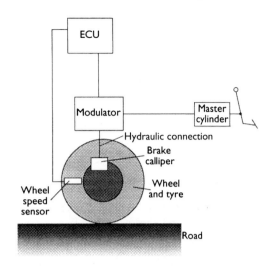

Figure 8.8 ABS block diagram

constant or allowed to increase. The maximum pressure is determined by the driver's pressure on the brake pedal.

From the wheel speed sensors the ECU calculates the following.

Vehicle reference speed	Determined from the combination of two diagonal wheel sensor signals. After the start of braking the ECU uses this value as its reference
Wheel acceleration or deceleration	This is a live measurement that is constantly changing
Brake slip	Although this cannot be measured directly a value can be calculated from the vehicle reference speed. This figure is then used to determine when/if ABS should take control of the brake pressure
Vehicle deceleration	During brake pressure control the ECU uses the vehicle reference speed as the starting point and decreases it in a linear manner. The rate of decrease is determined by the evaluation of all signals received from the wheel sensors driven and non-driven wheels on the vehicle must be treated in different ways as they behave differently when braking. A logical combination of wheel deceleration/acceleration and slip are used as the controlled variable. The actual strategy used for ABS control varies with the operating conditions

8.3.4 ABS components

There are a few variations between manufacturers involving a number of different components. For the majority of systems, however, there are three main components.

Wheel speed sensors

Most of these devices are simple inductance sensors and work in conjunction with a toothed wheel. They consist of a permanent magnet and a soft iron rod around which is wound a coil of wire. As the toothed wheel rotates the changes in inductance of the magnetic circuit generates a signal; the frequency and voltage of which are proportional to wheel speed. The frequency is the signal used by the ECU. The coil resistance is in the order of 800 to 1000 Ω. Coaxial cable is used to prevent interference affecting the signal. Some systems now use 'Hall effect' sensors.

Electronic control unit

The function of the ECU is to take in information from the wheel sensors and calculate the best course of action for the hydraulic modulator. The heart of a modern ECU consists of two microprocessors such as the Motorola 68HC11, which run the same programme independently of each other. This ensures greater security against any fault which could adversely affect braking performance, because the operation of each processor should be identical. If a fault is detected, the ABS disconnects itself and operates a warning light. Both processors have nonvolatile memory into which fault codes can be written for later service and diagnostic access. The ECU also has suitable input signal processing stages and output or driver stages for actuator control.

The ECU performs a self-test after the ignition is switched on. A failure will result in disconnection of the system. The following list forms the self-test procedure:

- current supply;
- exterior and interior interfaces;
- transmission of data;
- communication between the two microprocessors;
- operation of valves and relays;
- operation of fault memory control;
- reading and writing functions of the internal memory.

All this takes about 300 mS!

Hydraulic modulator

The hydraulic modulator as shown in Figure 8.9 has three operating positions:

- **pressure buildup** brake line open to the pump;
- **pressure holding** brake line closed;
- **pressure release** brake line open to the reservoir.

The valves are controlled by electrical solenoids, which have a low inductance so they react very quickly. The motor only runs when ABS is activated. Figure 8.10 shows an ABS hydraulic modulator with integrated ECU. This is the latest Bosch System version 8.1.

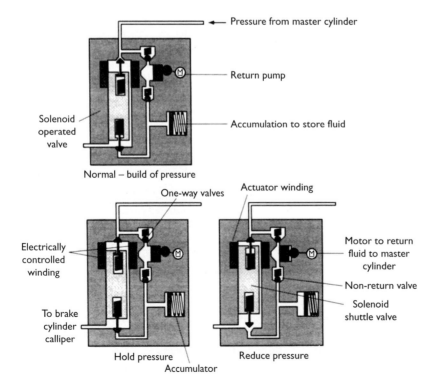

Figure 8.9 ABS hydraulic modulator operating positions

Chassis systems   177

8.4 Diagnostics – anti-lock brakes

8.4.1 Systematic testing procedure

> *Note:* ABS problems may require specialist attention – but don't be afraid to check the basics. An important note, however, is that some systems require special equipment to reinitialise the ECU if it has been disconnected (BMW for example).

Figure 8.10 ABS 8.1 modulator and ECU (*Source:* Bosch Press)

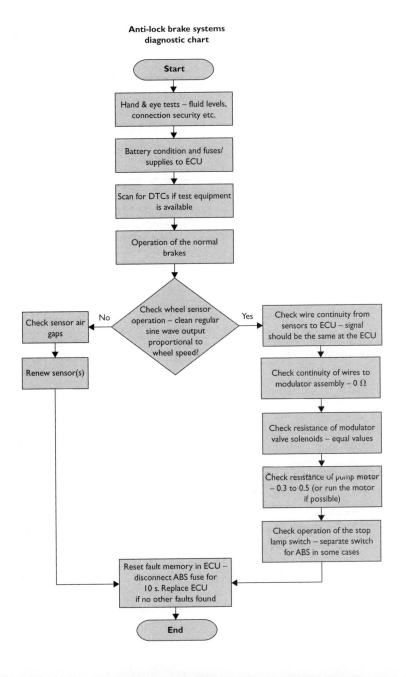

If the reported fault is the ABS warning light staying on proceed as follows (a scanner may be needed to reset the light).

8.4.2 Anti-lock brakes fault diagnosis table

Symptom	Possible cause
ABS not working and/or warning light on	Wheel sensor or associated wiring open circuit/high resistance Wheel sensor air gap incorrect Power supply/earth to ECU low or not present Connections to modulator open circuit No supply/earth connection to pump motor Modulator windings open circuit or high resistance
Warning light comes on intermittently	Wheel sensor air gap incorrect Wheel sensor air gap contaminated Loose wire connection

8.4.3 Bleeding anti-lock brakes

Special procedures may be required to bleed the hydraulic system when ABS is fitted. Refer to appropriate data for the particular vehicle. An example is reproduced as in Figure 8.11.

8.5 Traction control

8.5.1 Introduction

The steerability of a vehicle is not only lost when the wheels lock up on braking; the same effect arises if the wheels spin when driving off under severe acceleration. Electronic traction control has been developed as a supplement to ABS. This control system prevents the wheels from spinning when moving off or when accelerating sharply while on the move. In this way, an individual wheel which is spinning is braked in a controlled manner. If both or all of the wheels are spinning, the drive torque is reduced by means of an engine control function. Traction control has become known as ASR or TCR.

Traction control is not normally available as an independent system, but in combination with ABS. This is because many of the components required are the same as for the ABS. Traction control only requires a change in logic control in the ECU and a few extra control elements such as control of the throttle. Figure 8.12 shows a block diagram of a traction control system. Note the links with ABS and the engine control system.

Traction control will intervene to:

- maintain stability;
- reduction of yawing moment reactions;
- provide optimum propulsion at all speeds;
- reduce driver workload.

An automatic control system can intervene in many cases more quickly and precisely than the driver of the vehicle. This allows stability to be maintained at a time when the driver might not have been able to cope with the situation. Figure 8.13 shows an ABS and traction control modulator, complete with an ECU.

8.5.2 Control functions

Control of tractive force can be by a number of methods.

Throttle control	This can be via an actuator, which can move the throttle cable or, if the vehicle employs a drive by wire accelerator, then control will be in conjunction with the engine management ECU. This throttle control will be independent of the driver's throttle pedal position. This method alone is relatively slow to control engine torque
Ignition control	If ignition is retarded the engine torque can be reduced by up to 50% in a very short space of time. The timing is adjusted by a set ramp from the ignition map value
Braking effect	If the spinning wheel is restricted by brake pressure the reduction in torque at the affected wheel is very fast. Maximum brake pressure is not used to ensure that passenger comfort is maintained

Mercedes Benz C220 2.2 Sport D

ABS diagnostics

Removal/bleeding the system

Speed sensor
- The sensors are protected by tubes which are handed left and right for the vehicle
- Remove safety harness from nearest block connector
- Remove rear road wheel for access to sensor if needed
- Remove brake calliper
- Remove sensor retaining bolt(s) and remove sensor and shims if fitted
- Lightly lubricate sensor sleeve and push in sensor as far as it will go, do not rotate wheel
- Refit sensor retaining bolt and tighten to correct torque
- Check air gap and adjust if needed
- Reconnect harness

Hydraulic control unit
- Disconnect battery
- Remove fluid regulator
- Remove relay cover if fitted
- Remove earth lead
- Remove assistance cable from the centrifugal regulator
- Fit absorbent cloth underneath hydraulic control unit
- Undo hydraulic pipe(s)
- Disconnect retaining nuts/bolts/bracket/mountings
- Remove hydraulic control unit
- Refit in reverse order
- Tighten mounting nuts/bolts to correct torque
- Tighten hydraulic pipes to correct torque
- Bleed system—see bleed instructions

Electronic control unit
- Disconnect battery
- Remove battery, if needed and tray
- Remove seat/trim or mountings for access
- Remove wiring multi-plug/block
- Remove ECU securing screws/bolts or strap
- Remove ECU
- Refit in reverse order

G or acceleration sensor
- This operation only applies to four wheel drive models
- Disconnect electrical connections
- Drill out mounting bolts
- Remove mounting bolts/nuts and remove unit
- Fit unit with arrow pointing in direction of vehicle movement

Bleeding the brakes
- Do not switch on ignition before bleeding the system as this could cause air bubbles to form in the hydraulic unit
- Bleed system without the aid of the servo, on a level surface and without the wheels suspended
- If the hydraulic control unit is fitted with bleed screw(s), bleed these first in the conventional manner. Bleed screw first then second bleed screw (if fitted with two bleed screws)
- Cars fitted with delay valve must be bled before the remainder of the system
- To bleed delay valve release screw (below bleed screw) one turn
- Pump pedal 10 times and hold pressure
- Open bleed screw and close before pressure is completely lost
- For cars with diagonally split systems bleed each system in turn starting with right rear, left front, left rear, right front
- For cars with front/rear split systems bleed rear first then front
- Fill fluid reservoir
- Check operation of warning lights
- Low speed road tests and check ABS operation

Figure 8.11 Data re ABS bleeding, etc.

8.5.3 System operation

The description that follows is for a vehicle with an electronic accelerator (drive by wire). A simple sensor determines the position of the accelerator and, taking into account other variables such as engine temperature and speed for example, the throttle is set at the optimum position by a servo-motor. When accelerating the increase in engine torque leads to an increase in driving torque at the wheels. To achieve optimum acceleration the maximum possible driving torque must be transferred to the road. If driving torque exceeds that

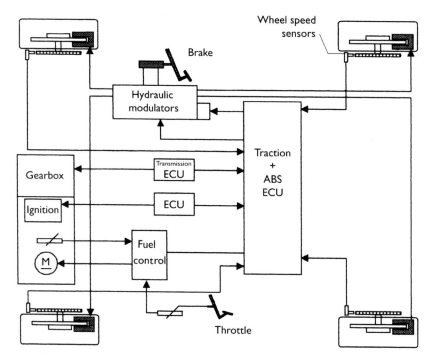

Figure 8.12 Traction control integrated with ABS and other systems

Figure 8.13 Bosch ABS version 8.1 hydraulic modulator

which can be transferred then wheel slip will occur on at least one wheel. The result of this is that the vehicle becomes unstable.

When wheel spin is detected the throttle position and ignition timing are adjusted but the best results are gained when the brakes are applied to the spinning wheel. This not only prevents the wheel from spinning but acts to provide a limited slip differential action. This is particularly good when on a road with varying braking force coefficients. When the brakes are applied a valve in the hydraulic modulator assembly moves over to allow traction control operation. This allows pressure from the pump to be applied to the brakes on the offending wheel. The valves, in the same way as with ABS, can provide pressure buildup, pressure hold and pressure reduction. This all takes place without the driver touching the brake pedal. The summary of this is that the braking force must be applied to the slipping wheel so as to equalise the combined braking coefficient for each driving wheel.

8.6 Diagnostics – traction control

8.6.1 Systematic testing

Note: Traction control (TCR or TCS or ASC) is usually linked with the ABS and problems may require specialist attention – but don't be afraid to check the basics. As with ABS, note that some systems require special equipment to reinitialise the ECU if it has been disconnected.

If the reported fault is the traction control system not working proceed as follows. Remember to check for the obvious – loose connections as in Figure 8.14, for example.

Traction control systems diagnostic chart

```
Start
  ↓
Hand & eye tests –
fluid levels, connection
security etc.
  ↓
Battery condition and
fuses/supplies to ECU.
Traction control switch
operation – 0 V drop
across or 0 ohms
  ↓
Scan for DTCs if test
equipment is available
  ↓
Normal operation of
the throttle
  ↓
Check wheel sensor operation – clean regular sine wave output proportional to wheel speed?
  No →  Check sensor air gaps → Renew sensor(s)
  Yes → Check wire continuity from sensors to ECU – signal should be the same at the ECU
        ↓
        Check continuity of wires to modulator and/or throttle control assembly – 0 Ω
        ↓
        Check resistance of controller/modulator valve solenoids – equal values
        ↓
        Replace ECU if no other faults found
  ↓
Reset fault memory in ECU
  ↓
End
```

8.6.2 Traction control fault diagnosis table

Symptom	Possible cause
Traction control inoperative	Wheel sensor or associated wiring open circuit/high resistance
	Wheel sensor air gap incorrect
	Power supply/earth to ECU low or not present
	Switch open circuit
	ABS system fault
	Throttle actuator inoperative or open circuit connections
	Communication link between ECUs open circuit
	ECU needs to be initialised

8.7 Steering and tyres

8.7.1 Tyres

The tyre performs two basic functions.

Figure 8.14 Check security of all connectors such as the one shown here

Figure 8.15 Tyres!

- It acts as the primary suspension, cushioning the vehicle from the effects of a rough surface.
- It provides frictional contact with the road surface. This allows the driving wheels to move the vehicle. The tyres also allow the front wheels to steer and the brakes to slow or stop the vehicle.

The tyre is a flexible casing which contains air. Tyres are manufactured from reinforced synthetic rubber. The tyre is made of an inner layer of fabric plies, which are wrapped around bead wires at the inner edges. The bead wires hold the tyre in position on the wheel rim. The fabric plies are coated with rubber, which is moulded to form the side walls and the tread of the tyre. Behind the tread is a reinforcing band, usually made of steel, rayon, or glass fibre. Modern tires are mostly tubeless, so they have a thin layer of rubber coating the inside to act as a seal.

8.7.2 Construction of a tubeless radial tyre

The wheel is made with a leak-proof rim and the valve is rubber mounted into a hole formed in the well of the rim. The tyre is made with an accurate bead, which fits tightly on to the rim. A thin rubber layer in the interior of the tyre makes an air tight seal.

The plies of a radial tyre pass from bead to bead at 90° to the circumference, or radially. There is a rigid belt band consisting of several layers of textile or metallic threads running round the tyre under the tread. Steel wire is often used in the construction of radial tyres. The radial tyre is flexible but retains high strength. It has good road holding and cornering power. In addition, radial tyres are economical due to their low 'rolling resistance'.

A major advantage of a radial tyre is its greatly improved grip even on wet roads. This is because the rigid belt band holds the tread flat on the road surface, when cornering. The rigid belt band also helps with the escape of water from under the tyre. Figure 8.15 shows some modern tyres.

8.7.3 Steering box and rack

Steering boxes contain a spiral gear known as a worm gear which rotates with the steering column. One form of design has a nut wrapped round the spiral and is therefore known as a worm and nut steering box. The grooves can be filled with re-circulating ball bearings, which reduce backlash or slack in the system and also reduce friction, making steering lighter. On vehicles with independent front suspension, an idler unit is needed together with a number of links and several joints. The basic weakness of the steering box system is in the number of swivelling joints and connections. If there is just slight wear at a number of points, the steering will not feel or be positive.

The steering rack (Figure 8.16) is now used almost without exception on light vehicles. This is because it is simple in design and very long lasting. The wheels turn on two large swivel joints. Another ball joint (often called a track rod end) is fitted on each swivel arm. A further ball joint to the ends of the rack connects the track rods. The rack is inside a lubricated tube and gaiters protect the inner ball joints. The pinion meshes with the teeth of the rack, and as it is turned by the steering

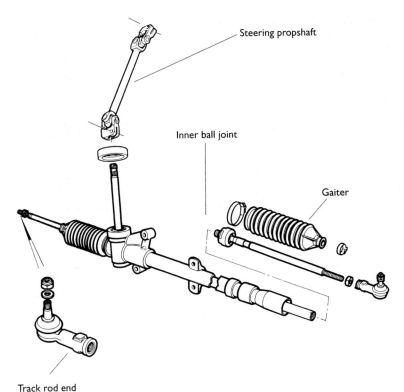

Track rod end

Figure 8.16 Steering rack

wheel, the rack is made to move back and forth, turning the front wheels on their swivel ball joints. On many vehicles now, the steering rack is augmented with hydraulic power assistance.

8.7.4 Power-assisted steering

Rack and pinion steering requires more turning effort than a steering box, although this is not too noticeable with smaller vehicles. However, heavier cars with larger engines or with wider tyres which scrub more, often benefit from power steering.

Many vehicles use a belt driven hydraulic pump to supply fluid under pressure for use in the system. Inside the rack and pinion housing is a hydraulic valve, which is operated as the pinion is turned for steering. The valve controls the flow of oil into a cylinder, which has a piston connected to the rack. This assists with the steering effort quite considerably.

A well designed system will retain 'feel' of road conditions for the driver to control the car. Steering a slow moving heavier vehicle when there is little room can be tiring, or even impossible for some drivers. This is where power steering has its best advantage. Many modern systems are able to make the power steering progressive. This means that as the speed of the vehicle increases, the assistance provided by the power steering reduces. This maintains better driver feel.

Very modern systems are starting to use electric power steering. This employs a very powerful electric motor as part of the steering linkage. There are two main types of electric power steering:

- Replacing the conventional hydraulic pump with an electric motor whilst the ram remains much the same.
- A drive motor, which directly assists with the steering and which has no hydraulic components.

The second system uses a small electric motor acting directly on the steering via an epicyclic gear train. This completely replaces the hydraulic pump and servo cylinder. It also eliminates the fuel penalty of the conventional pump and greatly simplifies the drive arrangements. Engine stall when the power steering is operated at idle speed is also eliminated. Figure 8.17 shows an electric power steering system.

An optical torque sensor is used to measure driver effort on the steering wheel. The sensor works by measuring light from an LED which is shining through holes which are aligned in discs at either end of a 50 mm torsion bar fitted into the steering column.

Figure 8.17 Electric power steering

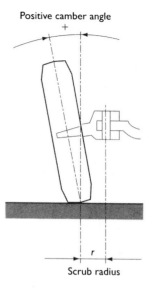

Figure 8.18 Camber angle

8.7.5 Steering characteristics

The steering characteristics of a vehicle, or in other words the way in which it reacts when cornering, can be described by one of three headings:

- oversteer;
- understeer;
- neutral.

Oversteer occurs when the rear of the vehicle tends to swing outward more than the front during cornering. This is because the slip angle on the rear axle is significantly greater than the front axle. This causes the vehicle to travel in a tighter circle, hence the term oversteer. If the steering angle is not reduced the vehicle will break away and all control will be lost. Turning the steering towards the opposite lock will reduce the front slip angle.

Understeer occurs when the front of the vehicle tends to swing outward more than the rear during cornering. This is because the slip angle on the rear axle is significantly smaller than the front axle. This causes the vehicle to travel in a greater circle, hence the term understeer. If the steering angle is not increased the vehicle will be carried out of the corner and all control will be lost. Turning the steering further into the bend will increase the front slip angle. Front engined vehicles tend to understeer because the centre of gravity is situated in front of the vehicle centre. The outward centrifugal force therefore has a greater effect on the front wheels than on the rear.

Neutral steering occurs when the centre of gravity is at the vehicle centre and the front and rear slip angles are equal. The cornering forces are therefore uniformly spread. Note, however, that understeer or oversteer can still occur if the cornering conditions change.

8.7.6 Camber

Note: Typical value is about 0.5° (values will vary so check specs).

On many cars, the front wheels are not mounted vertically to the road surface. Often they are tilted outwards at the top. This is called positive camber (Figure 8.18) and has the following effects:

- easier steering, less turning effort required;
- less wear on the steering linkages;
- less stress on main components;
- smaller scrub radius, which reduces the effect of wheel forces on the steering.

Negative camber has the effect of giving good cornering force. Some cars have rear wheels with negative camber. With independent suspension systems, wheels can change their camber from positive through neutral, to negative as the suspension is compressed. This varies, however, with the design and position of the suspension hinge points.

8.7.7 Castor

Note: Typical value is about 2 to 4° (values will vary so check specs).

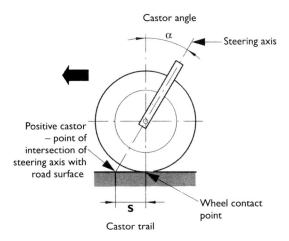

Figure 8.19 Positive castor

The front wheels tend to straighten themselves out after cornering. This is due to a castor action. Supermarket trolley wheels automatically run straight when pushed because the axle on which they rotate is behind the swivel mounting. Vehicle wheels get the same result by leaning the swivel pin mountings back so that the wheel axle is moved slightly behind the line of the swivel axis. The further the axle is behind the swivel, the stronger will be the straightening effect. The main effects of a positive castor angle (Figure 8.19) are:

- self-centring action;
- helps determine the steering torque when cornering.

Negative castor is used on some front wheel drive vehicles to reduce the return forces when cornering. Note that a combination of steering geometry angles is used to achieve the desired effect. This means that in some cases the swivel axis produces the desired self-centre action so the castor angle may need to be negative to reduce the return forces on corners.

8.7.8 Swivel axis inclination

> *Note:* Typical value is about 7 to 9° (values will vary so check specs).

The swivel axis is also known as the steering axis. Swivel axis inclination (Figure 8.20) means the angle compared to vertical made by the two swivel joints when viewed from the front or rear. On a strut type suspension system the angle is broadly that made by the strut. This angle always leans in towards the middle of the vehicle. The

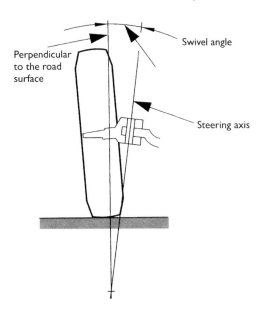

Figure 8.20 Swivel axis

swivel axis inclination (also called kingpin inclination) is mainly for:

- producing a self-centre action;
- improving steering control on corners;
- giving a lighter steering action.

Scrub radius, wheel camber and swivel axis inclination all have an effect on one another. The swivel axis inclination mainly affects the self-centring action, also known as the aligning torque. Because of the axis inclination the vehicle is raised slightly at the front as the wheels are turned. The weight of the vehicle therefore tries to force the wheels back into the straight-ahead position.

8.7.9 Tracking

As a front wheel drive car drives forward the tyres pull on the road surface taking up the small amount of free play in the mountings and joints. For this reason the tracking is often set toe-out so that the wheels point straight ahead when the vehicle is moving. Rear wheel drive tends to make the opposite happen because it pushes against the front wheels. The front wheels are therefore set toe-in. When the car moves, the front wheels are pushed out taking up the slack in the joints, so the wheels again end up straight ahead. The amount of toe-in or toe-out is very small, normally not exceeding 5 mm (the difference in the distance between the front and rear of the front wheels). Correctly set tracking ensures true rolling of the wheels and therefore reduced tyre wear. Figure 8.21 shows wheels set toe-in and toe-out.

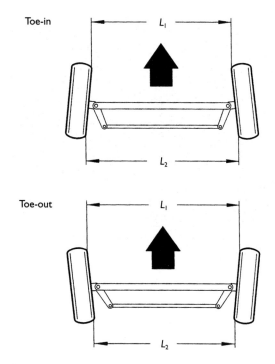

Figure 8.21 Tracking

8.7.10 Scrub radius

The scrub radius is the distance between the contact point of the steering axis with the road and the wheel centre contact point. The purpose of designing in a scrub radius is to reduce the steering force and to prevent steering shimmy. It also helps to stabilise the straight-ahead position.

It is possible to design the steering with a negative, positive or zero scrub radius as described below.

Scrub radius	Description	Properties
Negative	The contact point of the steering axis hits the road between the wheel centre and the outer edge of the wheel	Braking forces produce a torque which tends to make the wheel turn inwards. The result of this is that the wheel with the greatest braking force is turned in with greater torque. This steers the vehicle away from the side with the heaviest braking producing a built in counter steer action which has a stabilising effect
Positive	The contact point of the steering axis hits the road between the wheel centre and the inner edge of the wheel	A positive scrub radius makes turning the steering easier. However, braking forces produce a torque which tends to make the wheel turn outwards. The result of this is that the wheel with the greatest braking force is turned out with greater torque. Under different road conditions this can have the effect of producing an unwanted steering angle
Zero	The contact point of the steering axis hits the road at the same place as the wheel centre	This makes the steering heavy when the vehicle is at rest because the wheel cannot roll at the steering angle. However, no separate turning torque about the steering axis is created

From the information given you will realise that decisions about steering geometry are not clearcut. One change may have a particular advantage in one area but a disadvantage in another. To assist with fault diagnosis a good understanding of steering geometry is essential.

8.8 Diagnostics – steering and tyres

8.8.1 Systematic testing

If the reported fault is heavy steering proceed as follows.

1. Ask if the problem has just developed. Road test to confirm.
2. Check the obvious such as tyre pressures. Is the vehicle loaded to excess? Check geometry?
3. Assuming tyre pressure and condition is as it should be we must move on to further tests.
4. For example, jack up and support the front of the car. Operate the steering lock to lock. Disconnect one track rod end and move the wheel on that side, and so on.
5. If the fault is in the steering rack then this should be replaced and the tracking should be set.
6. Test the operation with a road test and inspect all other related components for security and safety.

8.8.2 Test equipment

Note: You should always refer to the manufacturer's instructions appropriate to the equipment you are using.

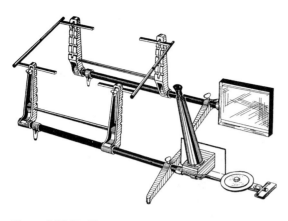

Figure 8.22 Tracking gauges

Tyre pressure gauge and tread depth gauge

Often under-rated pieces of test equipment. Correctly inflated tyres make the vehicle handle better, stop better and use less fuel. The correct depth of tread means the vehicle will be significantly safer to drive, particularly in wet conditions.

Tracking gauges (Figure 8.22)

The toe-in and toe-out of a vehicle's front wheels is very important. Many types of tracking gauges are available. One of the most common uses a frame placed against each wheel with a mirror on one side and a moveable viewer on the other. The viewer is moved until marks are lined up and the tracking can then be measured.

Wheel balancer (Figure 8.23)

This is a large fixed piece of equipment in most cases. The wheel is removed from the car, fixed on to the machine and spun at high speed. Sensors in the tester measure the balance of the wheel. The tester then indicates the amount of weight which should be added to a certain position. The weight is added by fitting small lead weights.

8.8.3 Four-wheel alignment

Standard front wheel alignment is simply a way of making sure the wheels are operating parallel with one another, and that the tyres meet the road at the correct angle. Four-wheel alignment makes sure that the rear wheels follow the front wheels in a parallel path.

Different manufacturers set different specifications for the angles created between the suspension, steering, wheels and the frame of the vehicle. When these angles are correct, the vehicle is properly aligned.

Figure 8.23 Wheel balancer

The main reasons for correct alignment are to ensure that the vehicle achieves:

- minimum rolling friction
- maximum tyre mileage
- stability on the road
- steering control for the driver.

Diagnosing incorrect alignments is usually just a matter of examining the:

- tyres for unusual wear
- wheels for damage
- steering wheel for position.

In addition a road test is usually necessary to check that the vehicle is not pulling to one side, wandering or weaving. Four basic wheel settings or angles determine whether a vehicle is properly aligned.

- Camber is the inward or outward tilt of a wheel compared to a vertical line. If the camber is out of adjustment, it will cause tyre wear on one side of the tyre's tread.
- Caster is the degree that the car's steering axis is tilted forward or backward from the vertical as viewed from the side of the car. If the caster is out of adjustment, it can cause problems with self-centring and wander. Caster has little effect on tyre wear.
- Toe refers to the directions in which two wheels point relative to each other. Incorrect toe causes rapid tyre wear to both tyres equally. Toe is the most common adjustment and it is always adjustable on the front wheels and is adjustable on the rear wheels of some cars.

Figure 8.24 Laser alignment digital readout

Figure 8.25 Laser alignment scale

- Offset is the amount that the rear wheels are out of line, or off set, with the front. Ideally, each rear wheel should be exactly in line with the corresponding front wheel.

Laser based equipment is now almost always used to set alignment (Figure 8.24). The front wheels are placed on turntables and measuring equipment connected to each of the four wheels. In simple terms, a laser is used to 'draw' a perfect box around the vehicle so that readouts on displays or scales show how much the wheel positions differ from this 'perfect' box. In addition, spirit levels are used for some measurements.

Before checking and adjusting any wheel alignments setting, it is good practice to check the suspension and vehicle ride height (Figure 8.25). Any bent or broken suspension component will mean that setting alignment will be at best difficult and at worst pointless!

8.8.4 Test results

Some of the information you may have to get from other sources such as data books or a workshop manual is listed in the following table.

Test carried out	Information required
Tracking	The data for tracking will be given as either an angle or a distance measurement. Ensure that you use the appropriate data for your type of test equipment. Distance will be a figure such as 3 mm toe-in, and angle one such as 50′ toe-in (50′ means 50 minutes). The angle of 1° is split into 60 minutes, so in this case the angle is 50/60 or 5/6 of a degree
Pressures	A simple measurement which should be in bars. You will find, however that many places still use PSI (pounds per square inch). As in all other cases only compare like with like.

Chassis systems **189**

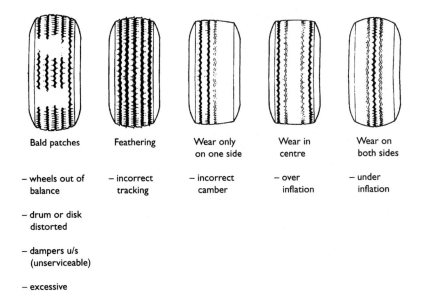

- wheels out of balance
- drum or disk distorted
- dampers u/s (unserviceable)
- excessive braking

— incorrect tracking

— incorrect camber

— over inflation

— under inflation

Figure 8.26 Tyre faults

| Tread depth | The minimum measurement (e.g. 1.6 mm over 75% of the tread area but please note the current local legal requirements) |

8.8.5 Tyres fault diagnosis table

The following table lists some of the faults which can occur if tyres and/or the vehicle are not maintained correctly. Figure 8.26 shows the same.

Symptom	Possible cause/fault
Wear on both outer edges of the tread	Under inflation
Wear in the centre of the tread all round the tyre	Over inflation
Wear just on one side of the tread	Incorrect camber
Feathering	Tracking not set correctly
Bald patches	Unbalanced wheels or unusual driving technique!

8.8.6 Tyre inflation pressures

The pressure at which the tyres should be set is determined by a number of factors such as:

- load to be carried;
- number of plies;
- operating conditions;
- section of the tyre.

Tyre pressures must be set at the values recommended by the manufacturers. Pressure will vary according to the temperature of the tyre – this is affected by operating conditions. Tyre pressure should always be adjusted when the tyre is cold and be checked at regular intervals.

8.8.7 Steering fault diagnosis table 1

Symptom	Possible faults	Suggested action
Excessive free play at steering wheel	Play between the rack and pinion or in the steering box	Renew in most cases but adjustment may be possible
	Ball joints or tie rod joints worn	Renew
	Column coupling loose or bushes worn	Secure or renew
Vehicle wanders, hard to keep in a straight line	As above	As above
	Alignment incorrect	Adjust to recommended setting
	Incorrect tyre pressure or mix of tyre types is not suitable	Adjust pressures or replace tyres as required
	Worn wheel bearings	Renew
Stiff steering	Wheel alignment incorrect	Adjust to recommended setting
	Tyre pressures too low	Adjust pressures
	Ball joints or rack seizing	Renew
Wheel wobble	Wheels out of balance	Balance or renew

	Wear in suspension linkages	Renew
	Alignment incorrect	Adjust to recommended setting
Understeer or oversteer	Tyre pressures incorrect	Adjust pressures
	Dangerous mix of tyre types	Replace tyres as required
	Excessive freeplay in suspension or steering system	Renew components as required

8.8.8 Steering, wheels and tyres fault diagnosis

Symptom	Possible cause
Wandering or instability	Incorrect wheel alignment Worn steering joints Wheels out of balance Wheel nuts or bolts loose
Wheel wobble	Front or rear Wheels out of balance Damaged or distorted wheels/tyres Worn steering joints
Pulling to one side	Defective tyre Excessively worn components Incorrect wheel alignment
Excessive tyre wear	Incorrect wheel alignment Worn steering joints Wheels out of balance Incorrect inflation pressures Driving style! Worn dampers
Excessive free play	Worn track rod end or swivel joints Steering column bushes worn Steering column universal joint worn
Stiff steering	Lack of steering gear lubrication Seized track rod end joint or suspension swivel joint Incorrect wheel alignment Damage to steering components

8.9 Suspension

8.9.1 Introduction

The purpose of a suspension system can best be summarised by the following requirements:

- cushion the car, passengers and load from road surface irregularities;
- resist the effects of steering, braking and acceleration, even on hills and when loads are carried;
- keep tyres in contact with the road at all times;
- work in conjunction with the tyres and seat springs to give an acceptable ride at all speeds.

The above list is difficult to achieve completely, so some sort of compromise has to be reached. Because of this many different methods have been tried, and many are still in use. Keep these four main requirements in mind and it will help you to understand why some systems are constructed in different ways.

A vehicle needs a suspension system to cushion and damp out road shocks so providing comfort to the passengers and preventing damage to the load and vehicle components. A spring between the wheel and the vehicle body allows the wheel to follow the road surface. The tyre plays an important role in absorbing small road shocks. It is often described as the primary form of suspension. The vehicle body is supported by springs located between the body and the wheel axles. Together with the damper these components are referred to as the suspension system.

As a wheel hits a bump in the road it is moved upwards with quite some force. An unsprung wheel is affected only by gravity, which will try to return the wheel to the road surface but most of the energy will be transferred to the body. When a spring is used between the wheel and the vehicle body, most of the energy in the bouncing wheel is stored in the spring and not passed to the vehicle body. The vehicle body will now only move upwards through a very small distance compared to the movement of the wheel.

8.9.2 Suspension system layouts

On older types of vehicle, a beam axle was used to support two stub axles. Beam axles are now rarely used in car suspension systems, although many commercial vehicles use them because of their greater strength and constant ground clearance.

The need for a better suspension system came from the demand for improved ride quality and improved handling. Independent front suspension (IFS) was developed to meet this need. The main advantages of independent front suspension are as follows:

- when one wheel is lifted or drops, it does not affect the opposite wheel;
- the unsprung mass is lower, therefore the road wheel stays in better contact with the road;
- problems with changing steering geometry are reduced;

Figure 8.27 Front wheel drive vehicle showing the suspension layout

- there is more space for the engine at the front;
- softer springing with larger wheel movement is possible.

There are a number of basic suspension systems in common use. Figure 8.27 shows a typical suspension layout.

8.9.3 Front axle suspensions

As with most design aspects of the vehicle, compromise often has to be reached between performance, body styling and cost. The following table compares the common front axle suspension systems.

Name	Description	Advantages	Disadvantages
Double transverse arms	Independently suspended wheels located by two arms perpendicular to direction of travel. The arms support stub axles	Low bonnet line Only slight changes of track and camber with suspension movements	A large number of pivot points is required High production costs
Transverse arms with leaf spring	A transverse arm and a leaf spring locate the wheel	The spring can act as an anti-roll bar, hence low cost	Harsh response when lightly loaded Major changes of camber as vehicle is loaded
Transverse arm with McPherson strut	A combination of the spring, damper, wheel hub, steering arm and axle joints in one unit	Only slight changes in track and camber with suspension movement Forces on the joints are reduced because of the long strut	The body must be strengthened around the upper mounting A low bonnet line is difficult to achieve
Double trailing arms	Two trailing arms support the stub axle. These can act on torsion bars often formed as a single assembly	No change in castor, camber or track with suspension movement Can be assembled and adjusted off the vehicle	Lots of space is required at the front of the vehicle Expensive to produce Acceleration and braking cause pitching movements which in turn changes the wheel base

Figure 8.28 shows a suspension system with front struts and rear trailing arms.

8.9.4 Rear axle suspensions

The following table compares the common rear axle suspension systems.

Name	Description	Advantages	Disadvantages
Rigid axle with leaf springs	The final drive, differential and axle shafts are all one unit	Rear track remains constant reducing tyre wear
Good directional stability because no camber change causes body roll on corners
Low cost
Strong design for load carrying	High unsprung mass
The interaction of the wheels causes lateral movement reducing tyre adhesion when the suspension is compressed on one side
Rigid axle with A-bracket	Solid axle with coil springs and a central joint supports the axle on the body	Rear of the vehicle pulls down on braking which stabilises the vehicle	High cost
Large unsprung mass
Rigid axle with compression/tension struts	Coil springs provide the springing and the axle is located by struts	Suspension extension is reduced when braking or accelerating
The springs are isolated from these forces	High loads on the welded joints
High weight overall
Large unsprung mass
Torsion beam trailing arm axle	Two links are used, connected by a 'U' section that has low torsional stiffness but high resistance to bending	Track and camber does not change
Low unsprung mass
Simple to produce
Space saving	Torsion bar springing on this system can be more expensive than coil springs
Torsion beam axle with Panhard rod	Two links are welded to an axle tube or 'U' section and lateral forces are taken by a Panhard rod	Track and camber does not change
Simple flexible joints to the bodywork	Torsion bar springing on this system can be more expensive than coil springs
Trailing arms	The pivot axis of the trailing arms is at 90° to the direction of vehicle travel	When braking the rear of the vehicle pulls down giving stable handling
Track and camber does not change
Space saving	Slight change of wheel base when the suspension is compressed
Semi-trailing arms – fixed length drive shafts	The trailing arms are pivoted at an angle to the direction of travel. Only one UJ is required because the radius of the suspension arm is the same as the driveshaft when the suspension is compressed	Only very small dive when braking
Lower cost than when variable length shafts are used	Sharp changes in track when the suspension is compressed resulting in tyre wear
Slight tendency to oversteer
Semi-trailing arms – variable length drive shafts	The final drive assembly is mounted to the body and two UJs are used on each shaft	The two arms are independent of each other
Only slight track changes	Large camber changes
High cost because of the drive shafts and joints

8.9.5 Anti-roll bar

The main purpose of an anti-roll bar (Figure 8.29) is to reduce body roll on corners. The anti-roll bar can be thought of as a torsion bar. The centre is pivoted on the body and each end bends to make connection with the suspension/wheel assembly. When the suspension is compressed on both sides, the anti-roll bar has no effect because it pivots on its mountings.

As the suspension is compressed on just one side, a twisting force is exerted on the anti-roll bar. The anti-roll bar is now under torsional load. Part of this load is transmitted to the opposite wheel, pulling it upwards. This reduces the amount

- ensure good tyre adhesion;
- support the weight of the vehicle;
- transmit gravity forces to the wheels.

There are a number of different types of spring in use on modern light vehicles. The following table lists these together with their main features.

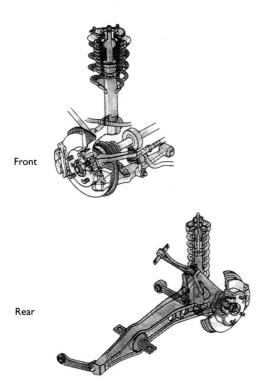

Figure 8.28 Suspension system with front struts and rear trailing arms

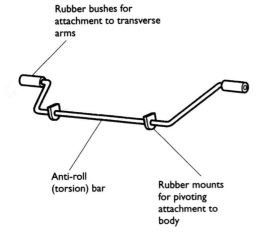

Figure 8.29 Anti-roll bar and mountings

of body roll on corners. The disadvantages are that some of the 'independence' is lost and the overall ride is harsher. Anti-roll bars can be fitted to both front and rear axles.

8.9.6 Springs

The requirements of the springs can be summarised as follows:

- absorb road shocks from uneven surfaces;
- control ground clearance and ride height;

Name	Comments	Characteristics
Coil springs	The most common spring currently in use on light vehicles. The coil spring is a torsion bar wound into a spiral	Can be progressive if the diameter of the spring is tapered conically Cannot transmit lateral or longitudinal forces, hence the need for links or arms Little internal damping Little or no maintenance High travel
Leaf springs	These springs can be single or multiple leaf. They are most often mounted longitudinally. Nowadays they are only ever used on commercial vehicles	Can transmit longitudinal and lateral forces Short travel High internal damping High load capacity Maintenance may be required Low height but high weight
Torsion bar springs	A torsion bar is a spring where twisting loads are applied to a steel bar. They can be round or square section, solid or hollow. Their surface must be finished accurately to eliminate pressure points, which may cause cracking and fatigue failure. They can be fitted longitudinally or laterally	Maintenance-free but can be adjusted Transmit longitudinal and lateral forces Limited self-damping Linear rate Low weight May have limited fatigue life
Rubber springs	Nowadays rubber springs are only used as a supplement to other forms of springs. They are, however, popular on trailers and caravans	Progressive rate Transmit longitudinal and lateral forces Short travel Low weight and low cost Their springing and damping properties can change with temperature
Air springs	Air springs can be thought of as being like a balloon or football on which	Expensive Good quality ride Electronic control can be used

194 Advanced automotive fault diagnosis

	the car is supported. The system involves compressors and air tanks. They are not normally used on light vehicles	Progressive spring rate High production cost
Hydro-pneumatic springs	A hydro-pneumatic spring is a gas spring with hydraulic force transmission. Nitrogen is usually used as the gas. The damper can be built in as part of the hydraulic system. The springs can be hydraulically connected together to reduce pitch or roll. Ride height control can be achieved by pumping oil into or out of the working chamber	Progressive rate Ride height control Damping built in Pressurised oil supply is required Expensive and complicated

8.9.7 Dampers (shock absorbers)

The functions of a damper can be summarised as follows:

- ensure directional stability;
- ensure good contact between the tyres and the road;
- prevent buildup of vertical movements;
- reduce oscillations;
- reduce wear on tyres and chassis components.

There are a number of different types of damper.

Friction damper	Not used on cars today but you will find this system used as part of caravan or trailer stabilisers
Lever type damper	Used on earlier vehicles, the lever operates a piston which forces oil into a chamber
Twin tube telescopic damper	This is a commonly used type of damper consisting of two tubes. An outer tube forms a reservoir space and contains the oil displaced from an inner tube. Oil is forced through a valve by the action of a piston as the damper moves up or down. The reservoir space is essential to make up for the changes in volume as the piston rod moves in and out
Single tube telescopic damper	This is often referred to as a gas damper. However, the damping action is still achieved by forcing oil through a restriction. The gas space behind a separator piston is to compensate for the changes in cylinder volume caused as the piston rod moves. It is at a pressure of about 25 bar
Twin tube gas damper (Figure 8.30)	The twin tube gas damper is an improvement on the well-used twin tube system. The gas cushion is used in this case to prevent oil foaming. The gas pressure on the oil prevents foaming which in turn ensures constant operation under all operating conditions. Gas pressure is lower than for a single tube damper at about 5 bar
Variable rate damper	This is a special variation of the twin tube gas damper. The damping characteristics vary depending on the load on the vehicle. Bypass grooves are machined in the upper half of the working chamber. With light loads the damper works in this area with a soft damping effect. When the load is increased the piston moves lower down the working chamber away from the grooves resulting in full damping effect
Electronically controlled dampers	These are dampers where the damping rate can be controlled by solenoid valves inside the units. With suitable electronic control, the characteristics can be changed within milliseconds to react to driving and/or load conditions

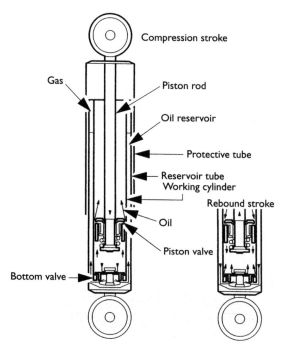

Figure 8.30 Twin tube gas damper

8.10 Diagnostics – suspension

8.10.1 Systematic testing

If the reported fault is poor handling proceed as follows.

1. Road test to confirm the fault.
2. With the vehicle on a lift inspect obvious items like tyres and dampers.
3. Consider if the problem is suspension related or in the steering for example. You may have decided this from road testing.
4. Inspect all the components of the system you suspect, for example dampers for correct operation and suspension bushes for condition and security. Let's assume the fault was one front damper not operating to the required standard.
5. Renew both of the dampers at the front to ensure balanced performance.
6. Road test again and check for correct operation of the suspension and other systems.

8.10.2 Test equipment

Note: You should always refer to the manufacturer's instructions appropriate to the equipment you are using.

Damper tester

This is a device that will draw a graph to show the response of the dampers. It may be useful for providing paper evidence of the operating condition but a physical examination is normally adequate.

The operating principle is shown as Figure 8.31 which indicates that the damper is not operating correctly in this case.

8.10.3 Test results

Some of the information you may have to get from other sources such as data books or a workshop manual is listed in the following table.

Test carried out	Information required
Damper operation	The vehicle body should move down as you press on it, bounce back just past the start point and then return to the rest position
Suspension bush condition	Simple levering if appropriate should not show excessive movement, cracks or separation of rubber bushes
Trim height	This is available from data books as a measurement from say the wheel centre to a point on the car body above

8.10.4 Suspension fault diagnosis table 1

Symptom	Possible faults	Suggested action
Excessive pitch or roll when driving	Dampers worn	Replace in pairs
Car sits lopsided	Broken spring Leak in hydraulic suspension	Replace in pairs Rectify by replacing unit or fitting new pipes
Knocking noises	Excessive free play in a suspension joint	Renew
Excessive tyre wear	Steering/suspension geometry incorrect (may be due to accident damage)	Check and adjust or replace any 'bent' or out of true components

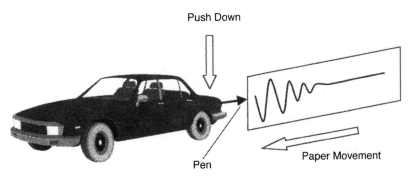

Figure 8.31 Damper testing procedure

8.10.5 Suspension fault diagnosis table 2

Symptom	Possible cause
Excessive pitching	Defective dampers
	Broken or weak spring
	Worn or damaged anti-roll bar mountings
Wandering or instability	Broken or weak spring
	Worn suspension joints
	Defective dampers
Wheel wobble	Worn suspension joints
Pulling to one side	Worn suspension joints
	Accident damage to suspension alignment
Excessive tyre wear	Worn suspension joints
	Accident damage to suspension alignment
	Incorrect trim height (particularly hydro-elastic systems)

> *Knowledge check questions*
>
> To use these questions, you should first try to answer them without help but if necessary, refer back to the content of the chapter. Use notes, lists and sketches to answer them. It is not necessary to write pages and pages of text!
>
> 1. Describe how to test the operation of an ABS wheel speed sensor.
> 2. List in a logical sequence, a series of tests to determine the cause of steering pulling to one side when braking.
> 3. Describe a method of testing a damper (shock absorber).
> 4. Make a sketch to show three different types of tyre wear and state for each a possible cause.
> 5. Explain why it may be necessary to check the run-out on a brake disk (rotor) and describe how this is done.

9
Electrical systems

9.1 Electronic components and circuits

9.1.1 Introduction

This section describing the principles and applications of various electronic components and circuits is not intended to explain their detailed operation. Overall, an understanding of basic electronic principles will help to show how electronic systems operate. These can range from a simple interior light delay unit to the most complicated engine management system. Testing individual electronic components is a useful diagnostic procedure.

9.1.2 Components

The symbols for the following electronic components are shown in Figure 9.1.

Resistors are probably the most widely used component in electronic circuits. Two factors must be considered when choosing a suitable resistor, the ohms value and the power rating. Resistors are used to limit current flow and provide fixed voltage drops. Most resistors used in electronic circuits are made from small carbon rods; the size of the rod determines the resistance. Carbon resistors have a negative temperature coefficient (NTC) and this must be considered for some applications. Thin-film resistors have more stable temperature properties and are constructed by depositing a layer of carbon on to an insulated former such as glass. The resistance value can be manufactured very accurately by spiral grooves cut into the carbon film. For higher power applications resistors are usually wire wound. Variable forms of most resistors are available. The resistance of a circuit is its opposition to current flow.

A capacitor is a device for storing an electric charge. In its simple form it consists of two plates separated by an insulating material. One plate can have excess electrons compared to the other. On vehicles its main uses are for reducing arcing across contracts and for radio interference suppression circuits as well as in ECUs. Capacitors are described as two plates separated by a dielectric. The area of the plates, the distance between them and the nature of the dielectric determine the value of capacitance. Metal foil sheets insulated by a type of paper are often used to construct capacitors. The sheets are rolled up together inside a tin can. To achieve higher values of capacitance it is necessary to reduce the distance between the plates in order to keep the overall size of the device manageable. This is achieved by immersing one plate in an electrolyte to deposit a layer of oxide typically 10^{-4} mm thick thus ensuring a higher capacitance value. The problem, however, is that this now makes the device polarity conscious and only able to withstand low voltages.

Diodes are often described as one-way valves and for most applications this is an acceptable description. A diode is a PN junction allowing electron flow from the N type material to the P type material. The materials are usually constructed from doped silicon. Diodes are not perfect devices and a voltage of about 0.6 V is required to switch the diode on in its forward biased direction.

Zener diodes are very similar in operation with the exception that they are designed to breakdown and conduct in the reverse direction at a predetermined voltage. They can be thought of as a type of pressure relief valve.

Transistors are the devices that have allowed the development of today's complex and small electronic systems. The transistor is used as either a solid state switch or as an amplifier. They are constructed from the same P and N type semiconductor materials as the diodes and can be made in either NPN or PNP format. The three terminals are known as the base, collector and emitter. When the base is supplied with the correct bias the circuit between the collector and emitter will conduct. The base current can be in the order of 50 to 200 times less than the emitter current. The ratio of the current flowing through the base compared to the current through the emitter is an indication of the amplification factor of the device.

198 Advanced automotive fault diagnosis

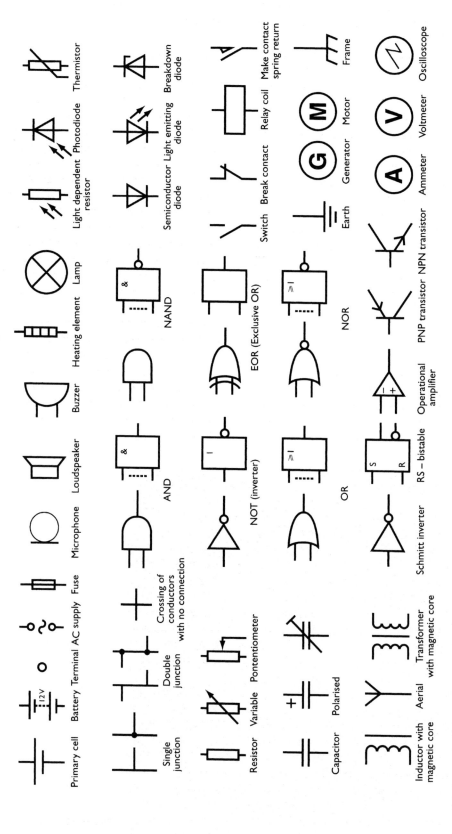

Figure 9.1 Electrical and electronic symbols

A Darlington pair is a simple combination of two transistors which will give a high current gain, typically several thousand. The transistors are usually mounted on a heat sink and overall the device will have three terminals marked as a single transistor – base, collector and emitter. The input impedance of this type of circuit is in the order of 1 MΩ; hence it will not load any previous part of a circuit connected to its input. The Darlington pair configuration is used for many switching applications. A common use of a Darlington pair is for the switching of coil primary current in the ignition circuit.

Another type of transistor is the FET or field effect transistor. This device has higher input resistance than the bipolar type described above. They are constructed in their basic form as n-channel or p-channel devices. The three terminals are known as the gate, source and drain. The voltage on the gate terminal controls the conductance of the circuit between the drain and the source.

Inductors are most often used as part of an oscillator or amplifier circuit. The basic construction of an inductor is a coil of wire wound on a former. It is the magnetic effect of the changes in current flow which gives this device the properties of inductance. Inductance is a difficult property to control particularly as the inductance value increases. This is due to magnetic coupling with other devices. Iron cores are used to increase the inductance value. This also allows for adjustable devices by moving the position of the core. Inductors, particularly of higher values, are often known as chokes and may be used in DC circuits to smooth the voltage.

9.1.3 Integrated circuits

Integrated circuits or ICs are constructed on a single slice of silicon often known as a substrate. Combinations of some of the components mentioned previously can be used to carry out various tasks such as switching, amplifying and logic functions. The components required for these circuits can be made directly on to the slice of silicon. The great advantage of this is not just the size of the ICs but the speed at which they can be made to work due to the short distances between components. Switching speed in excess of 1 MHz is typical.

The range and type of integrated circuits now available is so extensive that a chip is available for almost any application. The integration level of chips is now exceeding VLSI (very large scale integration). This means that there can be more than 100 000 active elements on one chip! Development in this area is moving so fast that often the science of electronics is now concerned mostly with choosing the correct combination of chips and discrete components are only used as final switching or power output stages. Figure 9.2 shows a highly magnified view of a typical IC.

Figure 9.2 Integrated circuit components

9.1.4 Digital circuits

With some practical problems, it is possible to express the outcome as a simple yes/no or true/false answer. Let's take a simple example: if the answer to the first or the second question is 'yes', then switch on the brake warning light; if both answers are 'no' then switch it off.

Is the handbrake on?
Is the level in the brake fluid reservoir low?

In this case, we need the output of an electrical circuit to be 'on' when *either one or both* of the inputs to the circuit are 'on'. The inputs will be via simple switches on the handbrake and in the brake reservoir. The digital device required to carry out the above task is an OR gate. An OR gate for use on this system would have two inputs (a and b) and one output (c). Only when 'a' OR 'b' is supplied will 'c' produce a voltage.

Once a problem can be described in logic states then a suitable digital or logic circuit can also determine the answer to the problem. Simple circuits can also be constructed to hold the logic state of their last input, these are in effect simple forms of 'memory'. By combining vast quantities of these basic digital building blocks, circuits can be constructed to carry out the most complex tasks in a fraction of a second. Due to IC technology, it is now possible to create hundreds of thousands if not millions of these basic circuits on one chip. This has given rise to the modern electronic control systems used for vehicle applications as well as all the countless other uses for a computer.

In electronic circuits true/false values are assigned voltage values. In one system known as TTL (transistor-transistor-logic), true or logic '1' is represented by a voltage of 3.5 V and false or logic '0', by 0 V.

9.1.5 Electronic component testing

Individual electronic components can be tested in a number of ways but a digital multimeter is normally the favourite option. The table below suggests some methods of testing components removed from the circuit.

Component	Test method
Resistor	Measure the resistance value with an ohmmeter and compare this to the value written or colour coded on the component
Capacitor	A capacitor can be difficult to test without specialist equipment but try this. Charge the capacitor up to 12 V and connect it to a digital voltmeter. As most digital meters have an internal resistance of about 10 MΩ, calculate the expected discharge time ($T = 5CR$) and see if the device complies! A capacitor from a contact breaker ignition system should take about five seconds to discharge in this way
Inductor	An inductor is a coil of wire so a resistance check is the best method to test for continuity
Diode	Many multimeters have a diode test function. If so the device should read open circuit in one direction and about 0.4 to 0.6 V in the other direction. This is its switch on voltage. If no meter is available with this function then wire the diode to a battery via a small bulb; it should light with the diode one way and not the other
LED	LEDs can be tested by connecting them to a 1.5 V battery. Note the polarity though; the longest leg or the flat side of the case is negative
Transistor (bipolar)	Some multimeters even have transistor testing connections but, if not available, the transistor can be connected into a simple circuit as in Figure 9.3 and voltage tests carried out as shown. This also illustrates a method of testing electronic circuits in general. It is fair to point out that without specific data it is difficult for the non-specialist to test unfamiliar circuit boards. It's always worth checking for obvious breaks and dry joints though!
Digital components	A logic probe can be used. This is a device with a very high internal resistance so it does not affect the circuit under test. Two different coloured lights are used; one glows for a 'logic 1' and the other for 'logic 0'. Specific data is required in most cases but basic tests can be carried out

9.2 Multiplexing

9.2.1 Limits of the conventional wiring system

The complexity of modern wiring systems has been increasing steadily over the last twenty-five years or so and recently has increased dramatically. It has now reached a point where the size and weight of the wiring harness is a major problem. The number of separate wires required on a top of the range vehicle can be in the region of 1200. The wiring loom required to control all functions in or from the driver's door can require up to 50 wires, the systems in the dashboard area alone can use over 100 wires and connections. This is clearly becoming a problem as apart from the obvious issue of size and weight, the number of connections and number of wires increase the possibility of faults developing. It has been estimated that the complexity of the vehicle wiring system doubles every 10 years.

The number of systems controlled by electronics is continually increasing. A number of these are already in common use and the others are becoming more widely adopted. Some examples of these systems are listed below.

- Engine management
- Stability control
- Anti-lock brakes
- Transmission management
- Active suspension
- Communications and multimedia.

All the systems listed above work in their own right but are also linked to each other. Many of the sensors that provide inputs to one electronic control unit are common to all or some of the others. One solution to this is to use one computer to control all systems. This however would be very expensive to produce in small numbers. A second solution is to use a common data bus. This would allow communication between modules and would make the information from the various vehicle sensors available to all of them.

9.2.2 Controller area networks (CAN)

Bosch has developed the protocol known as CAN or controller area network. This system is claimed to meet practically all requirements with a very small chip surface (easy to manufacture and, therefore, cheaper). CAN is suitable for transmitting data in the area of drive line components, chassis components and mobile communications. It is a compact system, which will make it practical for use in many areas. Two variations on the physical layer are available which suit different transmission rates. One for data transmission of between 100 K and 1 M baud (bits per second), to be used for rapid control devices. The other will transmit between 10 K and 100 K baud as a low speed bus for simple switching and control operations.

The CAN message signal consists of a sequence of binary digits (bits). Voltage (or light fibre optics) present indicates the value '1'; none present indicates '0'. The actual message can vary between 44 and 108 bits in length. This is made up of a start bit, name, control bits, the data itself, a cyclic redundancy check (CRC) for error detection, a confirmation signal and finally a number of stop bits.

The name portion of the signal identifies the message destination and also its priority. As the transmitter puts a message on the bus it also reads the name back from the bus. If the name is not the same as the one it sent then another transmitter must be in operation which has a higher priority. If this is the case it will stop

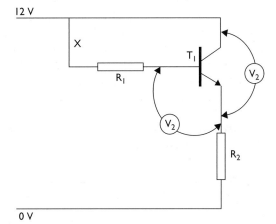

Use resistors $R_1 + R_2$ of about 1 kΩ when connected as shown V_1 should read 0.6 to 0.7 V and V_2 about 1 V. Disconnect wire X. V_1 should now read 0 V and V_2 12 V.

Figure 9.3 Transistor testing

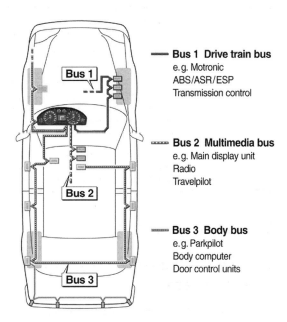

Bus 1 Drive train bus
e.g. Motronic
ABS/ASR/ESP
Transmission control

Bus 2 Multimedia bus
e.g. Main display unit
Radio
Travelpilot

Bus 3 Body bus
e.g. Parkpilot
Body computer
Door control units

Figure 9.4 CAN systems on a vehicle

transmission of its own message. This is very important in the case of motor vehicle data transmission.

All messages are sent to all units and each unit makes the decision whether the message should be acted upon or not. This means that further systems can be added to the bus at any time and can make use of data on the bus without affecting any of the other systems.

9.2.3 Summary

CAN is a shared broadcast bus which runs at speeds up to 1 Mbit/s. It is based around sending messages (or *frames*) which are of variable length, between 0 and 8 bytes. Each frame has an *identifier*, which must be unique (i.e. two nodes on the same bus must not send frames with the same identifier). The interface between the CAN bus and a CPU is usually called the *CAN controller*.

The Bosch CAN specification does not prescribe physical layer specifications. This resulted in two major physical layer designs. Both communicate using a differential voltage on a pair of wires and are often referred to as a high-speed and a low-speed physical layer. The low-speed architecture can change to a single-wire operating method (referenced to earth/ground) when one of the two wires is faulty because of a short or open circuit. Because of the nature of the circuitry required to perform this function, this architecture is very expensive to implement at bus speeds above 125 kbit/s. This is why 125 kbit/s is the division between high-speed and low-speed CAN.

The two wires operate in differential mode, in other words they carry inverted voltages (to reduce interference). The levels depend on which standard is being used. The voltage on the two wires, known as CAN-High and CAN-Low are as follows.

Table A ISO 11898 (CAN High Speed) standard

Signal	Recessive state			Dominant state		
	Min	Nominal	Max	Min	Nominal	Max
CAN-High	2.0 V	2.5 V	3.0 V	2.75 V	3.5 V	4.5 V
CAN-Low	2.0 V	2.5 V	3.0 V	0.5 V	1.5 V	2.25 V

Table B ISO 11519 (CAN Low Speed) standard

Signal	Recessive state			Dominant state		
	Min	Nominal	Max	Min	Nominal	Max
CAN-High	1.6 V	1.75 V	1.9 V	3.85 V	4.0 V	5.0 V
CAN-Low	3.1 V	3.25 V	3.4 V	0 V	1.0 V	1.15 V

For the recessive state the nominal voltage for the two wires is the same to decrease the power drawn from the nodes.

The voltage level on the CAN bus is recessive when the bus is idle.

Benefits of in-vehicle networking can be summarised as follows.

- A smaller number of wires is required for each function. This reduces the size and cost of the wiring harness as well as its weight. Reliability, serviceability, and installation issues are improved.
- General sensor data, such as vehicle speed, engine temperature and air temperature can be shared. This eliminates the need for redundant sensors.
- Functions can be added through software changes unlike existing systems, which require an additional module or input/output pins for each function added.
- New features can be enabled by networking, for example, each driver's preference for ride firmness, seat position, steering assist effort, mirror position and radio station presets can be stored in a memory profile.

9.2.4 CAN diagnostics

The integrity of the signal on the controller area network can be checked in two ways. The first way is to examine the signal on a dual channel scope connected to the CAN-High and CAN-Low lines (Figure 9.5).

In this display, it is possible to verify that:

- data is being continuously exchanged along the CAN bus;
- the voltage levels are correct;
- a signal is present on both CAN lines.

CAN uses a differential signal so the signal on one line should be a coincident mirror image of the data on the other line. The usual reasons for examining the CAN signals is where a CAN fault has been indicated by OBD, or to check the CAN connection to a suspected faulty CAN node. Manufacturer's data should be referred to for precise waveform parameters.

The CAN data shown in Figure 9.6 is captured on a much faster timebase and allows the individual state changes to be examined. This enables the mirror image nature of the signals, and the coincidence of the edges to be verified.

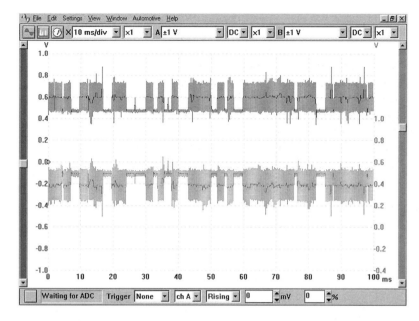

Figure 9.5 CAN signals

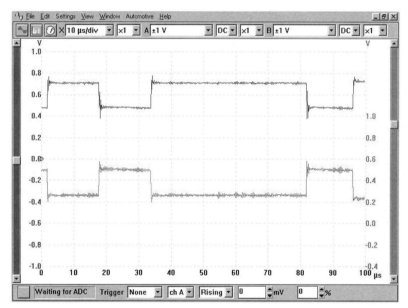

Figure 9.6 CAN signals on a fast timebase

The signals are equal and opposite and they are of the same amplitude (voltage). The edges are clean and coincident with each other. This shows that the vehicle data bus (CANbus) is enabling communication between the nodes and the CAN controller unit. This test effectively verifies the integrity of the bus at this point in the network. If a particular node is not responding correctly, the fault is likely to be the node itself. The rest of the bus should work correctly.

It is usually recommended to check the condition of the signals present at the connector of each of the ECUs on the network. The data at each node will always be the same on the same bus. Remember that much of the data on the bus is safety critical, so do **not** use insulation piercing probes!

The second way of checking the CAN signals is to use a suitable reader or scanner. The AutoTap scanner discussed in Chapter 3 will do this with suitable software.

9.3 Lighting

9.3.1 External lights

Figure 9.7 shows the rear lights of a Mazda 323. Note how in common with many manufacturers, the lenses are almost smooth. This is because the

Figure 9.7 Rear lights of the Mazda 323

reflectors now carry out diffusion of the light. Regulations exist relating to external lights. The following is a simplified interpretation of current rules.

Side lights	A vehicle must have two side lights each with wattage of less than 7 W. Most vehicles have the side light incorporated as part of the headlight assembly
Rear lights	Again two must be fitted each with wattage not less than 5 W. Lights used in Europe must be 'E' marked and show a diffused light. Position must be within 400 mm from the vehicle edge and over 500 mm apart, and between 350 and 1500 mm above the ground
Brake lights	Two lights often combined with the rear lights. They must be between 15 and 36 W each, with diffused light, and must operate when any form of first line brake is applied. Brake lights must be between 350 and 1500 mm above the ground and at least 500 mm apart in a symmetrical position. High level brake lights are now allowed and if fitted must operate with the primary brake lights
Reverse lights	No more than two lights may be fitted with a maximum wattage each of 24 W. The light must not dazzle and either be switched automatically from the gearbox or with a switch incorporating a warning light. Safety reversing 'beepers' are now often fitted in conjunction with this circuit, particularly on larger vehicles
Day running lights	Volvo use day running lights as these are in fact required in Sweden and Finland. These lights come on with the ignition and must only work in conjunction with the rear lights. Their function is to indicate that the vehicle is moving or about to move. They switch off when parking or headlights are selected
Rear fog lights	One or two may be fitted but if only one it must be on the off side or centre line of the vehicle. They must be between 250 and 1000 mm above the ground and over 100 mm from any brake light. The wattage is normally 21 W and they must only operate when either the side lights, headlights or front fog lights are in use
Front spot and fog lights	If front spot lights are fitted (auxiliary driving lights), they must be between 500 and 1200 mm above the ground and more than 400 mm from the side of the vehicle. If the lights are non-dipping then they must only operate when the headlights are on main beam. Front fog lamps are fitted below 500 mm from the ground and may only be used in fog or falling snow. Spot lamps are designed to produce a long beam of light to illuminate the road in the distance. Fog lights are designed to produce a sharp cut-off line such as to illuminate the road just in front of the vehicle but without reflecting back or causing glare

9.3.2 Lighting circuits

Figure 9.8 shows a simplified lighting circuit. Whilst this representation helps to demonstrate the way in which a lighting circuit operates, it is not now used in this simple form. The circuit does, however, help to show in a simple way how various lights in and around the vehicle operate with respect to each other. For example fog lights can be wired to work only when the side lights are on. Another example is how the headlights cannot

9.3.3 Gas discharge lighting

Gas discharge headlamps (GDL) are now fitted to some vehicles. They have the potential to provide more effective illumination and new design possibilities for the front of a vehicle. The conflict between aerodynamic styling and suitable lighting positions is an economy/safety trade off, which is undesirable. The new headlamps make a significant contribution towards improving this situation because they can be relatively small. The GDL system consists of three main components.

- **Lamp** this operates in a very different way from conventional incandescent bulbs. A much higher voltage is needed.
- **Ballast system** this contains an ignition and control unit and converts the electrical system voltage into the operating voltage required by the lamp. It controls the ignition stage and run up as well as regulating during continuous use and finally monitors operation as a safety aspect.
- **Headlamp** the design of the headlamp is broadly similar to conventional units. However, in order to meet the limits set for dazzle, a more accurate finish is needed and hence more production costs are involved.

Figure 9.10 shows a xenon lamp system.

9.3.4 LED lighting

The advantages of LED lighting are clear, the greatest being reliability. LEDs have a typical rated life of over 50 000 hours compared to just a few thousand for incandescent lamps. The environment in which vehicle lights have to survive is hostile to say the least. Extreme variations in temperature and humidity as well as serious shocks and vibration have to be endured. Figure 9.11 shows some lights employing LEDs.

LEDs are more expensive than bulbs but the potential savings in design costs due to sealed units being used and greater freedom of design could outweigh the extra expense. A further advantage is that they turn on quicker than ordinary bulbs. This time is approximately the difference between 130 mS for the LEDs, and 200 mS for bulbs. If this is related to a vehicle brake light at motorway speeds, then the increased reaction time equates to about a car length. This is potentially a major contribution to road safety. LEDs as high level brake lights are becoming popular because of the shock resistance, which will allow them to be mounted on the boot lid.

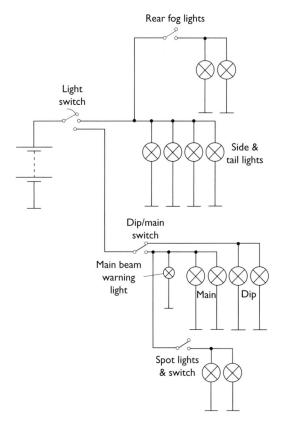

Figure 9.8 Simplified lighting circuit

be operated without the side lights first being switched on.

Dim dip headlights are an attempt to stop drivers just using side lights in semi-dark or poor visibility conditions. The circuit is such that when side lights and ignition are on together, then the headlights will come on automatically at about one sixth of normal power.

> *Note:* If there is any doubt as to the visibility or conditions, switch on dipped headlights. If your vehicle is in good order it will not discharge the battery.

Dim dip lights are achieved in one of two ways. The first uses a simple resistor in series with the headlight bulb and the second is to use a 'chopper' module which switches the power to the headlights on and off rapidly. In either case the 'dimmer' is bypassed when the driver selects normal headlights. The most cost effective method is using a resistor but this has the problem that the resistor (about 1 Ω) gets quite hot and hence has to be positioned appropriately. Figure 9.9 shows a typical modern vehicle lighting circuit.

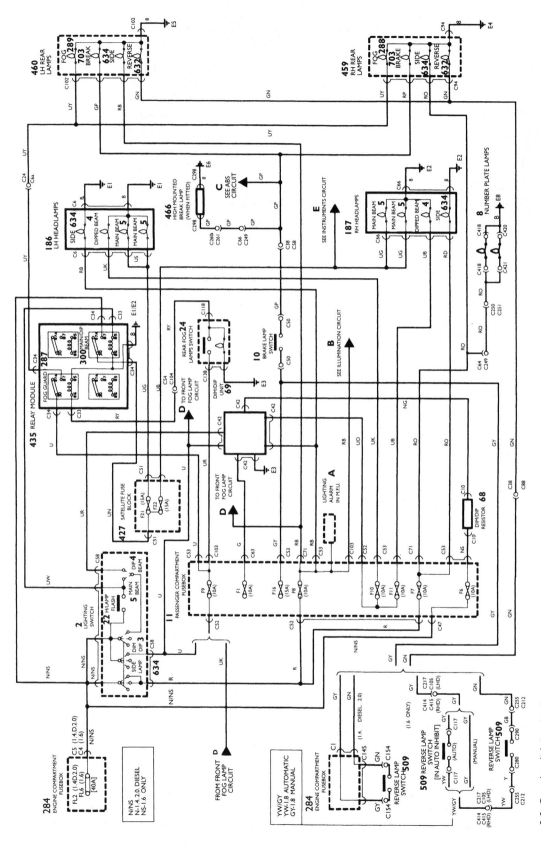

Figure 9.9 Complete lighting circuit

Electrical systems 207

Figure 9.10 Xenon lighting

Figure 9.11 Adaptive lighting using LEDs

Heavy vehicle side marker lights are an area of use where LEDs have proved popular. Many lighting manufacturers are already producing lights for the after market. Being able to use sealed units will greatly increase the life expectancy. Side indicator repeaters are a similar issue due to the harsh environmental conditions.

9.4 Diagnosing lighting system faults

9.4.1 Testing procedure

The process of checking a lighting system circuit is broadly shown in the chart on next page.

Figure 9.12 shows a simplified dim dip lighting circuit with meters connected for testing. Refer to Chapter 2 for further details about 'volt drop testing'. A simple principle to keep in mind is that the circuit should be able to supply all the available battery voltage to the consumers (bulbs, etc.). A loss of 5 to 10% may be acceptable. With all the switches in the 'on' position appropriate to where the meters are connected the following readings should be obtained:

- V_1 12.6 V (if less check battery condition);
- V_2 0 to 0.2 V (if more the ignition switch contacts have a high resistance);
- V_3 0 to 0.2 V (if more the dim dip relay contacts have a high resistance);
- V_4 0 to 0.2 V (if more there is a high resistance in the circuit between the output of the light switch and the junction for the tail lights);

208 Advanced automotive fault diagnosis

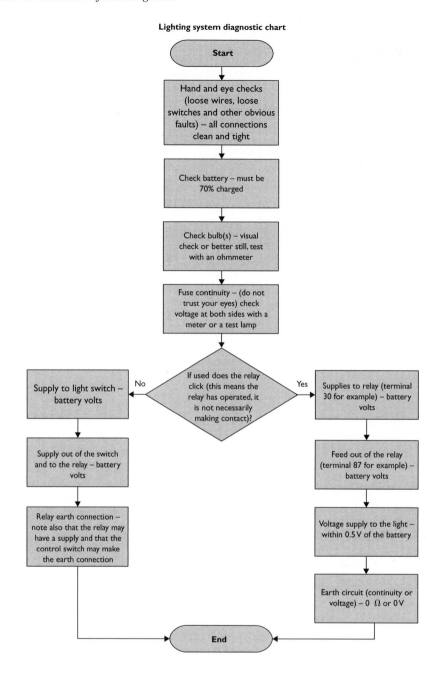

- V_5 0 to 0.2 V (if more then there is a high resistance in the light switch contacts, the dip switch contacts or the circuit – connect the meter across a smaller section to narrow down the fault);
- V_6 0 to 0.2 V (if more then there is a high resistance in the earth for that bulb);
- V_7 12 to 12.6 V if on normal dip or about 6 V if on dim dip (if less then there is a high resistance in the circuit – check other readings, etc. to narrow down the fault).

9.4.2 Lighting fault diagnosis table

Symptom	Possible fault
Lights dim	High resistance in the circuit
	Low alternator output
	Discoloured lenses or reflectors
Headlights out of adjustment	Suspension fault
	Loose fittings
	Damage to body panels
	Adjustment incorrect

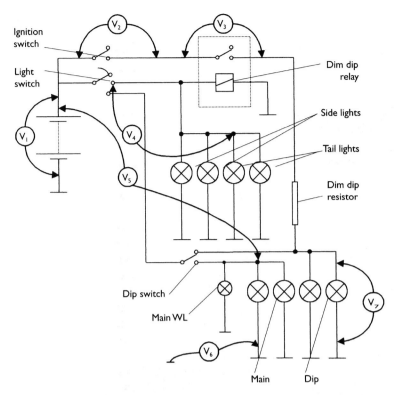

Figure 9.12 Lighting circuit testing

Lights do not work
- Bulbs blown
- Fuse blown
- Loose or broken wiring/connections/fuse
- Relay not working
- Corrosion in light units
- Switch not making contact

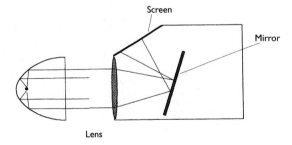

Figure 9.13 Principle of a beam setter

9.4.3 Headlight beam setting

Many types of beam setting equipment are available and most work on the same principle. This is represented by Figure 9.13. The method is the same as using an aiming board but is more convenient and accurate due to easier working and because less room is required. To set the headlights of a car using an aiming board the following procedure should be adopted.

1. Park the car on level ground square on to a vertical aiming board at a distance of 10 m if possible. The car should be unladen except for the driver.
2. Mark out the aiming board as shown in Figure 9.14.
3. Bounce the suspension to ensure it is level.
4. With the lights set on dip beam, adjust the cut-off line to the horizontal mark, which will be 1 cm below the height of the headlight centre for every 1 m the car is away from the board*. The break-off point should be adjusted to the centre line of each light in turn.

Note: If the required dip is 1% then 1 cm per 1 m. If 1.2% is required then 1.2 cm per 1 m, etc.

9.5 Auxiliaries

9.5.1 Wiper motors and linkages

Most wiper linkages consist of series or parallel mechanisms. Some older types use a flexible rack and wheel boxes similar to the operating mechanism of many sunroofs. One of the main

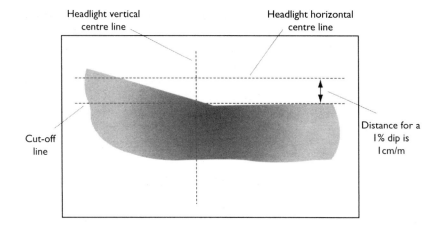

Figure 9.14 Headlight aiming board (Europe RHD)

Figure 9.15 Interesting headlights on the new Jaguar S-type

considerations for the design of a wiper linkage is the point at which the blades must reverse. This is because of the high forces on the motor and linkage at this time. If the reverse point is set so that the linkage is at its maximum force transmission angle then the reverse action of the blades puts less strain on the system. This also ensures smoother operation. Figure 9.16 shows a typical wiper linkage layout.

Most if not all wiper motors now in use are permanent magnet motors. The drive is taken via a worm gear to increase torque and reduce speed. Three brushes may be used to allow two-speed operation. The normal speed operates through two brushes placed in the usual positions opposite to each other. For a fast speed the third brush is placed closer to the earth brush. This reduces the number of armature windings between them, which reduces resistance hence increasing current and therefore speed. Figure 9.17 shows two typical wiper motors. Typical specifications for wiper motor speed and hence wipe frequency are 45 rev/min at normal speed and 65 rev/min at fast speed. The motor must be able to overcome the starting friction of each blade at a minimum speed of 5 rev/min.

The wiper motor or the associated circuit often has some kind of short circuit protection. This is to protect the motor in the event of stalling, if frozen to the screen for example. A thermal trip

Electrical systems 211

of some type is often used or a current sensing circuit in the wiper ECU if fitted. The maximum time a motor can withstand stalled current is normally specified. This is usually in the region of about 15 minutes.

The windscreen washer system usually consists of a simple DC permanent magnet motor driving a centrifugal water pump. The water, preferably with a cleaning additive, is directed onto an appropriate part of the screen by two or more jets. A non-return valve is often fitted in the line to the jets to prevent water siphoning back to the reservoir. This also allows 'instant' operation when the washer button is pressed. The washer circuit is normally linked in to the wiper circuit such that when the washers are operated the wipers start automatically and will continue for several more sweeps after the washers have stopped.

9.5.2 Wiper circuits

Figure 9.18 shows a circuit for fast, slow and intermittent wiper control. The switches are shown in the off position and the motor is stopped and in its park position. Note that the two main brushes of the motor are connected together via the limit switch, delay unit contacts and the wiper switch. This causes regenerative braking because of the current generated by the motor due to its momentum after the power is switched off. Being connected to a very low resistance loads up the

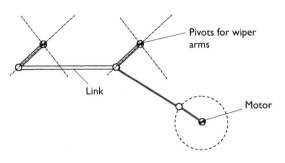

Figure 9.16 Wiper linkage

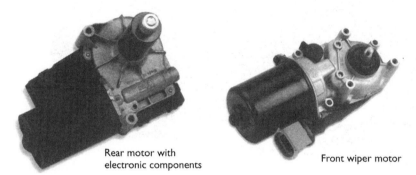

Figure 9.17 Wiper motors

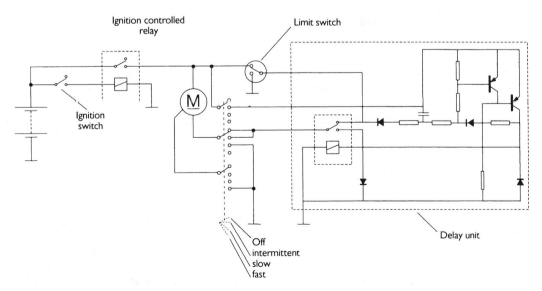

Figure 9.18 Wiper circuit with intermittent/delay operation as well as slow and fast speed

'generator' and it stops instantly when the park limit switch closes.

When either the delay contacts or the main switch contacts are operated the motor will run at slow speed. When fast speed is selected the third brush on the motor is used. On switching off, the motor will continue to run until the park limit switch changes over to the position shown. This switch is only in the position shown when the blades are in the parked position.

9.5.3 Headlight wipers and washers

There are two ways in which headlights are cleaned, first by high-pressure jets, and second by small wiper blades with low-pressure water supply. The second method is in fact much the same as windscreen cleaning but on a smaller scale. The high-pressure system tends to be favoured but can suffer in very cold conditions due to the fluid freezing. It is expected that the wash system should be capable of about 50 operations before refilling of the reservoir is necessary. Headlight cleaners are often combined with the windscreen washers. They operate each time the windscreen washers are activated, if the headlights are also switched on.

A retractable nozzle for headlight cleaners is often used. When the water pressure is pumped to the nozzle it is pushed from its retracted position, flush with the bodywork. When the washing is completed the jet is then retracted back into the housing.

9.5.4 Indicators and hazard lights

Direction indicators have a number of statutory requirements. The light produced must be amber, but they may be grouped with other lamps. The flashing rate must be between one and two per second with a relative 'on' time of between 30 and 57%. If a fault develops this must be apparent to the driver by the operation of a warning light on the dashboard. The fault can be indicated by a distinct change in frequency of operation or the warning light remaining on. If one of the main bulbs fails then the remaining lights should continue to flash perceptibly.

Legislation as to the mounting position of the exterior lamps exists such that the rear indicator lights must be within a set distance of the tail lights and within a set height. The wattage of indicator light bulbs is normally 21 W at 6, 12 or 24 V as appropriate. Figure 9.19 shows a typical indicator and hazard circuit.

Flasher units are rated by the number of bulbs they are capable of operating. When towing a trailer or caravan the unit must be able to operate at a higher wattage. Most units use a relay for the actual switching as this is not susceptible to voltage spikes and also provides an audible signal. Figure 9.20 shows the rear lights on a Ford Mondeo!

9.5.5 Brake lights

Figure 9.21 shows a typical brake light circuit. Most incorporate a relay to switch the lights, which is in turn operated by a spring-loaded switch on the brake pedal. Links from this circuit to cruise control may be found. This is to cause the cruise control to switch off as the brakes are operated.

9.5.6 Electric horns

Regulations in most countries state that the horn (or audible warning device) should produce a uniform sound. This makes sirens and melody type fanfare horns illegal! Most horns draw a large current so are switched by a suitable relay.

The standard horn operates by simple electromagnetic switching. Current flow causes an armature to which is attached a tone disc, to be attracted towards a stop. This opens a set of contacts which disconnects the current allowing the armature and disc to return under spring tension. The whole process keeps repeating when the horn switch is on. The frequency of movement and hence the fundamental tone is arranged to lie between 1.8 and 3.5 kHz. This note gives good penetration through traffic noise. Twin horn systems, which have a high and low tone horn, are often used. This produces a more pleasing sound but is still very audible in both town and higher speed conditions. Figure 9.22 shows a typical horn together with its associated circuit.

9.5.7 Engine cooling fan motors

Most engine cooling fan motors (radiator cooling) are simple PM types. The fans used often

Electrical systems **213**

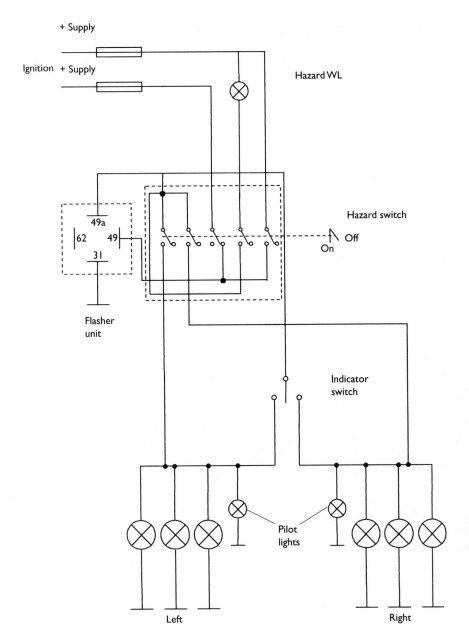

Figure 9.19 Indicator and hazard circuit

Figure 9.20 'His Rear Lights' (old Desmond Decker song)

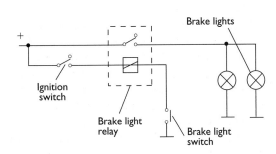

Figure 9.21 Brake light circuit

214 *Advanced automotive fault diagnosis*

have the blades placed asymmetrically (balanced but not in a regular pattern) to reduce noise when operating. Figure 9.23 shows a motor in position.

When twin cooling fans and motors are fitted, they can be run in series or parallel. This is often the case when air conditioning is used as the condenser is usually placed in front of the radiator and extra cooling air speed may be needed. A circuit for series or parallel operation of cooling fans is shown in Figure 9.24.

9.6 Diagnosing auxiliary system faults

9.6.1 Testing procedure

The process of checking an auxiliary system circuit is broadly as follows.

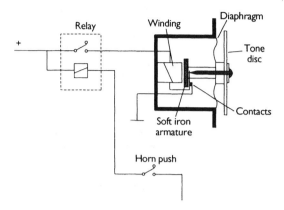

Figure 9.22 Horn and circuit

Figure 9.23 Cooling fan motor

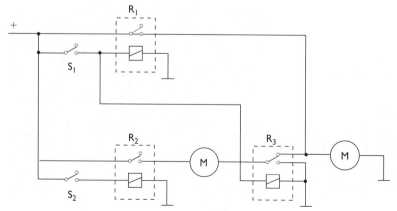

Slow speed – relay R_2 only (S_2 on)
Full speed – relays R_1, R_2 and R_3 (S_1 and S_2 on)

Figure 9.24 Two-speed twin cooling fan circuit

Electrical systems 215

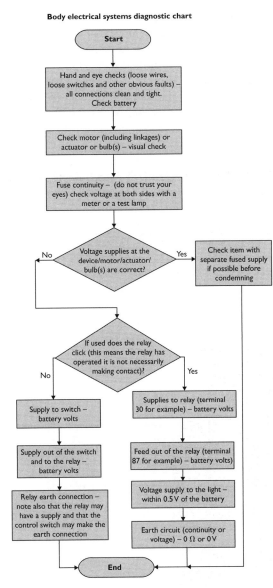

Body electrical systems diagnostic chart

9.6.2 Auxiliaries fault diagnosis table

Symptom	Possible fault
Horn not working or poor sound quality	Loose or broken wiring/connections/fuse Corrosion in horn connections Switch not making contact High resistance contact on switch or wiring Relay not working
Wipers not working or poor operation	Loose or broken wiring/connections/fuse Corrosion in wiper connections Switch not making contact High resistance contact on switch or wiring Relay/timer not working Motor brushes or slip ring connections worn Limit switch contacts open circuit or high resistance Blades and/or arm springs in poor condition
Washers not working or poor operation	Loose or broken wiring/connections/fuse Corrosion in washer motor connections Switch not making contact Pump motor poor or not working Blocked pipes or jets Incorrect fluid additive used
Indicators not working or incorrect operating speed	Bulb(s) blown Loose or broken wiring/connections/fuse Corrosion in horn connections Switch not making contact High resistance contact on switch or wiring Relay not working
Heater blower not working or poor operation	Loose or broken wiring/connections/fuse Switch not making contact Motor brushes worn Speed selection resistors open circuit

Figure 9.26 shows the zones of a windscreen as well as a non-circular wiping pattern. The zones are important for assessing if damage to the screen is critical or not.

9.6.3 Wiper motor and circuit testing

Figure 9.27 shows a procedure recommended by Lucas for testing a wiper motor. The expected reading on the ammeter should not be more that about 5 A. Several types of wiper motor are in current use so take care to make the appropriate connections for this test. Remember to use a fused jumper lead as a precaution.

Figure 9.25 Fuse box

9.7 In car entertainment (ICE) security and communications

9.7.1 ICE

Controls on most sets will include volume, treble, bass, balance and fade. Cassette tape options will include Dolby® filters to reduce hiss and other tape selections such as chrome or metal. A digital display of course, will provide a visual output of operating condition. This is also linked into the vehicle lighting to prevent glare at night. Track selection and programming for one or several compact discs is possible. Figure 9.28 shows a modern ICE system display and control panel together with a top quality speaker (woofer).

Many ICE systems are coded to deter theft. The code is activated if the main supply is disconnected and will not allow the set to work until the correct code has been re-entered. Some systems now include a plug in electronic 'key card', which makes the set worthless when removed.

Good ICE systems include at least six speakers, two larger speakers in the rear parcel shelf to produce good low frequency reproduction, two front door speakers for mid range and two front door tweeters for high frequency notes. Speakers are a very important part of a sound system. No matter how good the receiver or CD player is,

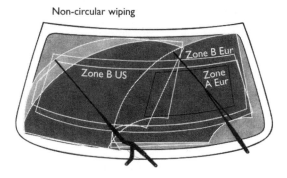

Figure 9.26 Windscreen zones

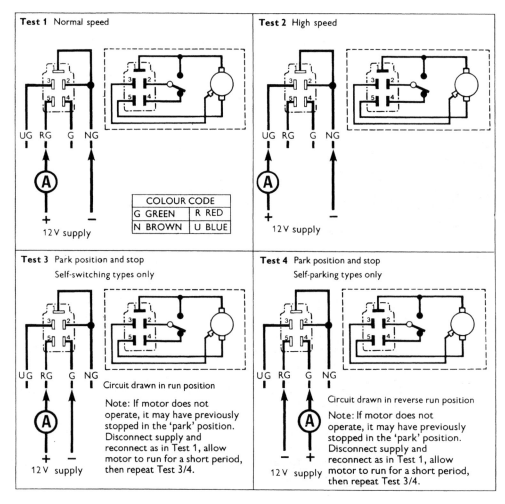

Figure 9.27 Wiper motor tests

the sound quality will be reduced if inferior speakers are used. Equally if the speakers are of a lower power output rating than the set, distortion will result at best and damage to the speakers at worst. Speakers fall generally into the following categories:

- **tweeters** high frequency reproduction;
- **mid range** middle range frequency reproduction (treble);
- **woofers** low frequency reproduction (bass);
- **sub-woofers** very low frequency reproduction.

The radio data system RDS has become a standard on many radio sets. It is an extra inaudible digital signal which is sent with FM broadcasts in a similar way to how teletext is sent with TV signals. RDS provides information so a receiver can appear to act intelligently. The possibilities available when RDS is used are as follows.

- The station name can be displayed in place of the frequency.
- There can be automatic tuning to the best available signal for the chosen radio station. For example, in the UK a journey from the south of England to Scotland would mean the radio would have to be retuned up to ten times. RDS will do this without the driver even knowing.
- Traffic information broadcasts can be identified and a setting made so that whatever you are listening to at the time can be interrupted.

The radio broadcast data system (RBDS) is an extension of RDS which has been in use in Europe since 1984. The system allows the broadcaster to transmit text information at the rate of about 1200 bits per second. The information is transmitted on a 57 kHz suppressed sub-carrier as part of the FM MPX signal.

RBDS was developed for the North American market by the National Radio Systems Committee (NRSC), a joint committee composed of the Electronic Industries Association (EIA) and the National Association of Broadcasters (NAB). The applications for the transmission of text to the vehicle are interesting and include:

- song title and artist;
- traffic, accident and road hazard information;
- stock information;
- weather.

In emergency situations, the audio system can be enabled to interrupt the cassette, CD, or normal radio broadcast to alert the user.

Figure 9.29 shows a circuit typical of many ICE systems.

Figure 9.28 ICE display and speaker

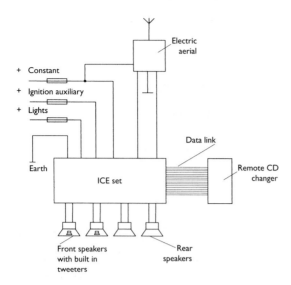

Figure 9.29 ICE circuit

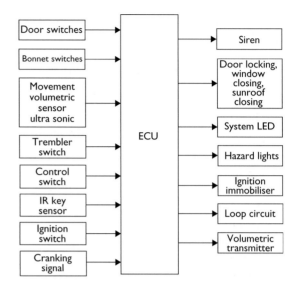

Figure 9.30 Block diagram of a complex alarm system

9.7.2 Security systems

Car and alarm manufacturers are constantly fighting to improve security. Building the alarm system as an integral part of the vehicle electronics has made significant improvements. Even so, retro fit systems can still be very effective. Three main types of intruder alarm are used:

- switch operated on all entry points;
- battery voltage sensed;
- volumetric sensing.

There are three main ways to disable the vehicle:

- ignition circuit cut off;
- starter circuit cut off;
- engine ECU code lock.

Most alarm systems are made for 12 V, negative earth vehicles. They have electronic sirens and give an audible signal when arming and disarming. They are all triggered when the car door opens and will automatically reset after a period of time, often one or two minutes. The alarms are triggered instantly when entry point is breached. Most systems are two pieces, with separate control unit and siren; most will have the control unit in the passenger compartment and the siren under the bonnet.

Most systems now come with two infrared (IR) remote 'keys' that use small button type batteries and have an LED that shows when the signal is being sent. They operate with one vehicle only. Intrusion sensors such as car movement and volumetric sensing can be adjusted for sensitivity.

When operating with flashing lights most systems draw about 5 A. Without flashing lights (siren only) the current draw is less than 1 A. The sirens produce a sound level of about 95 dB, when measured 2 m in front of the vehicle.

Figure 9.30 shows a block diagram of a complex alarm system. The system, as is usual, can be considered as a series of inputs and outputs. This is particularly useful for diagnosing faults. Most factory fitted alarms are combined with the central door locking system. This allows the facility mentioned in a previous section known as lazy lock. Pressing the button on the remote unit, as well as setting the alarm, closes the windows and sunroof and locks the doors.

A security code in the engine ECU is a powerful deterrent. This can only be 'unlocked' to allow the engine to start when it receives a coded signal. Ford and other manufacturers use a special ignition key which is programmed with the required information. Even the correct 'cut' key will not start the engine. Citroen, for example, have used a similar idea but the code has to be entered via a numerical keypad.

Of course nothing will stop the car being lifted on to a lorry and driven away, but this technique will mean a new engine control ECU will be needed. The cost will be high and also questions may be asked as to why a new ECU is required.

9.7.3 Mobile communications

If the success of the cellular industry is any indication of how much use we can make of the telephone, the future promises an even greater expansion. Cellular technology started to become useful in the 1980s and has continued to develop from then – very quickly.

The need and desire we perceive to keep in touch with each other is so great that an increasing number of business people now have up to five phone numbers: home, office, pager, fax and cellular. But within the foreseeable future, high tech digital radio technology and sophisticated telecommunications systems will enable all communications to be processed through a single number.

With personal numbering, a person carrying a pocket-sized phone will need only one phone number. Instead of people calling places, people will call people; we will not be tied to any particular place. Personal numbering will make business people more productive because they will be able to reach and be reached by colleagues and clients, anywhere and anytime, indoors or outdoors. When travelling from home to office or from one meeting to the next, it will be possible to communicate with anyone, whenever the need arises.

But where does this leave communication systems relating to the vehicle? It is my opinion that 'in vehicle' communication equipment for normal business and personal use will be by the simple pocket-sized mobile phone and that there is no further market for the car telephone. Hands free conversions will still be important.

CB radios and short range two-way systems such as used by taxi firms and service industries will still have a place for the time being. Even these may decline as the cellular network becomes cheaper and more convenient to use.

9.8 Diagnosing ICE, security and communication system faults

9.8.1 Testing procedure

The process of checking an ICE system circuit is broadly as follows.

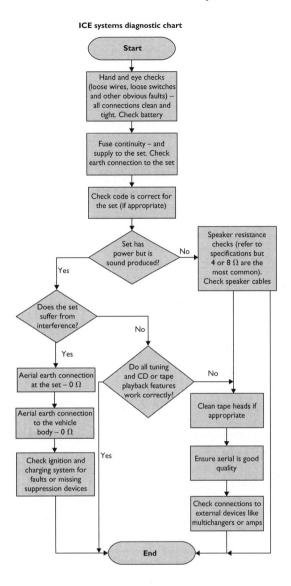

ICE systems diagnostic chart

ICE, security and communication system fault diagnosis table

Symptom	Possible fault
Alarm does not operate	Fuse blown
	Not set correctly
	Remote key battery discharged
	Open circuit connection to alarm unit
	ECU fault
	Receiver/transmitter fault
	Volumetric transmitter/receiver fault
Alarm goes off for no apparent reason	Drain on battery
	Loose connection
	Vibration/trembler/movement detection circuit set too sensitive
	Self-discharge in the battery
	Window left open allowing wind or even a bird or insect to cause interior movement
	Somebody really is trying to steal the car ...

Radio interference	Tracking HT components Static buildup on isolated body panels High resistance or open circuit aerial earth Suppression device open circuit
ICE system does not produce sound	Set not switched on! Loose or open circuit connections Trapped wires Connections to separate unit (amplifier, equaliser, etc.) incorrect Fuse blown
Unbalanced sound	Fade or balance controls not set correctly Speakers not wired correctly (front right, front left, rear right, rear left, etc.) Speaker open circuit or reduced output
Phasing	Speaker polarity incorrect. This should be marked but if not use a small battery to check all speakers are connected the same way. A small DC voltage will move the speaker cone in one direction
Speaker rattle	Insecure speaker(s) Trim not secure Inadequate baffles
Crackling noises	If one speaker – then try substitution If one channel – swap connections at the set to isolate the fault If all channels but only the radio then check interference Radio set circuit fault
Vibration Hum	Incorrect or loose mounting Speaker cables routed next to power supply wires Set fault
Distortion Poor radio reception	Incorrect power rating speakers Incorrect tuning 'Dark' spot/area. FM signals can be affected by tall buildings, etc. Aerial not fully extended Aerial earth loose or high resistance Tuner not trimmed to the aerial (older sets generally) Aerial sections not clean
Telephone reception poor	Low battery power Poor reception area Interference from the vehicle Loose connections on hands free circuit

9.8.2 Interference suppression

The process of interference suppression on a vehicle is to reduce the amount of unwanted noise produced from the speakers of an ICE system. This can, however, be quite difficult. To aid the discussion it is necessary to first understand the different types of interference. Figure 9.31 shows two signals, one clean and the other suffering from interference. The amount of interference can be stated as a signal to noise ratio. This is the useful field strength compared to the interference field strength at the receiver.

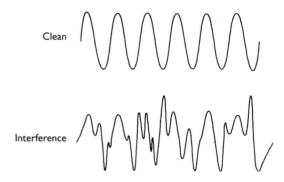

Figure 9.31 Radio signals

There are two overall issues to be considered relating to suppression of interference on a vehicle. These are as follows.

- **Short range** the effect of interference on the vehicle's radio system.
- **Long range** the effect of the vehicle on external receivers such as domestic televisions. This is covered by legislation making it illegal to cause disturbance to radios or televisions when using a vehicle.

Interference can propagate in one of four ways:

- line borne conducted through the wires;
- air borne radiated through the air to the aerial;
- capacitive coupling by an electric field;
- inductive coupling magnetic linking.

The sources of interference in the motor vehicle can be summarised quite simply as any circuit which is switched or interrupted suddenly. This includes the action of a switch and the commutation process in a motor, both of which produce rapidly increasing signals. The secret of suppression is to slow down this increase. Interference is produced from four main areas of the vehicle:

- ignition system;
- charging system;
- motors and switches;
- static discharges.

The ignition system of a vehicle is the largest source of interference, in particular the high tension side. Voltages up to 30 kV are now common and the current for a fraction of a second when the plug fires can peak in excess of 100 A. The interference caused by the ignition system is mostly above 30 MHz and the energy can peak for fractions of a second in the order of 500 kW.

The charging system produces noise because of the sparking at the brushes. Electronic regulators produce few problems but regulators with vibrating contacts can cause trouble.

Any motor or switch including relays is likely to produce some interference. The most popular sources are the wiper motor and heater motor. The starter is not considered due to its short usage time.

Buildup of static electricity is due to friction between the vehicle and the air, and the tyres and the road. If the static on say the bonnet builds up more than on the wing then a spark can be discharged. Using bonding straps to ensure all panels stay at the same potential easily prevents this. Due to the action of the tyres a potential can build up between the wheel rims and the chassis unless suitable bonding straps are fitted. The arc to ground can be as much as 10 kV.

There are five main techniques for suppressing radio interference:

- resistors;
- bonding;
- screening;
- capacitors;
- inductors.

Resistance is used exclusively in the ignition HT circuit, up to a maximum of about 20 kΩ per lead. This has the effect of limiting the peak current, which in turn limits the peak electromagnetic radiation. Providing excessive resistance is not used, the spark quality is not affected. These resistors effectively damp down the interference waves.

Bonding has been mentioned earlier. It is simply to ensure that all parts of the vehicle are at the same electrical potential to prevent sparking due to the buildup of static.

Screening is generally only used for specialist applications such as emergency services and the military. It involves completely enclosing the ignition system and other major sources of noise in a conductive screen, which is connected to the vehicle's chassis earth. This prevents interference waves escaping, it is a very effective technique but expensive. Often a limited amount of screening, metal covers on the plugs for example, can be used to good effect.

Capacitors and inductors are used to act as filters. This is achieved by using the changing value of 'resistance' to alternating signals as the frequency increases. The correct term for this resistance is either capacitive or inductive reactance. By choosing suitable values of capacitor in parallel and/or inductor in series it is possible to filter out unwanted signals of certain frequencies.

The aerial is worth a mention at this stage. Several types are in use; the most popular still being the rod aerial which is often telescopic. The advantage of a rod aerial is that it extends beyond the interference field of the vehicle. For reception in the AM bands the aerial represents a capacitance of 80 pF with a shunt resistance of about 1 MΩ. The set will often incorporate a trimmer to ensure that the aerial is matched to the set. Contact resistance between all parts of the aerial should be less than 20 mΩ. This is particularly important for the earth connection.

When receiving in the FM range the length of the aerial is very important. The ideal length of a rod aerial for FM reception is one quarter of the wavelength. In the middle of the FM band (94 MHz) this is about 80 cm. Due to the magnetic and electrical field of the vehicle and the effect of the coaxial cable, the most practical length is about 1 m. Some smaller aerials are available but whilst these may be more practical the signal strength is reduced. Aerials embedded into the vehicle windows or using the heated rear window element are good from the damage prevention aspect and for insensitivity to moisture, but produce a weaker signal often requiring an aerial amplifier to be included. Note that this will also amplify interference. Some top range vehicles use a rod aerial and a screen aerial, the set being able to detect and use the strongest signal. This reduces the effect of reflected signals and causes less flutter.

Consideration must be given to the position of an external aerial. This has to be a compromise taking into account the following factors:

- **rod length**   1 m if possible;
- **coaxial cable length**   longer cable reduces the signal strength;
- **position**   as far away as reasonably possible from the ignition system;
- **potential for vandalism**   out of easy reach;
- **aesthetic appearance**   does it fit with the style of the vehicle;
- **angle of fitting**   vertical is best for AM horizontal for FM!

Most quality sets also include a system known as interference absorption. This is a circuit built in to the set consisting of high quality filters and is not adjustable.

9.9 Body electrical systems

9.9.1 Electric seat adjustment

Adjustment of the seat is achieved by using a number of motors to allow positioning of different parts of the seat. A typical motor reverse circuit is shown in Figure 9.32. When the switch is moved *one* of the relays will operate and this changes the polarity of the supply to *one* side of the motor. Movement is possible in the following ways:

- front to rear;
- cushion height rear;
- cushion height front;
- backrest tilt;
- headrest height;
- lumbar support.

When seat position is set some vehicles have position memories to allow automatic repositioning, if the seat has been moved. This is often combined with electric mirror adjustment. Figure 9.33 shows how the circuit is constructed to allow position memory. As the seat is moved a variable resistor, mechanically linked to the motor, is also moved. The resistance value provides feedback to an ECU. This can be 'remembered' in a number of ways; the best technique is to supply the resistor with a fixed voltage such that the output relative to the seat position is proportional to position. This voltage can then be 'analogue to digital' converted, which produces a simple 'number' to store in a digital memory. When the driver presses a memory recall switch the motor relays are activated by the ECU until the number in memory and the number fed back from the seat are equal. This facility is often isolated when the engine is running to prevent the seat moving into a dangerous position as the car is being driven. Position of the seats can still be adjusted by operating the switches as normal.

9.9.2 Electric mirrors

Many vehicles have electrical adjustment of mirrors, particularly on the passenger side. The system used is much the same as has been discussed above in relation to seat movement. Two small motors are used to move the mirror vertically or horizontally. Many mirrors also contain a small heating element on the rear of the glass. This is operated for a few minutes when the ignition is first switched on and can also be linked to the heated rear window circuit. Figure 9.34 shows an electrically operated mirror circuit, which includes feedback resistors for positional memory.

9.9.3 Electric sunroof operation

The operation of an electric sunroof is similar to the motor reverse circuit discussed earlier in this

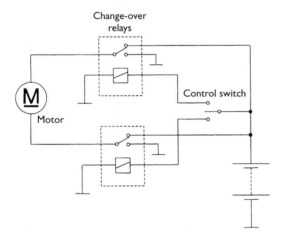

Figure 9.32 Motor reverse circuit

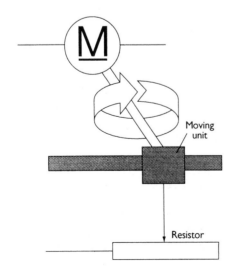

Figure 9.33 Position memory for electric seats

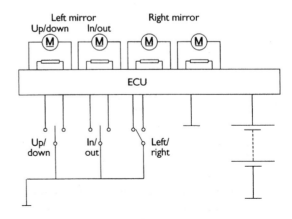

Figure 9.34 Feedback resistors for positional memory and circuit

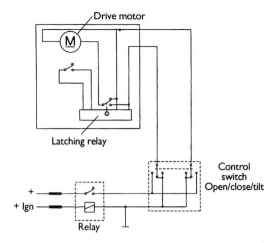

Figure 9.35 Sunroof circuit

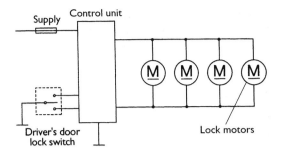

Figure 9.36 Door lock circuit

chapter. However, further components and circuitry are needed to allow the roof to slide, tilt and stop in the closed position. The extra components used are a micro switch and a latching relay. A latching relay works in much the same way as a normal relay except that it locks into position each time it is energised. The mechanism used to achieve this is much like that used in ball point pens that use a button on top.

The micro switch is mechanically positioned such as to operate when the roof is in its closed position. A rocker switch allows the driver to adjust the roof. The circuit for an electrically operated sunroof is shown in Figure 9.35. The switch provides the supply to the motor to run it in the chosen direction. The roof will be caused to open or tilt. When the switch is operated to close the roof the motor is run in the appropriate direction until the micro switch closes when the roof is in its closed position. This causes the latching relay to change over which stops the motor. The control switch has now to be released. If the switch is pressed again the latching relay will once more change over and the motor will be allowed to run.

9.9.4 Door locking circuit

When the key is turned in the driver's door lock all the other doors on the vehicle should also lock. Motors or solenoids in each door achieve this. If the system can only be operated from the driver's door key then an actuator is not required in this door. If the system can be operated from either front door or by remote control then all the doors need an actuator. Vehicles with sophisticated alarm systems often lock all the doors as the alarm is set.

Figure 9.36 shows a door locking circuit. The main control unit contains two change-over relays, which are actuated by either the door lock switch or, if fitted, the remote infrared key. The motors for each door lock are simply wired in parallel and all operate at the same time.

Most door actuators are now small motors which, via suitable gear reduction, operate a linear rod in either direction to lock or unlock the doors. A simple motor reverse circuit is used to achieve the required action.

Infrared central door locking is controlled by a small hand-held transmitter and an infrared sensor receiver unit as well as a decoder in the main control unit. This layout will vary slightly between different manufacturers. When the infrared key is operated by pressing a small switch a complex code is transmitted. The number of codes used is well in excess of 50 000. The infrared sensor picks up this code and sends it in an electrical form to the main control unit. If the received code is correct the relays are triggered and the door locks are either locked or unlocked. If an incorrect code is received on three consecutive occasions when attempting to unlock the doors, then the infrared system will switch itself off until the door is opened by the key. This will also reset the system and allow the correct code to again operate the locks. This technique prevents a scanning type transmitter unit from being used to open the doors.

9.9.5 Electric window operation

The basic form of electric window operation is similar to many of the systems discussed so far in this chapter, that is a motor reversing system either by relays or directly by a switch. More sophisticated systems are now becoming more popular for reasons of safety as well as improved comfort. The following features are now available from many manufacturers:

- one-shot up or down;
- inch up or down;
- lazy lock;
- back-off.

When a window is operated in one-shot or one-touch mode the window is driven in the chosen direction until either the switch position is reversed, the motor stalls or the ECU receives a signal from the door lock circuit. The problem with one-shot operation is that if a child, for example, should become trapped in the window there is a serious risk of injury. To prevent this, the back-off feature is used. An extra commutator is fitted to the motor armature and produces a signal via two brushes, proportional to the motor speed. If the rate of change of speed of the motor is detected as being below a certain threshold when closing, then the ECU will reverse the motor until the window is fully open.

By counting the number of pulses received the ECU can also determine the window position. This is important, as the window must not reverse when it stalls in the closed position. In order for the ECU to know the window position it must be initialised. This is often done simply by operating the motor to drive the window first fully open, and then fully closed. If this is not done then the one-shot close will not operate. On some systems Hall effect sensors are used to detect motor speed. Other systems sense the current being drawn by the motor and use this as an indication of speed.

Lazy lock feature allows the car to be fully secured by one operation of a remote infrared key. This is done by the link between the door lock ECU and the window and sunroof ECUs. A signal is supplied which causes all the windows to close in turn, then the sunroof and finally locks the doors. The alarm will also be set if required. The windows close in turn to prevent the excessive current demand which would occur if they all tried to operate at the same time.

A circuit for electric windows is shown in Figure 9.37. Note the connections to other systems such as door locking and the rear window isolation switch. This is commonly fitted to allow

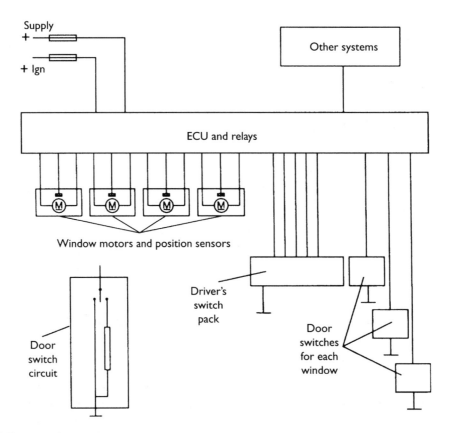

Figure 9.37 Electric window circuit

the driver to prevent rear window operation for child safety for example.

9.10 Diagnosing body electrical system faults

9.10.1 Testing procedure

The following procedure is very generic but with a little adaptation can be applied to any electrical system. Refer to manufacturer's recommendations if in any doubt. The process of checking any system circuit is broadly as follows.

9.10.2 Body electrical systems fault diagnosis

Symptom	Possible fault
Electric units not operating	If *all* units not operating: open circuit in main supply main fuse blown relay coil or contacts open circuit or high resistance
(unit = window, door lock, mirror etc.)	If *one* unit is not operating: fuse blown control switch open circuit motor seized or open circuit back-off safety circuit signal incorrect (windows)

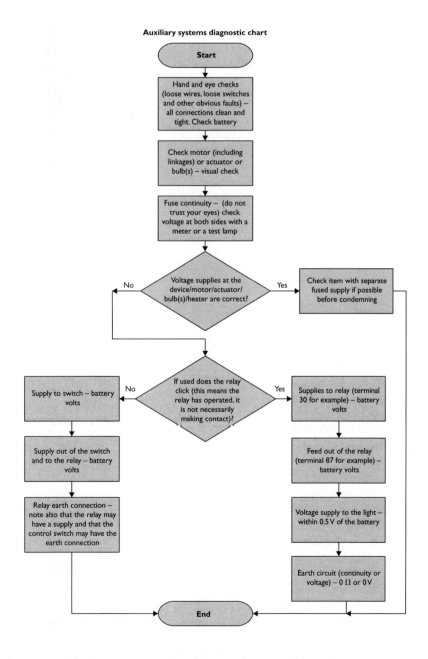

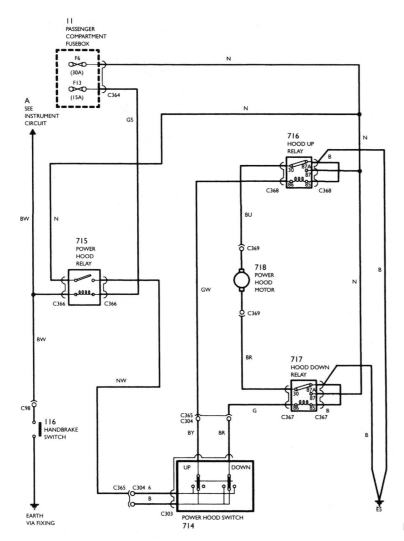

Figure 9.38 Power hood circuit

9.10.3 Circuit systematic testing

The circuit shown in Figure 9.38 is for a power hood on a vehicle. The following faultfinding guide is an example of how to approach a problem with a system such as this in a logical manner.

If the power hood will not operate with the ignition switch at the correct position and the handbrake applied proceed as follows.

1. Check fuses 6 and 13.
2. Check 12 V supply on N wire from fuse 6.
3. Check for 12 V on GS wire at power hood relay.
4. Check continuity from power hood relay to earth on BW wire.
5. Check power hood relay.
6. Check for 12 V on NW wire at hood switch. Check for 12 V on N wire at hood up and down relays.
7. Check continuity from hood up and down relays to earth on B wire.
8. Check switch operation.
9. Check pump motor operation.

If the power hood will operate in one direction only proceed as follows.

1. Check for 12 V on N wire at hood up or down relay as appropriate.
2. Check continuity from hood up or down relay to earth on B wire.
3. Check relay.

9.11 Instrumentation

9.11.1 Gauges

Thermal gauges, which are ideal for fuel and engine temperature indication, have been in use

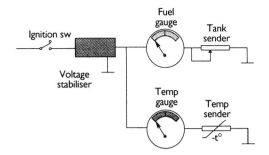

Figure 9.39 Thermal gauge circuit

for many years. This will continue because of their simple design and inherent 'thermal' damping. The gauge works by utilizing the heating effect of electricity and the widely adopted benefit of the bimetal strip. As a current flows through a simple heating coil wound on a bimetal strip, heat causes the strip to bend. The bimetal strip is connected to a pointer on a suitable scale. The amount of bend is proportional to the heat, which in turn is proportional to the current flowing. Providing the sensor can vary its resistance in proportion to the measurement and (e.g. fuel level), the gauge will indicate a suitable representation as long as it has been calibrated for the particular task. Figure 9.39 shows a representation of a typical thermal gauge circuit.

Thermal type gauges are used with a variable resistor and float in a fuel tank or with a thermistor in the engine water jacket. The resistance of the fuel tank sender can be made non-linear to counteract any non-linear response of the gauge. The sender resistance is at a maximum when the tank is empty.

A constant voltage supply is required to prevent changes in the vehicle system voltage affecting the reading. This is because if system voltage increased the current flowing would increase and hence the gauges would read higher. Most voltage stabilisers are simple zener diode circuits.

Air cored gauges work on the same principle as a compass needle lining up with a magnetic field. The needle of the display is attached to a very small permanent magnet. Three coils of wire are used and each produces a magnetic field. The magnet will line up with the resultant of the three fields. The current flowing and the number of turns (ampere-turns) determine the strength of the magnetic flux produced by each coil. As the number of turns remains constant the current is the key factor. Figure 9.40 shows the principle of the air cored gauge together with the circuit for

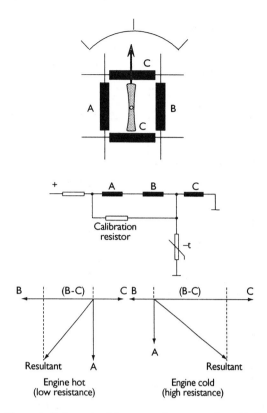

Figure 9.40 Principle of the air cored gauge together with the circuit when used as a fuel lever or temperature indicator and the resultant magnetic fields

use as a temperature indicator. The ballast resistor on the left is used to limit maximum current and the calibration resistor is used for calibration! The thermistor is the temperature sender. As the thermistor resistance is increased the current in all three coils will change. Current through C will be increased but the current in coils A and B will decrease.

The air cored gauge has a number of advantages. It has almost instant response and as the needle is held in a magnetic field it will not move as the vehicle changes position. The gauge can be arranged to continue to register the last position even when switched off or, if a small 'pull off' magnet is used, it will return to its zero position. As a system voltage change would affect the current flowing in all three coils variations are cancelled out negating the need for voltage stabilisation. Note that the operation is similar to the moving iron gauge.

9.11.2 Digital instrumentation

The block diagram shown in Figure 9.41 is typical of a digital instrumentation system. All signal conditioning and logic functions are carried out in

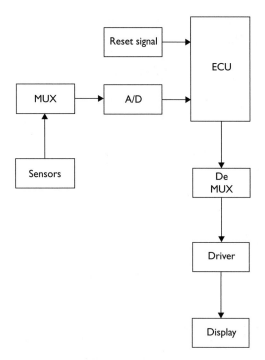

Figure 9.41 Digital instrumentation

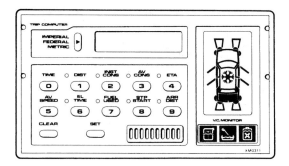

Figure 9.42 Trip computer display and a vehicle 'map'

the ECU. This will often form part of the dashboard assembly. Standard sensors provide information to the ECU, which in turn will drive suitable displays. The ECU contains a ROM (read only memory) section, which allows it to be programmed to a specific vehicle. The gauges used are as described in the above sections. Some of the extra functions available with this system are described briefly as follows.

- **Low fuel warning light** Can be made to illuminate at a particular resistance reading from the fuel tank sender unit.
- **High engine temperature warning light** Can be made to operate at a set resistance of the thermistor.
- **Steady reading of the temperature gauge** To prevent the gauge fluctuating as the cooling system thermostat operates, the gauge can be made to read only at say five set figures. For example, if the input resistance varies from 240 to 200 Ω as the thermostat operates, the ECU will output just one reading corresponding to 'normal' on the gauge. If the resistance is much higher or lower the gauge will read to one of the five higher or lower positions. This gives a low resolution but high readability for the driver.
- **Oil pressure or other warning lights can be made to flash** This is more likely to catch the driver's attention.

- **Service or inspection interval warning lights can be used** The warning lights are operated broadly as a function of time but, for example, the service interval is reduced if the engine experiences high speeds and/or high temperatures. Oil condition sensors are also used to help determine service intervals.
- **Alternator warning light** Works as normal but the same or an extra light can be made to operate if the output is reduced or if the drive belt slips. This is achieved by a wire from one phase of the alternator providing a pulsed signal, which is compared to a pulsed signal from the ignition. If the ratio of the pulses changed this would indicate a slipping belt.

9.11.3 Vehicle condition monitoring

VCM or vehicle condition monitoring is a form of instrumentation. It has now become difficult to separate it from the more normal instrumentation system discussed in the first part of this chapter. Figure 9.42 shows a typical display unit, which also incorporates the vehicle map (see next section). The complete VCM system can include driver information relating to the following list:

- high engine temperature;
- low fuel;
- low brake fluid;
- worn brake pads;
- low coolant level;
- low oil level;
- low screen washer fluid;
- low outside temperature;
- bulb failure;
- doors, bonnet or boot open warning.

A circuit is shown in Figure 9.43, which can be used to operate bulb failure warning lights for whatever particular circuit it is monitoring. The simple principle is that the reed relay is only

Electrical systems   229

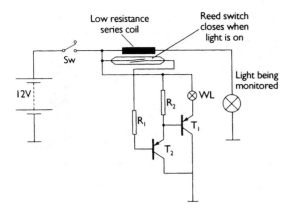

Figure 9.43 Bulb failure warning circuit

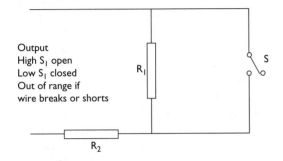

Figure 9.44 Equivalent circuit of a dual resistance self-testing system

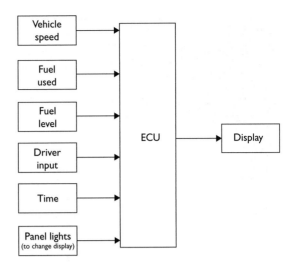

Figure 9.45 Trip computer block diagram

operated when the bulb being monitored is drawing current. The fluid and temperature level monitoring systems work in a similar way to the systems described earlier but in some cases the level of a fluid is by a float and switch.

Oil level can be monitored by measuring the resistance of a heated wire on the end of the dip stick. A small current is passed through the wire to heat it. How much of the wire is covered by oil will determine its temperature and therefore its resistance.

Many of the circuits monitored use a dual resistance system so that the circuit itself is also checked. Figure 9.44 shows the equivalent circuit for this technique. In effect it will produce one of three possible outputs: high resistance, low resistance or an out of range reading. The high or low resistance readings are used to indicate say correct fluid level and low fluid level. A figure outside these limits would indicate a circuit fault of either a short or open circuit connection.

The display is often just a collection of LEDs or a back lit liquid crystal display (LCD). These are arranged into suitable patterns and shapes such as to represent the circuit or system being monitored. A door open will illuminate a symbol which looks as if the door of the vehicle map (plan view of the car) is open. Low outside temperature or ice warning is often a large snowflake!

9.11.4 Trip computer

The trip computer used on many top range vehicles is arguably an expensive novelty, but is popular none-the-less. The display and key pad of a typical trip computer is shown above. The functions available on most systems are:

- time and date;
- elapsed time or a stop watch;
- estimated time of arrival;
- average fuel consumption;
- range on remaining fuel;
- trip distance.

The above details can usually be displayed in imperial, US or metric units as required. Figure 9.45 shows a block diagram of a trip computer system. Note that several systems use the same inputs and that several systems 'communicate' with each other. This makes the overall wiring very bulky – if not complicated.

9.11.5 Displays

If the junction of a diode is manufactured in a certain way, light will be emitted from the junction when a current is made to pass in the forward biased direction. This is an LED and will produce red, yellow or green light with slight changes in the manufacturing process. LEDs are used extensively as indicators on electronic

Figure 9.46 Display techniques

equipment and in digital displays. They last for a very long time (50 000 hours) and draw only a small current.

LED displays are tending to be replaced for automobile use by the liquid crystal type, which can be back lit to make it easier to read in the daylight. However, LEDs are still popular for many applications.

The actual display will normally consist of a number of LEDs arranged into a suitable pattern for the required output. This can range from the standard seven segment display to show numbers, to a custom designed speedometer display.

Liquid crystals are substances that do not melt directly from a solid to the liquid phase, but first pass through a paracrystalline stage in which the molecules are partially ordered. In this stage a liquid crystal is a cloudy or translucent fluid but still has some of the optical properties of a solid crystal.

Mechanical stress, electric and magnetic fields, pressure and temperature can alter the molecular structure of liquid crystals. A liquid crystal also scatters light that shines on it. Because of these properties, liquid crystals are used to display letters and numbers on calculators, digital watches and automobile instrument displays. LCDs are also used for portable computer screens and even television screens. The LCD has many more areas of potential use and developments are on going. In particular this type of display is now good enough to reproduce pictures and text on computer screens.

When a voltage of about 10 V at 50 Hz is applied to the crystal it becomes disorganised and the light passing through it is no longer twisted by 90°. This means that the light polarised by the first polariser will not pass through the second, and will therefore not be reflected. This will show as a dark area on the display. These areas are constructed into suitable segments in much the same way as with LEDs to provide whatever type of display is required. The size of each individual area can be very small such as to form one pixel of a TV or computer screen if appropriate.

A vacuum fluorescent display works in much the same way as a television tube and screen. It is becoming increasingly popular for vehicle use because it produces a bright light (which is adjustable), and a wider choice of colours than LED or LCD displays. The glass front of the display can be coloured to improve the readability and aesthetic value. This type of display has many advantages but the main problem for automobile use is its susceptibility to shock and vibration. This can be overcome, however, with suitable mountings. Figure 9.46 shows some instrument and other displays using a number of the previously discussed techniques.

9.12 Diagnosing instruments system faults

9.12.1 Testing procedure

The process of checking a thermal gauge fuel or temperature instrument system is broadly as follows.

Electrical systems 231

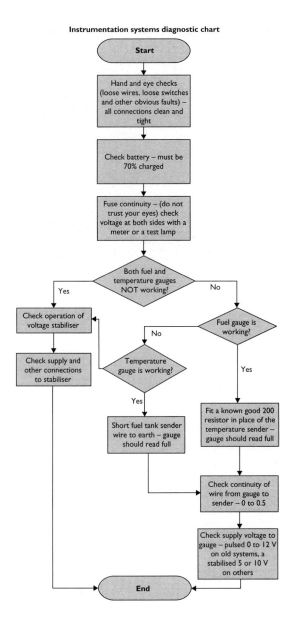

Instrumentation systems diagnostic chart

9.12.2 Instrumentation fault diagnosis table

Symptom	Possible fault
Fuel and temperature gauges both read high or low	Voltage stabiliser
Gauges read full/hot or empty/cold all the time	Short/open circuit sensors Short or open circuit wiring
Instruments do not work	Loose or broken wiring/connections/fuse Inoperative instrument voltage stabiliser Sender units (sensor) faulty Gauge unit fault (not very common)

9.12.3 Black box technique for instrumentation

Instrumentation systems, like most others, now revolve around an ECU. The ECU is considered to be a 'black box'; in other words we know what it should do but how it does it is irrelevant! Figure 9.47 shows an instrumentation system where the instrument pack could be considered as a black box. Normal faultfinding or testing techniques can now be applied to the sensors and supply circuits.

Remember also the 'sensor to ECU method' of testing described in Section 2.5.8. A resistance test carried out on a component such as the tank unit (lower right) would give a direct measure of its resistance. A second reading at the instrument pack between the GB and BP wires, if the same as the first, would confirm that the circuit is in good order.

> **Warning**
> The circuit supply must always be off when carrying out ohmmeter tests.

9.13 Heating, ventilation and air conditioning (HVAC)

9.13.1 Ventilation and heating

To allow fresh air from outside the vehicle to be circulated inside the cabin, a pressure difference must be created. This is achieved by using a plenum chamber. A plenum chamber by definition holds a gas (in this case air) at a pressure higher than the ambient pressure. The plenum chamber on a vehicle is usually situated just below the windscreen, behind the bonnet. When the vehicle is moving the airflow over the vehicle will cause a higher pressure in this area. Suitable flaps and drains are utilised to prevent water entering the car through this opening.

By means of distribution trunking, control flaps and suitable 'nozzles', the air can be directed as required. This system is enhanced with the addition of a variable speed blower motor. Figure 9.48 shows a typical ventilation and heating system layout.

When extra air is forced into a vehicle cabin the interior pressure would increase if no outlet was available. Most passenger cars have the outlet grilles on each side of the vehicle above or near the rear quarter panels or doors.

232 Advanced automotive fault diagnosis

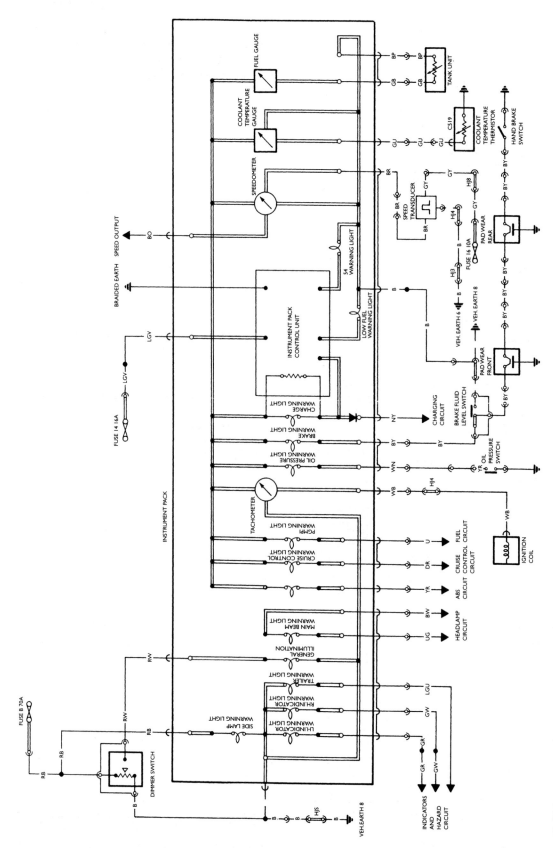

Figure 9.47 Instrument pack warning circuit

Electrical systems 233

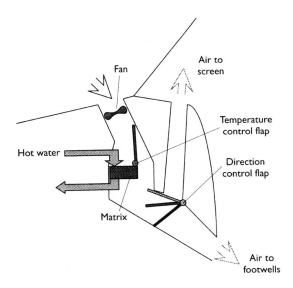

Figure 9.48 Heating and ventilation system

Figure 9.49 Heating and ventilation motor and fans

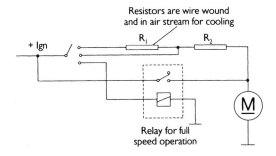

Figure 9.50 Circuit diagram of a three-speed control system

9.13.2 Heating system – water-cooled engine

Heat from the engine is utilised to increase the temperature of the car interior. This is achieved by use of a heat exchanger, called the heater matrix. Due to the action of the thermostat in the engine cooling system the water temperature remains broadly constant. This allows for the air being passed over the heater matrix to be heated by a set amount depending on the outside air temperature and the rate of airflow. A source of hot air is therefore available for heating the vehicle interior. However, some form of control is required over how much heat (if any) is required. The method used on most modern vehicles is the blending technique. This is simply a control flap, which determines how much of the air being passed into the vehicle is directed over the heater matrix. The main drawback of this system is the change in air-flow with vehicle speed. Some systems use a valve to control the hot coolant flowing to the heater matrix.

By a suitable arrangement of flaps it is possible to direct air of the chosen temperature to selected areas of the vehicle interior. In general, basic systems allow the warm air to be adjusted between the inside of the windscreen and the driver and passenger footwells. Most vehicles also have small vents directing warm air at the driver's and front passenger's side windows. Fresh cool air outlets with directional nozzles are also fitted.

One final facility, which is available on many vehicles, is the choice between fresh or recirculated air. The main reason for this is to decrease the time it takes to demist or defrost the vehicle windows, and simply to heat the car interior more quickly to a higher temperature. The other reason is that, for example in heavy congested traffic, the outside air may not be very clean.

9.13.3 Heater blower motors

The motors used to increase airflow are simple permanent magnet two brush motors. The blower fan is often the centrifugal type and in many cases the blades are positioned asymmetrically to reduce resonant noise. Figure 9.49 shows a typical motor and fan arrangement. Varying the voltage supplied controls motor speed. This is achieved by using dropping resistors. The speed in some cases is made 'infinitely' variable, by the use of a variable resistor. In most cases the motor is controlled to three or four set speeds.

Figure 9.50 shows a circuit diagram typical of a three-speed control system. The resistors are usually wire wound and are placed in the air stream to prevent overheating. These resistors will have low values in the region of 1 Ω or less.

9.13.4 Electronic heating control

Most vehicles that have electronic control of the heating system also include air conditioning,

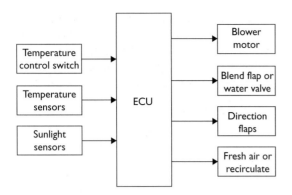

Figure 9.51 An electronically controlled vehicle heating system

which is covered in the next section. However, a short description at this stage will help to lead into the more complex systems. Figure 9.51 shows a block diagram representing an electronically controlled vehicle heating system.

This system requires control of the blower motor, blend flap, direction flaps and the fresh or recirculated air flap. The technique involves one or a number of temperature sensors suitably positioned in the vehicle interior, to provide information for the ECU. The ECU responds to information received from these sensors and sets the controls to their optimum positions. The whole arrangement is in fact a simple closed loop feedback system with the air temperature closing the loop. The ECU has to compare the position of the temperature control switch with the information that is supplied by the sensors and either cool or heat the care interior as required.

9.13.5 Air conditioning introduction

A vehicle fitted with air conditioning allows the temperature of the cabin to be controlled to the ideal or most comfortable value determined by the ambient conditions. The system as a whole still utilises the standard heating and ventilation components, but with the important addition of an evaporator, which both cools and dehumidifies the air.

Air conditioning can be manually controlled or, as is not often the case, combined with some form of electronic control. The system as a whole can be thought of as a type of refrigerator or heat exchanger. Heat is removed from the car interior and dispersed to the outside air.

To understand the principle of refrigeration the following terms and definitions will be useful.

- Heat is a form of energy.
- Temperature means the degree of heat of an object.
- Heat will only flow from a higher to a lower temperature.
- Heat quantity is measured in 'calories' (more often kcal).
- 1 kcal heat quantity changes the temperature of 1 kg of liquid water by 1°C.
- Change of state is a term used to describe the changing of a solid to liquid, a liquid to a gas, a gas to a liquid or a liquid to a solid.
- Evaporation is used to describe the change of state from a liquid to a gas.
- Condensation is used to describe the change of state from gas to liquid.
- Latent heat describes the energy required to evaporate a liquid without changing its temperature (breaking of molecular bonds), or the amount of heat given off when a gas condenses back into a liquid without changing temperature (making of molecular bonds).

Latent heat in the change of state of a refrigerant is the key to air conditioning. A simple example of this is that if you put a liquid such as methylated spirits on your hand it feels cold. This is because it evaporates and the change of state (liquid to gas) uses heat from your body. This is why the process is often thought of as 'unheating' rather than cooling.

The refrigerant used in many air conditioning systems is known as R134A. This substance changes state from liquid to gas at −26.3°C. R134A is HFC based rather than CFC, due to the problems with atmospheric ozone depletion associated with CFC based refrigerants. A key to understanding refrigeration is to remember that low pressure refrigerant will have low temperature, and high-pressure refrigerant will have a high temperature.

Figure 9.52 shows the basic principle of an air conditioning – or refrigeration – system. The basic components are the evaporator, condenser and pump or compressor. The evaporator is situated in the car; the condenser is outside the car, usually in the air stream; and the compressor is driven by the engine.

As the pump operates it will cause the pressure on its intake side to fall, which will allow the refrigerant in the evaporator to evaporate and draw heat from the vehicle interior. The high

Electrical systems   **235**

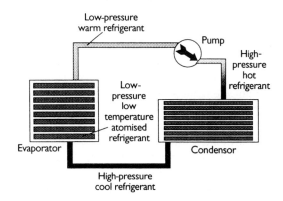

Figure 9.52 Basic principles of air conditioning

pressure or output of the pump is connected to the condenser. The pressure causes the refrigerant to condense (in the condenser), thus giving off heat outside the vehicle as it changes state. Figure 9.53 shows some typical components of an air conditioning system.

9.13.6 Air conditioning overview

The operation of the system is continuous cycle. The compressor pumps low pressure but heat laden vapour from the evaporator, compresses it and pumps it as a super heated vapour under high pressure to the condenser. The temperature of the refrigerant at this stage is much higher than the outside air temperature hence it gives up its heat via the fins on the condenser as it changes state back to a liquid.

This high-pressure liquid is then passed to the receiver drier where any vapour which has not yet turned back to a liquid is stored, and a desiccant bag removes any moisture (water) that is contaminating the refrigerant. The high-pressure liquid is now passed through the thermostatic expansion valve and is converted back to a low-pressure liquid as it passes through a restriction in the valve into the evaporator. This valve is the element of the system that controls the refrigerant flow and hence the amount of cooling provided. As the liquid changes state to a gas in the evaporator, it takes up heat from its surroundings, thus cooling or 'unheating' the air that is forced over the fins. The low-pressure vapour leaves the evaporator returning to the pump thus completing the cycle.

If the temperature of the refrigerant increases beyond certain limits, condenser cooling fans can be switched in to supplement the ram air effect. A safety switch is fitted in the high-pressure side

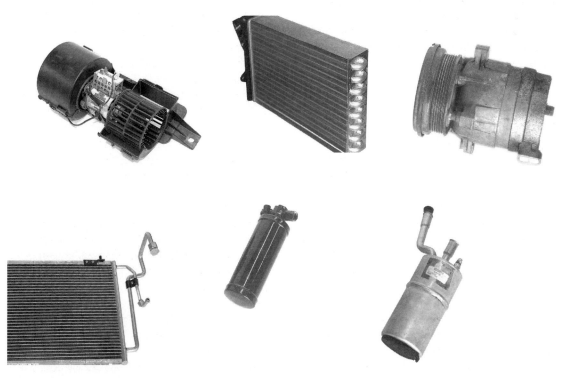

Figure 9.53 Heating, ventilation and air conditioning HVAC components

of most systems. It is often known as a high-low pressure switch, as it will switch off the compressor if the pressure is too high due to a component fault, or if the pressure is too low due to a leakage, thus protecting the compressor.

9.13.7 Automatic temperature control

Full temperature control systems provide a comfortable interior temperature in line with the passenger controlled input. The ECU has full control of fan speed, air distribution, air temperature, fresh or recirculated air and the air conditioning pump. These systems will soon be able to control automatic demist or defrost when reliable sensors are available. A single button currently will set the system to full defrost or demist.

A number of sensors are used to provide input to the ECU.

- Ambient temperature sensor mounted outside the vehicle to allow compensation for extreme temperature variation. This device is usually a thermistor.
- Solar light sensor mounted on the fascia panel. This device is a photodiode and allows a measurement of direct sunlight from which the ECU can determine whether to increase the air to the face vents.
- The in car temperature sensors are simple thermistors but to allow for an accurate reading a small motor and fan can be used to take a sample of interior air and direct it over the sensing elements.
- A coolant temperature sensor is used to monitor the temperature of the coolant supplied to the heater matrix. This sensor is used to prevent operation of the system until coolant temperature is high enough to heat the vehicle interior.
- Driver input control switches.

The ECU takes information from all of the above sources and will set the system in the most appropriate manner as determined by the software. Control of the flaps can be either by solenoid controlled vacuum actuators or by small motors. The main blower motor is controlled by a heavy-duty power transistor and is constantly variable. These systems are able to provide a comfortable interior temperature in exterior conditions ranging from −10 to +35°C even in extreme sunlight.

9.13.8 Seat heating

The concept of seat heating is very simple. A heating element is placed in the seat, together with an on-off switch and a control to regulate the heat. However the design of these heaters is more complex than first appears. The heater must meet the following criteria.

- The heater must only supply the heat loss experienced by the person's body.
- Heat is to be supplied only at the major contact points.
- Leather and fabric seats require different systems due to their different thermal properties.
- Heating elements must fit the design of the seat.
- The elements must pass the same rigorous tests as the seat, such as squirm, jounce and bump tests.

Figure 9.54 shows a seat containing heating elements.

In order for the passengers (including the driver) to be comfortable rigorous tests have been

Figure 9.54 Heated seat – great on a cold morning!

carried out to find the optimum heat settings and the best position for the heating elements. Many tests are carried out on new designs, using manikin with sensors attached, to measure the temperature and heat flow.

The cable used for most heating elements consists of multi-strand alloyed copper. This cable may be coated with tin or insulated as the application demands. The heating element is laminated and bonded between layers of polyurethane foam.

9.13.9 Screen heating

Heating of the rear screen involves a very simple circuit as shown in Figure 9.55. The heating elements consist of a thin metallic strip bonded to the glass. When a current is passed through the elements, heat is generated and the window will defrost or demist. This circuit can draw high current, 10 to 15 A being typical. Because of this, the circuit will often contain a timer relay to prevent the heater being left on too long. The timer will switch off after 10 to 15 minutes. The elements are usually positioned to defrost the main area of the screen and the rest position of the rear wiper blade if fitted.

Front windscreen heating is being introduced on many vehicles. This of course presents more problems than the rear screen, as vision must not be obscured. The technology, drawn from the aircraft industry, involves very thin wires cast in to the glass. As with the heated rear window this device can consume a large current and is operated by timer relay.

9.14 Diagnostics – HVAC

9.14.1 Testing procedure

Warning
Do not work on the refrigerant side of air conditioning systems unless you have been trained and have access to suitable equipment.

The process of checking an air conditioning system is broadly as follows.

9.14.2 Air conditioning fault diagnosis table

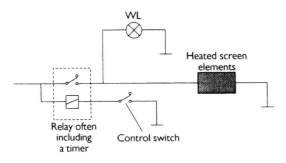

Figure 9.55 Screen heating circuit

Figure 9.56 Circuit voltage testing

Symptom	Possible fault
After stopping the compressor, pressure falls quickly to about 195 kPa and then falls gradually	Air in the system or if no bubbles are seen in the sight glass as the condenser is cooled with water then excessive refrigerant may be the fault
Discharge pressure low	Fault with the compressor or, if bubbles are seen, low refrigerant
Discharge temperature is lower than normal	Frozen evaporator
Suction pressure too high	High pressure valve fault, excessive refrigerant or expansion valve open too long
Suction and discharge pressure too high	Excessive refrigerant in the system or condenser not working due to fan fault or clogged fins
Suction and discharge pressure too low	Clogged or kinked pipes
Refrigerant loss	Oily marks (from the lubricant in the refrigerant) near joints or seals indicate leaks

9.14.3 Heating and ventilation fault diagnosis table

Symptom	Possible fault
Booster fan not operating at any speed	Open circuit fuse/supply/earth Motor inoperative/seized Dropping resistor(s) open circuit Switch open circuit Electronic speed controller not working
Booster fan only works on full speed	Dropping resistor(s) open circuit Switch open circuit Electronic speed controller not working
Control flap(s) will not move	Check vacuum connections (many work by vacuum operated actuators) Inspect cables
No hot air	Matrix blocked Blend flap stuck
No cold air	Blend flap stuck Blocked intake
Reduced temperature when set to 'hot'	Cooling system thermostat stuck open Heater matrix partially blocked Control flap not moving correctly

9.14.4 Air conditioning receiver drier sight glass

A very useful guide to diagnostics is the receiver drier sight glass. Figure 9.57 shows four possible symptoms and suggestions as to the possible fault.

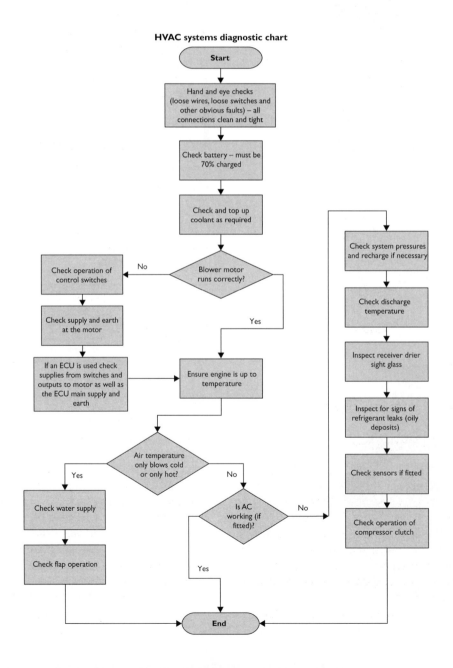

Electrical systems **239**

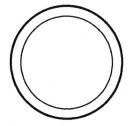

Clear – system OK *or* completely empty

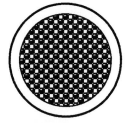

Foamy/bubbles – low on refrigerant and air in system

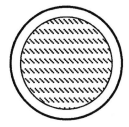

Streaky – Low on refrigerant and compressor oil circulation

Cloudy – Desiccant in receiver dryer contaminating the system

Figure 9.57 Air conditioning receiver drier sight glass

9.15 Cruise control

9.15.1 Introduction

Cruise control is the ideal example of a closed loop control system as shown in Figure 9.58. The purpose of cruise control is to allow the driver to set the vehicle speed and let the system maintain it automatically. The system reacts to the measured speed of the vehicle and adjusts the throttle accordingly. The reaction time is important so that the vehicle's speed does not feel as if it is surging up and down.

Other facilities are included such as allowing the speed to be gradually increased or decreased

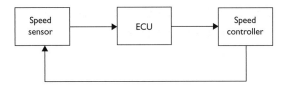

Figure 9.58 Cruise control – closed control loop

at the touch of a button. Most systems also remember the last set speed and will resume this again at the touch of a button.

To summarise and to add further refinements, the following is the list of functional requirements for a good cruise control system:

- hold the vehicle speed at the selected value;
- hold the speed with minimum surging;
- allow the vehicle to change speed;
- relinquish control immediately the brakes are applied;
- store the last set speed;
- contain built in safety features.

9.15.2 System description

The main switch switches on the cruise control; this in turn is ignition controlled. Most systems do not retain the speed setting in memory when the main switch has been turned off. Operating the 'set' switch programs the memory but this will normally work only if conditions similar to the following are met:

- vehicle speed is greater than 40 km/h;
- vehicle speed is less than 12 km/h;
- change of speed is less than 8 km/h/s;
- automatics must be in 'drive';
- brakes or clutch are not being operated;
- engine speed is stable.

Once the system is set, the speed is maintained to within about 3–4 km/h until it is deactivated by pressing the brake or clutch pedal, pressing the 'resume' switch or turning off the main control switch. The last 'set' speed is retained in memory except when the main switch is turned off.

If the cruise control system is required again then either the 'set' button will hold the vehicle at its current speed or the 'resume' button will accelerate the vehicle to the previous 'set' speed. When cruising at a set speed, the driver can press and hold the 'set' button to accelerate the vehicle until the desired speed is reached when the button is released. If the driver accelerates from the set speed, to overtake for example, then, when the throttle is released, the vehicle will slow down until it reaches the last set position.

9.15.3 Components

The main components of a typical cruise control system are as follows.

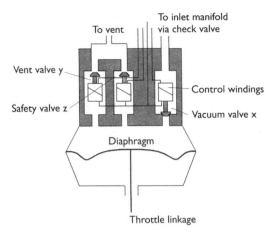

Figure 9.59 Vacuum operated cruise control actuator

Actuator	A number of methods are used to control the throttle position. Vehicles fitted with drive by wire systems allow the cruise control to operate the same actuator. A motor can be used to control the throttle cable or in many cases a vacuum operated diaphragm is used which is controlled by three simple valves. This technique is shown in Figure 9.59. When the speed needs to be increased valve 'x' is opened allowing low pressure from the inlet manifold to one side of the diaphragm. The atmospheric pressure on the other side will move the diaphragm and hence the throttle. To move the other way valve 'x' is closed and valve 'y' is opened allowing atmospheric pressure to enter the chamber. The spring moves the diaphragm back. If both valves are closed then the throttle position is held. Valve 'x' is normally closed and valve 'y' normally open; thus in the event of electrical failure cruise will not remain engaged and the manifold vacuum will not be disturbed. Valve 'z' provides extra safety and is controlled by the brake and clutch pedals
Main switch and warning lamp	This is a simple on/off switch located in easy reach of the driver on the dashboard. The warning lamp can be part of this switch or part of the main instrument display as long as it is in the driver's field of vision
'Set' and 'Resume' switches	These are fitted either on the steering wheel or on a stalk from the steering column. When part of the steering wheel slip rings are needed to transfer the connection. The 'set' button programmes the speed into memory and can also be used to increase the vehicle and memory speed. The 'resume' button allows the vehicle to reach its last set speed or to temporarily deactivates the control
Brake switch	This switch is very important, as braking would be dangerous if the cruise control system was trying to maintain the vehicle speed. This switch is normally of superior quality and is fitted in place or as a supplement to the brake light switch activated by the brake pedal. Adjustment of this switch is important
Clutch or automatic gearbox switch	The clutch switch is fitted in a similar manner to the brake switch. It deactivates the cruise system to prevent the engine speed increasing if the clutch is pressed. The automatic gearbox switch will only allow the cruise to be engaged when it

Figure 9.60 Headway sensor and control electronics

	is in the 'drive' position. This is again to prevent the engine overspeeding if the cruise tried to accelerate to a high road speed with the gear selector in position '1' or '2'. The gearbox will still change gear if accelerating back up to a set speed as long as it 'knows' top gear is available
Speed sensor	This will often be the same sensor that is used for the speedometer. If not several types are available, the most common producing a pulsed signal the frequency of which is proportional to the vehicle speed
Headway sensor (Figure 9.60)	Only on the very latest and top of the range systems, this device uses radar or light to sense the distance from the vehicle in front

9.16 Diagnostics – cruise control

9.16.1 Systematic testing

If the cruise control system will not operate then considering the ECU as a black box, the following procedure should be followed.

Electrical systems **241**

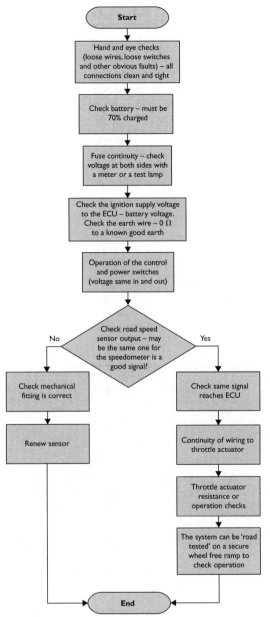

Cruise control systems diagnostic chart

9.16.2 Cruise control fault diagnosis table

Symptom	Possible fault
Cruise control will not set	Brake switch sticking on Safety valve/circuit fault Diaphragm holed Actuating motor open circuit or seized Steering wheel slip ring open circuit Supply/earth/fuse open circuit General wiring fault
Surging or uneven speed	Actuator cable out of adjustment ECU fault Engine/engine management fault

9.17 Air bags and belt tensioners

9.17.1 Introduction

Seat belt, seat belt tensioner and an air bag are at present the most effective restraint system in the event of a serious accident. At speeds in excess of 40 km/h the seat belt alone is no longer adequate. The method becoming most popular for an air bag system is that of building most of the required components into one unit. This reduces the amount of wiring and connections, thus improving reliability. An important aspect is that some form of system monitoring must be built in, as the operation cannot be tested – it only ever works once.

The sequence of events in the case of a frontal impact at about 35 km/h, as shown in Figure 9.62, is as follows.

1. Driver in normal seating position prior to impact.
2. About 15 ms after the impact the vehicle is strongly decelerated and the threshold for triggering the air bag is reached. The igniter ignites the fuel tablets in the inflater.
3. After about 30 ms the air bag unfolds and the driver will have moved forwards as the vehicle's crumple zones collapse. The seat belt will have locked or been tensioned depending on the system.
4. At 40 ms after impact the air bag will be fully inflated and the driver's momentum will be absorbed by the air bag.
5. About 120 ms after impact the driver will be moved back into the seat and the air bag will have almost deflated through the side vents allowing driver visibility.

Passenger air bag events are similar to the above description. A number of arrangements are used with the mounting of all components in the steering wheel centre becoming the most popular. None-the-less the basic principle of operation is the same.

9.17.2 Components and circuit

The main components of a basic air bag system are as follows:

- driver and passenger air bags;
- warning light;
- passenger seat switches;
- pyrotechnic inflater;

242 Advanced automotive fault diagnosis

- igniter;
- crash sensor(s);
- ECU.

The air bag is made of a nylon fabric with a coating on the inside. Prior to inflation the air bag is folded up under suitable padding which has specially designed break lines built in. Holes are provided in the side of the air bag to allow rapid deflation after deployment. The driver's air has a volume of about 60 litres and the passenger air bag about 160 litres. Figure 9.63 shows a steering wheel with an air bag fitted in the centre.

A warning light is used as part of the system monitoring circuit. This gives an indication of a potential malfunction and is an important part of the circuit. Some manufacturers use two bulbs for added reliability.

Consideration is being given to the use of a seat switch on the passenger side to prevent deployment when not occupied. This may be more appropriate to side-impact air bags.

The pyrotechnic inflater and the igniter can be considered together. The inflater in the case of the driver is located in the centre of the steering wheel. It contains a number of fuel tablets in a combustion chamber. The igniter consists of charged capacitors, which produce the ignition spark. The fuel tablets burn very rapidly and produce a given quantity of nitrogen gas at a given pressure. This gas is forced into the air bag through a filter and the bag inflates breaking through the padding in the wheel centre. After deployment a small amount of sodium hydroxide will be present in the air bag and vehicle interior. Personal protection equipment must be used when removing the old system and cleaning the vehicle interior.

The crash sensor can take a number of forms; these can be described as mechanical or electronic. The mechanical system works by a spring holding

Figure 9.61 Battery earth connection – so easy to forget the obvious

Figure 9.63 Supplementary restraint system (SRS) – an air bag is incorporated in this steering wheel

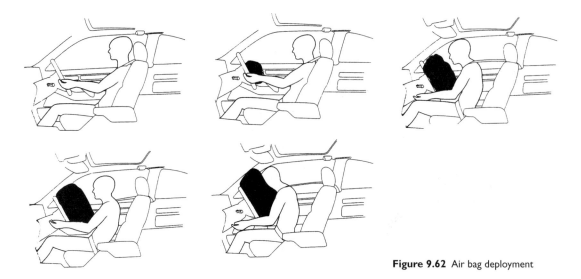

Figure 9.62 Air bag deployment

a roller in a set position until an impact above a predetermined limit provides enough force to overcome the spring and the roller moves, triggering a micro switch. The switch is normally open with a resistor in parallel to allow the system to be monitored. Two switches similar to this may be used to ensure that the bag is deployed only in the case of sufficient frontal impact. Note the air bag is not deployed in the event of a roll over. The other main type of crash sensor can be described as an accelerometer. This will sense deceleration, which is negative acceleration.

The final component to be considered is the ECU or diagnostic control unit. When a mechanical type crash sensor is used, in theory no electronic unit would be required. A simple circuit could be used to deploy the air bag when the sensor switch operated. However, it is the system monitoring or diagnostic part of the ECU which is most important. If a failure is detected in any part of the circuit then the warning light will be operated. Up to five or more faults can be stored in the ECU memory, which can be accessed by blink code or serial fault readers. Conventional testing of the system with a multimeter and jump wires is not to be recommended as it might cause the air bag to deploy!

A block diagram of an air bag circuit is shown in Figure 9.64. Note the 'safing' circuit, which is a crash sensor that prevents deployment in the event of a faulty main sensor. A digital based system using electronic sensors has about 10 ms at a vehicle speed of 50 km/h to decide if the supplementary restraint systems (SRS) should be activated. In this time about 10 000 computing operations are necessary. Data for the development of these algorithms is based on computer simulations but digital systems can also remember the events during a crash allowing real data to be collected.

9.17.3 Seat belt tensioners

Taking the 'slack' out of a seat belt in the event of an impact is a good contribution to vehicle passenger safety. The decision to take this action is the same as for the air bag. The two main types are:

- spring tension;
- pyrotechnic.

The mechanism used by one type of seat belt tensioner works by explosives. When the explosive charge is fired, the cable pulls a lever on the seat belt reel, which in turn tightens the belt. The unit must be replaced once deployed. This feature is sometimes described as anti-submarining. Figure 9.65 shows a full seat belt, air bag system and other safety features.

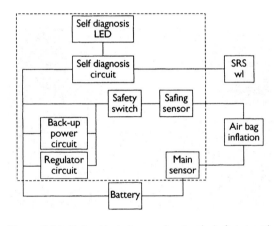

Figure 9.64 Air bag block diagram showing the 'safing circuit'

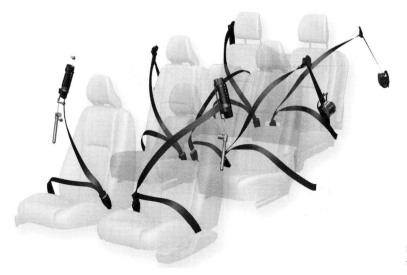

Figure 9.65 Seat belts and tensioners (*Source*: Volvo Media)

244 Advanced automotive fault diagnosis

Figure 9.66 The reason for SRS! (*Source*: Saab Media)

9.18 Diagnostics – air bags and belt tensioners

9.18.1 Systematic testing

Warning
Careless or incorrect diagnostic work could deploy the air bag causing serious injury. Leave well alone if in any doubt.

The only reported fault for air bags should be that the warning light is staying on! If an air bag has been deployed then all the major components should be replaced. Some basic tests that can be carried out are as follows.

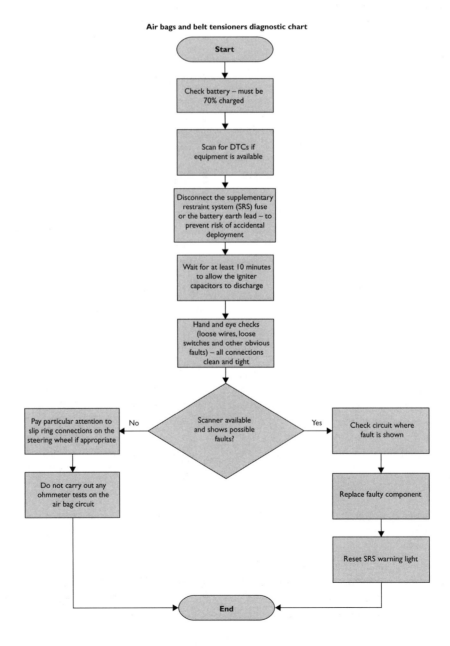

> **Warning**
> Do not carry out any electrical tests on the air bag circuit.

9.18.2 Air bags and belt tensioners fault diagnosis table

Symptom	Possible cause
Warning light on	Wiring fault
	Fuse blown on removed
	ECU fault
	Crash sensor fault
	Igniter fault

9.18.3 Deactivation and activation procedures

> **Warning**
> Do not work on air bag circuits unless fully trained.

Air bag simulators are required to carry out diagnosis and testing of the air bag system. For the frontal air bag(s) this tool is a 2.5 Ω resistor, used to simulate an air bag module connection to the system. Do not short-circuit the air bag module connections with a jumper wire. If a jumper wire is used to short-circuit the air bag module connections, a lamp fault code will be displayed and a diagnostic trouble code logged by the air bag control module.

Ford recommend the following procedure for the air bag system fitted to the Ford Focus.

Deactivation procedure

> **Warning**
> The backup power supply must be depleted, before any work is carried out on the supplementary restraint system. Wait at least one minute after disconnecting the battery ground cable. Failure to follow this instruction could cause accidental air bag deployment and may cause personal injury.

1. Disconnect the battery ground cable.
2. Wait one minute for the backup power supply in the Air Bag Control Module to deplete its stored energy.

> **Warning**
> Place the air bag module on a ground wired bench, with the trim cover facing up to avoid accidental deployment. Failure to follow this instruction may result in personal injury.

3. Remove the driver air bag module from the vehicle.
4. Connect the air bag simulator to the sub-harness in place of the driver air bag module at the top of the steering column.
5. Remove the passenger air bag module.
6. Connect the air bag simulator to the harness in place of the passenger air bag module.
7. Disconnect the driver five-way under seat connector.
8. Connect the air bag simulator to driver five-way under seat floor harness in place of the seat belt pretensioner and side air bag.
9. Disconnect the passenger five-way under seat connector.
10. Connect the air bag simulator to the passenger five-way under seat floor harness in place of the seat belt pretensioner and side air bag.
11. Reconnect the battery ground cable.

Reactivation procedure

> **Warning**
> The air bag simulators must be removed and the air bag modules reconnected when reactivated to avoid non-deployment in a collision. Failure to follow this instruction may result in personal injury.

1. Disconnect the battery ground cable.
2. Wait one minute for the backup power supply in the Air Bag Control Module to deplete its stored energy.
3. Remove the driver air bag simulator from the sub-harness at the top of the steering column.
4. Reconnect and install the driver air bag module.
5. Remove the passenger air bag simulator from the passenger air bag module harness.
6. Reconnect and install the passenger air bag module.
7. Remove the air bag simulator from the driver five-way under seat connector.
8. Reconnect the driver five-way under seat connector.
9. Remove the air bag simulator from the passenger five-way under seat connector.
10. Reconnect the passenger five-way under seat connector.
11. Reconnect the battery ground cable.
12. Prove out the system, repeat the self test and clear the fault codes.

> *Note:* This section is included as general guidance; however, do not assume it is relevant to all vehicles.

> ### Knowledge check questions
>
> To use these questions, you should first try to answer them without help but if necessary, refer back to the content of the chapter. Use notes, lists and sketches to answer them. It is not necessary to write pages and pages of text!
>
> 1. Explain what is meant by 'controller area network (CAN)' and why it is used to connect ECUs or nodes together.
> 2. List in a logical sequence, a series of steps to diagnose why a wiper motor operates on slow speed but not on fast speed.
> 3. Describe how to check that an on/off relay is operating correctly.
> 4. Describe how to test the output of a road speed sensor used as part of a cruise control system.
> 5. Make a sketch of a fuel gauge circuit and describe how to check it for correct operation.

10
Transmission systems

10.1 Manual transmission

10.1.1 Clutch

A clutch is a device for disconnecting and connecting rotating shafts. In a vehicle with a manual gearbox, the driver pushes down the clutch when changing gear to disconnect the engine from the gearbox. It also allows a temporary neutral position for, say, waiting at traffic lights and a gradual way of taking up drive from rest.

The clutch is made of two main parts, a pressure plate and a driven plate. The driven plate, often termed the clutch disc, is fitted on the shaft, which takes the drive into the gearbox. When the clutch is engaged, the pressure plate presses the driven plate against the engine flywheel. This allows drive to be passed to the gearbox. Pushing down the clutch springs the pressure plate away, which frees the driven plate. Figure 10.1 shows some typical clutch components including an electromechanical clutch actuator. The diaphragm type clutch replaced an earlier type with coil springs as it has a number of advantages when used on light vehicles:

- not affected by high speeds (coil springs can be thrown outwards);
- low pedal force making for easy operation;
- light and compact;
- clamping force increases or at least remains constant as the friction lining wears.

The method of controlling the clutch is quite simple. The mechanism consists of either a cable or hydraulic system.

10.1.2 Manual gearbox

The driver changes the gears of a manual gearbox by moving a hand-operated lever called a

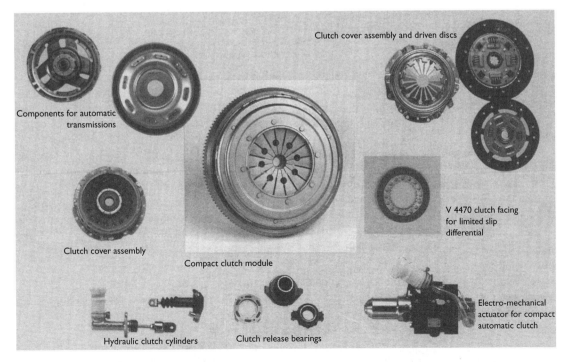

Figure 10.1 Clutch components (*Source*: Valeo)

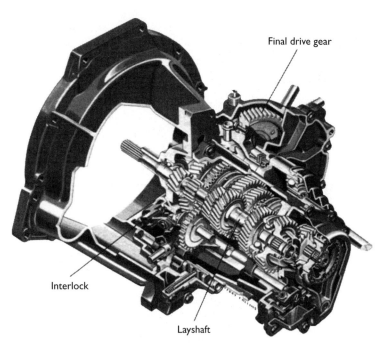

Figure 10.2 Five-speed gearbox (FWD) (*Source*: Ford)

gear stick! All manual gearboxes have a neutral position; three, four, or five forward gears; and a reverse gear. A few even have six forward gears now! The driver puts the gearbox into neutral as the engine is being started, or when a car is parked with the engine left running.

Power travels in to the gearbox via the input shaft. A gear at the end of this shaft drives a gear on another shaft called the countershaft or layshaft. A number of gears of various sizes are mounted on the layshaft. These gears drive other gears on a third motion shaft also known as the output shaft.

The gearbox produces various gear ratios by engaging different combinations of gears. For reverse, an extra gear called an idler operates between the countershaft and the output shaft. It turns the output shaft in the opposite direction to the input shaft. Figure 10.2 shows a front wheel drive gearbox.

Figure 10.3 shows the power flows through this box in each of the different gears. Note how in each case (with the exception of reverse) the gears do not move. This is why this type of gearbox has become known as constant mesh. In other words the gears are running in mesh with each other at all times. Dog clutches are used to select which gears will be locked to the output shaft. These clutches which are moved by selector levers, incorporate synchromesh mechanisms.

A synchromesh mechanism is needed because the teeth of the dog clutches would clash if they met at different speeds. The system works like a friction type cone clutch. The collar is in two parts and contains an outer toothed ring that is spring-loaded to sit centrally on the synchromesh hub. When the outer ring (synchroniser sleeve) is made to move by the action of the selector mechanism the cone clutch is also moved because of the locking keys. The gear speeds up as the cones touch, thus allowing the dog clutches to engage smoothly. A baulking ring is fitted between the cone on the gear wheel and the synchroniser hub. This is to prevent engagement until the speeds are synchronised.

A detent mechanism is used to hold the selected gear in mesh. In most cases this is just a simple ball and spring acting on the selector shaft(s). Figure 10.4 shows a rear wheel drive gearbox with the detent mechanism marked. Gear selection interlocks are a vital part of a gearbox. These are to prevent more than one gear from being engaged at any one time. On the single rail (one rod to change the gears) gearbox shown in the figure, the interlock mechanism is shown at the rear. As the rod is turned (side to side movement of the gear stick) towards first–second, third–fourth or fifth gear positions, the interlock will only engage with either the first–second, third–fourth or fifth gear selectors as appropriate. Equally when any selector clutch is in mesh the interlock will not allow the remaining selectors to change position.

10.1.3 Driveshafts and wheel bearings

Light vehicle driveshafts now fall into one of two main categories, the first being by far the most popular.

Transmission systems **249**

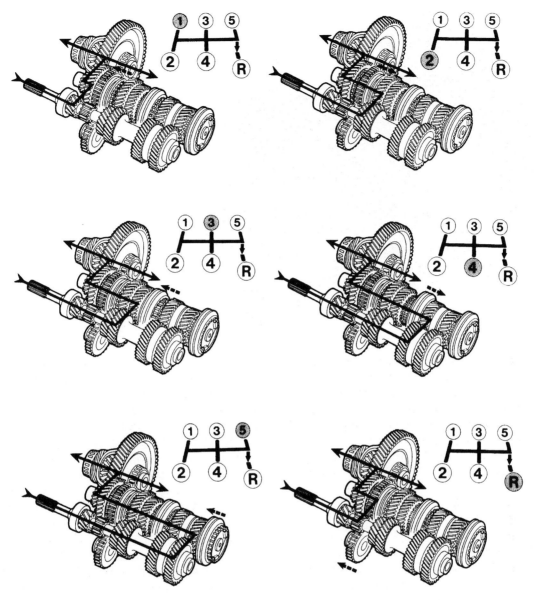

Figure 10.3 Gearbox power flows (*Source:* Ford)

- **Driveshafts with constant velocity joints (FWD)** transmit drive from the output of the final drive to each front wheel. They must also allow for suspension and steering movements.
- **Propshaft with universal joints (RWD)** transmits drive from the gearbox output to the final drive in the rear axle. Drive then continues through the final drive and differential, via two half shafts to each rear wheel. The propshaft must also allow for suspension movements.

Wheel bearings are also very important. They allow smooth rotation of the wheel but must also be able to withstand high stresses such as from load in the vehicle and when cornering.

Front bearings

Seal	Keeps out dirt and water, and keeps in the grease lubrication
Spacer	Ensures the correct positioning of the seal
Inner bearing	Supports the weight of the vehicle at the front, when still or moving. Ball bearings are used for most vehicles with specially shaped tracks for the balls. This is so the bearings can stand side loads when cornering
Swivel hub	Attachment for the suspension and steering as well as supporting the bearings
Outer bearing	As for inner bearing
Drive flange	Runs inside the centre race of the bearings. The wheel is bolted to this flange

250 Advanced automotive fault diagnosis

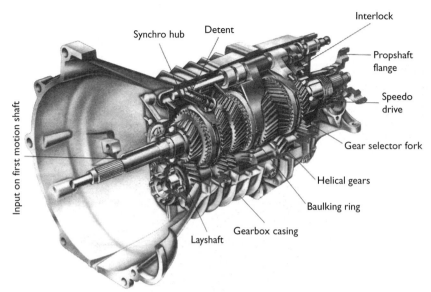

Figure 10.4 Rear wheel drive gearbox

Rear bearings

Stub axle	Solid mounted to the suspension arm, this stub axle fits in the centre of the two bearings
Seal	Keeps out dirt and water, and keeps in the grease lubrication
Inner bearing	Supports the weight of the vehicle at the rear, when still or moving. Ball bearings are used for most vehicles with specially shaped tracks for the balls. This is so the bearings can stand side loads when cornering
Spacer	To ensure the correct spacing and pressure between the two bearings
Drum	For the brakes and attachment of the wheel
Outer bearing	As for inner bearing
Washer	The heavy washer acts as a face for the nut to screw against
Castle nut and split pin	Holds all parts in position securely. With this type of bearing, no adjustment is made because both bearings are clamped on to the spacer. Some older cars use tapered bearings and adjustment is very important
Grease retainer cap	Retains grease, but should not be over packed. Also keeps out the dirt and water

10.1.4 Final drive and differential

Because of the speed at which an engine runs, and in order to produce enough torque at the road wheels, a fixed gear reduction is required. This is known as the final drive, and consists of just two gears. These are fitted after the output of the gearbox, on front wheel drive, or in the rear axle after the propshaft on rear wheel drive vehicles. The gears also turn the drive through ninety degrees on rear wheel drive vehicles. The ratio is normally about 4:1; in other words when the gearbox output is turning at 4000 rev/min the wheels will turn at 1000 rev/min.

Many cars now have a transverse engine, which drives the front wheels. The power of the engine therefore does not have to be carried through a right angle to the drive wheels. The final drive contains ordinary reducing gears rather than bevel gears.

The differential is a set of gears that divide the torque evenly between the two drive wheels. The differential also allows one wheel to rotate faster than the other does when necessary. When a car goes around a corner, the outside drive wheel travels further than the inside one. The outside wheel must therefore rotate faster than the inside one to cover the greater distance in the same time.

Some higher performance vehicles use limited slip differentials. The clutch plates are connected to the two output shafts and hence if controlled will in turn control the amount of slip. This can be used to counteract the effect of one wheel losing traction when high power is applied.

Differential locks are used on many off-road type vehicles. A simple dog clutch or similar device prevents the differential action. This allows far better traction on slippery surfaces.

10.1.5 Four-wheel drive systems

Four-wheel drive (4 WD) provides good traction on rough or slippery surfaces. Many cars are now available with 4 WD. In some vehicles, the driver can switch between four-wheel drive and two-wheel drive (2 WD). A vehicle with four-wheel drive delivers power to all four wheels. A transfer box is used to distribute the power between

Figure 10.5 Ford Maverick 4WD transmission

the front and rear wheels. Figure 10.5 shows the layout of the system used on the Ford Maverick. Note the extra components compared to a normal two-wheel drive system:

- **transfer gearbox** to provide an extra drive output;
- **differential on each axle** to allow cornering speed variations;
- **centre differential** to prevent wind up between the front and rear axles;
- **extra drive shafts** to supply drive to the extra axle.

The transfer gearbox on some vehicles may also contain extra reduction gears for low ratio drive.

One problem to overcome, however, with 4WD is that if three differentials are used then the chance of one wheel slipping actually increases. This is because the drive will always be transferred to the wheel with least traction – like running a 2WD car with one driving wheel jacked up. To overcome this problem and take advantage of the extra traction available a viscous coupling is combined with an epicyclic gear train to form the centre differential.

The drive can now be distributed proportionally. A typical value is about 35% to the front and 65% to the rear wheels. However, the viscous clutch coupling acts so that if a wheel starts to slip, the greater difference in speed across the coupling will cause more friction and hence more drive will pass through the coupling. This tends to act so that the drive is automatically distributed to the most effective driving axle. A 'Hyvo' or silent chain drive is often used to drive from the transfer box.

10.2 Diagnostics – manual transmission

10.2.1 Systematic testing

If the reported fault is a slipping clutch proceed as follows.

1. Road test to confirm when the fault occurs.
2. Look for oil leaking from the bell housing or general area of the clutch. Check adjustment if possible.
3. If adjustment is correct then the clutch must be examined.
4. In this example the clutch assembly must be removed for visual examination.
5. Replace parts as necessary; this is often done as a kit comprising the clutch plate and cover as well as a bearing in some cases.
6. Road test and check operation of all the transmission.

10.2.2 Test equipment

Note: You should always refer to the manufacturer's instructions appropriate to the equipment you are using.

Stethoscope (Figure 10.6)

This is a useful device that can be used in a number of diagnostic situations. In its basic form it is a long screwdriver! The probe (or screwdriver blade) is placed near the suspected component such as a bearing. The ear piece (or screwdriver handle placed next to the ear) amplifies the sound. Take care though; even a good bearing can sound rough using this method. Compare a known good noise with the suspected one.

10.2.3 Test results

Some of the information you may have to get from other sources such as data books or a workshop manual is listed in the following table.

Test carried out	Information required
Backlash or freeplay	Backlash data is often given, as the distance component will move. The backlash between two gears for example should be very small
Overdrive operation	Which gears the overdrive is meant to operate in!

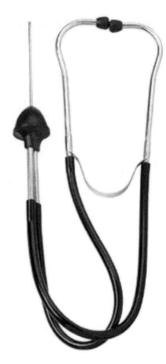

Figure 10.6 Stethoscope

10.2.4 Manual transmission fault diagnosis table 1

Symptom	Possible causes or faults	Suggested action
Clutch slipping	Clutch worn out	Renew
	Adjustment incorrect	Adjust or check auto-adjuster
	Oil contamination	Rectify oil leak – clutch may also need to be renewed
Jumps out of gear	Gearbox detent fault	Gearbox may require overhaul
Noisy when changing gear	Synchromesh worn	Gearbox may require overhaul
Rapid knocking noise when cornering	Driveshaft CV joints worn or without lubrication	Renew or lubricate joint
		Ensure gaiter is in place and in good condition
Whining noise	Wheel bearing worn	Renew
	Other bearings	Investigate and renew if possible
Difficult to change gear	Clutch out of adjustment	Adjust or check auto-adjuster
Clutch hydraulic fault		Check system for air and/or leaks
Gearbox selectors worn		Gearbox may require overhaul

10.2.5 Manual gearbox fault diagnosis table 2

Symptom	Possible cause
Noisy in a particular gear (with engine running)	Damaged gear Worn bearing
Noisy in neutral (with engine running)	Input shaft bearings worn (goes away when clutch is pushed down?) Lack of lubricating oil Clutch release bearing worn (gets worse when clutch is pushed down?)
Difficult to engage gears	Clutch problem Gear linkage worn or not adjusted correctly Work synchromesh units Lack of lubrication
Jumps out of gear	Gear linkage worn or not adjusted correctly Worn selector forks Detent not working Weak synchromesh units
Vibration	Lack of lubrication Worn bearings Mountings loose
Oil leaks	Gaskets leaking Worn seals

10.2.6 Clutch faults diagnosis table

Symptom	Possible cause
No pedal resistance	Broken cable Air in hydraulic system Hydraulic seals worn Release bearing or fork broken Diaphragm spring broken
Clutch does not disengage	As above Disc sticking in gearbox splines Disc sticking to flywheel Faulty pressure plate
Clutch slip	Incorrect adjustment Worn disc linings Contaminated linings (oil or grease) Faulty pressure plate
Judder when engaging	Contaminated linings (oil or grease) Worn disc linings Distorted or worn pressure plate Engine mountings worn, loose or broken Clutch disc hub splines worn
Noisy operation	Broken components Release bearing seized Disc cushioning springs broken
Snatching	Disc cushioning springs broken Operating mechanism sticking (lubrication?)

10.2.7 Drive shafts fault diagnosis table

Symptom	Possible cause
Vibration	Incorrect alignment of propshaft joints Worn universal or CV joints Bent shaft Mountings worn
Grease leaking	Gaiters split or clips loose
Knocking noises	Dry joints Worn CV joints (gets worse on tight corners)

10.2.8 Final drive fault diagnosis table

Symptom	Possible cause
Oil leaks	Gaskets split Driveshaft oil seals Final drive output bearings worn (drive-shafts drop and cause leaks)
Noisy operation	Low oil level Incorrect preload adjustment Bearings worn
Whining noise	Low oil level Worn differential gears

10.3 Automatic transmission

10.3.1 Introduction

An automatic gearbox contains special devices that automatically provide various gear ratios, as they are needed. Most automatic gearboxes have three or four forward gears and reverse. Instead of a gearstick, the driver moves a lever called a selector. Most automatic gearboxes now have selector positions for park, neutral, reverse, drive, 2 and 1 (or 3, 2 and 1 in some cases). The engine will only start if the selector is in either the park or neutral position. In park, the drive shaft is locked so that the drive wheels cannot move. It is now quite common when the engine is running to be able to move the selector out of park only if you are pressing the brake pedal. This is a very good safety feature as it prevents sudden movement of the vehicle.

For ordinary driving, the driver moves the selector to the drive position. The transmission starts out in the lowest gear and automatically shifts into higher gears as the car picks up speed. The driver can use the lower positions of the gearbox for going up or down steep hills or driving through mud or snow. When in position 3, 2, or 1, the gearbox will not change above the lowest gear specified.

10.3.2 Torque converter operation

The torque converter is a device that almost all automatic transmissions now use. It delivers power from the engine to the gearbox like a basic fluid flywheel but also increases the torque when the car begins to move. The torque converter resembles a large doughnut sliced in half. One half, called the pump impeller, is bolted to the drive plate or flywheel. The other half, called the turbine, is connected to the gearbox-input shaft. Each half is lined with vanes or blades. The pump and the turbine face each other in a case filled with oil. A bladed wheel called a stator is fitted between them.

The engine causes the pump to rotate and throw oil against the vanes of the turbine. The force of the oil makes the turbine rotate and send power to the transmission. After striking the turbine vanes, the oil passes through the stator and returns to the pump. When the pump reaches a specific rate of rotation, a reaction between the oil and the stator increases the torque. In a fluid flywheel oil returning to the impeller tends to slow it

Figure 10.7 This cut-away gearbox shows the torque converter components

down. In a torque converter the stator or reactor diverts the oil towards the centre of the impeller for extra thrust. Figure 10.7 shows a gearbox with a cut-away torque converter.

When the engine is running slowly, the oil may not have enough force to rotate the turbine. But when the driver presses the accelerator pedal, the engine runs faster and so does the impeller. The action of the impeller increases the force of the oil. This force gradually becomes strong enough to rotate the turbine and move the vehicle. A torque converter can double the applied torque when moving off from rest. As engine speed increases the torque multiplication tapers off until at cruising speed there is no increase in torque. The reactor or stator then freewheels on its one-way clutch at the same speed as the turbine.

The fluid flywheel action reduces efficiency because the pump tends to rotate faster than the turbine. In other words some slip will occur (about 2%). To improve efficiency, many transmissions now include a lock-up clutch. When the pump reaches a specific rate of rotation, this clutch locks the pump and turbine together, allowing them to rotate as one.

10.3.3 Epicyclic gearbox operation

Epicyclic gears are a special set of gears that are part of most automatic gearboxes. They consist of three elements:

- a sun gear, located in the centre;
- the carrier that holds two, three, or four planet gears, which mesh with the sun gear and revolve around it;
- an internal gear or annulus is a ring with internal teeth, which surrounds the planet gears and meshes with them.

Any part of a set of planetary gears can be held stationary or locked to one of the others. This will produce different gear ratios. Most of the automatic gearboxes of the type shown earlier have two sets of planetary gears that are arranged in line as shown in Figure 10.8. This provides the necessary number of gear ratios. The appropriate elements in the gear train are held stationary by a system of hydraulically operated brake bands and clutches. These are worked by a series of hydraulically operated valves in the lower part of the gearbox. Oil pressure to operate the clutches and brake bands is supplied by a pump. The supply for this is the oil in the sump of the gearbox.

Unless the driver moves the gear selector to operate the valves, automatic gear changes are made depending on just two factors:

- **throttle opening** a cable is connected from the throttle to the gearbox;
- **road speed** when the vehicle reaches a set speed a governor allows pump pressure to take over from the throttle.

The cable from the throttle also allows a facility known as 'kick down'. This allows the driver to change down a gear, such as for overtaking, by pressing the throttle all the way down.

Many modern automatic gearboxes now use gears the same as in manual boxes. The changing of ratios is similar to the manual operation except that hydraulic clutches and valves are used.

10.3.4 Constantly variable transmission

Figure 10.9 shows the Ford CTX (Constantly variable TransaXle) transmission. This kind of automatic transmission called a continuously variable transmission uses two pairs of cone-shaped pulleys connected by a metal belt. The key to this system is the high friction drive belt. The belt is made from high performance steel and transmits drive by thrust rather than tension. The ratio of the rotations, often called the gear ratio, is determined by how far the belt rides from the centres of the pulleys. The transmission can produce an unlimited number of ratios. As the car changes speed the ratio is continuously adjusted. Cars with this system are said to use fuel more efficiently than cars with set gear ratios. Within the gearbox hydraulic control is used to move the pulleys and hence change the drive ratio. An epicyclic gear set is used to provide a reverse gear as well as a fixed ratio.

Transmission systems 255

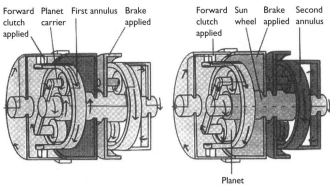

First gear: the line shows power flow through the two epicyclic gears. Driving the front annulus and braking the rear planet carrier gives two gear reductions

Second gear: one gear reduction is achieved by driving the first annulus and braking the common sun wheel

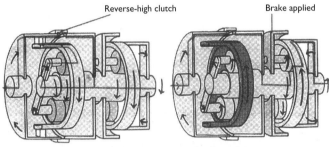

Top gear: all gears are held stationary in relation to each other, and the complete assembly rotates to provide direct drive

Reverse: braking the second planet carrier causes the rear annulus and output shaft to turn in the opposite direction

Figure 10.8 Two epicyclic gear sets can produce three forward gears and reverse

Figure 10.9 Ford CTX transmission

10.3.5 Electronic control of transmission

The main aim of electronically controlled automatic transmission (ECAT) is to improve on conventional automatic transmission in the following ways:

- gear changes should be smoother and quieter;
- improved performance;
- reduced fuel consumption;
- reduction of characteristic changes over system life;
- increased reliability.

The important points to remember are that gear changes and lock-up of the torque converter are controlled by hydraulic pressure. In an ECAT system electrically controlled solenoid valves can influence this hydraulic pressure. Most ECAT systems now have a transmission ECU that is in communication with the engine control ECU.

With an ECAT system the actual point of gearshift is determined from pre-programmed memory within the ECU. Data from other sensors is also taken into consideration. Actual gearshifts are initiated by changes in hydraulic pressure, which is controlled by solenoid valves.

The two main control functions of this system are hydraulic pressure and engine torque. A temporary reduction in engine torque during gear shifting allows smooth operation. This is because the peak of gearbox output torque which causes the characteristic surge during gear changes on conventional automatics is suppressed. Because of these control functions smooth gearshifts are possible and, due to the learning ability of some ECUs, the characteristics remain throughout the life of the system.

The ability to lock-up the torque converter has been used for some time even on vehicles with more conventional automatic transmission. This gives better fuel economy, quietness and improved driveability. Lock-up is carried out using a hydraulic valve, which can be operated gradually to produce a smooth transition. The timing of lock-up is determined from ECU memory in terms of the vehicle speed and acceleration.

10.3.6 Semi-automatic transmission

A number of different types of semi-automatic transmission are either in use or under development. An interesting system is the electronically controlled clutch. This technology is a spin off from formula 1 and top rally cars. The electronic clutch was developed for these racing vehicles to improve the get away performance and speed of gear changes.

For production vehicles a system has been developed which can interpret the driver's intention. With greater throttle openings the clutch operation changes to prevent abuse and driveline damage. Electrical control of the clutch release bearing position is by a solenoid actuator or electric motor with worm gear reduction. This allows the time to reach the ideal take off position to be reduced and the ability of the clutch to transmit torque to be improved.

This technique has now been developed to the stage where it will operate the clutch automatically as the stick is moved to change gear. Sensors are fitted to monitor engine and road speed as well as gear position. The Ferrari Mondiale uses a system similar to the one described here.

10.4 Diagnostics – automatic transmission

10.4.1 Systematic testing

If the reported fault is that the kick down does not operate proceed as follows.

1. Road test to confirm the problem.
2. Is the problem worse when the engine is hot? Check the transmission fluid level! Has work been done to the engine?
3. If fluid level is correct then you must investigate further. Work on the engine may have disturbed the kick down cable.
4. Check the adjustment/fitting of the kick down cable.
5. Adjust if incorrect.
6. Run and repeat road test.

10.4.2 Test equipment

Note: You should always refer to the manufacturer's instructions appropriate to the equipment you are using.

Revcounter

A revcounter may be used during a stall test to check the operation of the torque converter and the automatic gearbox.

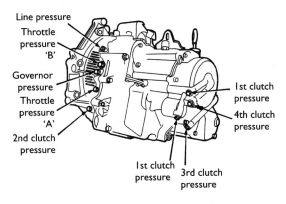

Figure 10.10 Transmission pressure testing

Pressure gauge

This is a standard type of gauge but with suitable adapters for connection to a gearbox. Figure 10.10 shows where various tests can be carried out on an automatic gearbox.

10.4.3 Test results

Some of the information you may have to get from other sources such as data books or a workshop manual is listed in the following table.

Test carried out	Information required
Stall test	Highest revs expected and the recommended duration of the test
Kick down test	Rev range in which the kick down should operate. For example, above a certain number of engine revs it may not be allowed to work

10.4.4 Automatic gearbox fault diagnosis table 1

Symptom	Possible faults	Suggested action
Slip, rough shifts, noisy operation or no drive	There are numerous faults that can cause these symptoms	Check the obvious such as fluid levels and condition. Carry out a stall test. Refer to a specialist if necessary

10.4.5 Automatic gearbox fault diagnosis table 2

Symptom	Possible cause
Fluid leaks	Gaskets or seals broken or worn Dip stick tube seal Oil cooler or pipes leaking
Discoloured and/or burnt smell to fluid	Low fluid level Slipping clutches and/or brake bands in the gearbox Fluid requires changing
Gear selection fault	Incorrect selector adjustment Low fluid level Incorrect kick down cable adjustment Load sensor fault (maybe vacuum pipe, etc.)
No kick down	Incorrect kick down cable adjustment Kick down cable broken Low fluid level
Engine will not start or starts in gear	Inhibitor switch adjustment incorrect Faulty inhibitor switch Incorrect selector adjustment
Transmission slip, no drive or poor quality shifts	Low fluid level Internal automatic gearbox faults often require the attention of a specialist

10.4.6 ECAT fault diagnosis

Carrying out a test on an electronically controlled automatic transmission (ECAT) is like any other 'black box' system. As discussed previously, do consider all the mechanical aspects of the system before looking for electrical faults.

Looking at the ECAT block diagram (Figure 10.11) it can be seen that the system can be considered as a series of inputs and outputs. Many of the sensors (inputs) can be tested using normal equipment such as a scope or multimeter.

The difficulty is often that the sensors and actuator are not easily accessible. Figure 10.12 shows an electrohydraulic valve block, sensors and ECU combines and illustrates this quite well. The diagnostic link is almost essential for diagnosing these systems. Of course if the complete unit is replaced then the fault will more than likely be fixed – as long as it was an electrical problem of course!

A typical test procedure for an ECAT would be as follows.

1. Check fluid condition and level.
2. Check for leaks.
3. Look for other symptoms on the vehicle (an engine problem may be connected).
4. Road test and check operation of all gears.
5. Pressure test the system pump pressure*.
6. Pressure test the converter pressure*.
7. Interrogate fault codes if possible.
8. Check as indicated by the DTC.

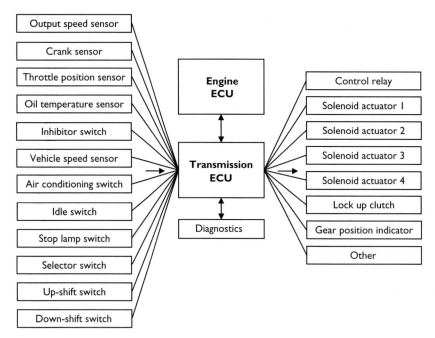

Figure 10.11 ECAT block diagram showing inputs and outputs

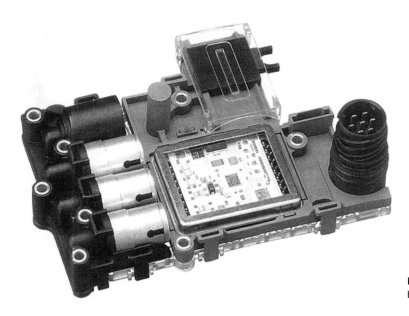

Figure 10.12 Electrohydraulic valve block, ECU and sensors combined

9. Carry out a stall test.
10. Check sensor signals using a scope.
11. Check sensors such as control switches.
12. Check resistance of solenoid actuators.

The above list is very generic but will serve as a useful reminder of the process.

*When carrying out pressure testing, manufacturers data is essential. However, as a guide the following figures are typical.

	Selector position	Gear	Rev/min (approx.)	Bar (approx.)
Pump	D	1	800	7
		2, 3, 4 (as per the system)	4000	5
	R	R	700–4000	12
Converter	D	4 (or top)	With clutch locked	0.7 max

10.4.7 Automatic transmission – stall test

To assist with the diagnosis of automatic transmission faults a stall test is often used. The duration of a stall test must not be more than about seven seconds. You should also allow at least two minutes before repeating the test. Refer to manufacturer's recommendations if necessary.

> **Warning**
> If the precautions mentioned are not observed, the gearbox will overheat.

The function of this test is to determine the correct operation of the torque converter and that there is no transmission clutch slip. Proceed as follows.

1. Run engine up to normal operating temperature by road test if possible.
2. Check transmission fluid level and adjust if necessary.
3. Connect a revcounter to the engine.
4. Apply handbrake and chock the wheels.
5. Apply foot brake, select 'D' and fully press down the throttle for about seven seconds.
6. Note the highest rev/min obtained (2500 to 2750 is a typically acceptable range).
7. Allow two minutes for cooling and then repeat the test in '2' and 'R'.

> *Knowledge check questions*
>
> To use these questions, you should first try to answer them without help but if necessary, refer back to the content of the chapter. Use notes, lists and sketches to answer them. It is not necessary to write pages and pages of text!
>
> 1. Describe how to use a road test to diagnose a suspected CV joint fault.
> 2. Explain why a stall test may be used to diagnose automatic transmission faults.
> 3. List in a logical sequence, a procedure for checking the operation of an electronically controlled automatic transmission.
> 4. Describe a procedure used to test for a slipping clutch.
> 5. Describe a series of steps that could be used to diagnose the source of a 'rumbling' noise from a transmission system.

11
Conclusion, web resources and developments

11.1 Introduction

11.1.1 Overview

If like me you continue to be fascinated by the technology of the automobile, then you will also be interested in the future technology. Figure 11.1 is an illustration of how technology has developed and perhaps how it will continue to do so.

To keep up to date you will have to keep learning! But remember, the way in which you apply your knowledge is just as important as the knowledge itself. Logic is the key to diagnostics.

This final chapter includes some general information about web resources and software but also gives an overview of diagnostics to date – and possible future developments.

11.1.2 The future of diagnostics

What does the pace of technology change mean for the diagnostic technician?

On-board diagnostics will continue to develop but there will always be a need for the highly skilled diagnostic technician. In the recent past, due to the cost and complexity of some equipment, only the large main dealers could afford it. However, it is now possible to buy excellent scanning or code reading equipment for just a few hundred pounds/dollars. If you continue to keep up to date with the technology, learn how to use good quality basic test and scanning equipment, and apply the diagnostic processes highlighted in this book, you will never be short of work!

11.2 Web contacts

11.2.1 Introduction

Like almost everybody, you are no doubt already interested in the Internet and the World Wide Web. In this short section I have highlighted some of the resources available to you on the net.

All the major vehicle manufacturers have web pages as do most of the parts suppliers. There are user groups, forums for discussions and many other sources of information, or lets admit it – sites that are just good fun! So go ahead and surf the net. I would be most pleased to meet you at my site where you will find up to date information about my books, new technologies and links to just about everywhere (Figure 11.3). Just point your browser to: www.automotive-technology.co.uk

All the links in the next section can be accessed from here as well as many more. It will also be kept up-to-date as I find more interesting places to visit. Please let me know if you have any suggestions for useful links – I am mostly interested in sites that

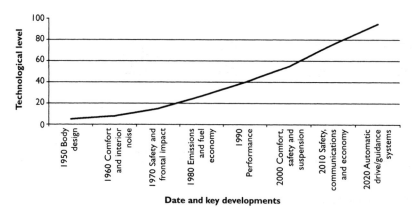

Figure 11.1 Technological developments

provide good technical information, not just advertising material! My email address is: tom.denton@automotive-technology.co.uk

11.2.2 Web site addresses

Here are just a few selected web addresses that I have found useful and interesting. Remember that there are lots more on my home page: www.automotive-technology.co.uk

www.auto-training.co.uk
Excellent multimedia training on this site designed to match UK qualifications

www.ase-autotraining.com
Excellent multimedia training available on this site designed to match USA qualifications

www.onboarddiagnostics.co.uk
This site is a great reference source for OBD information and equipment

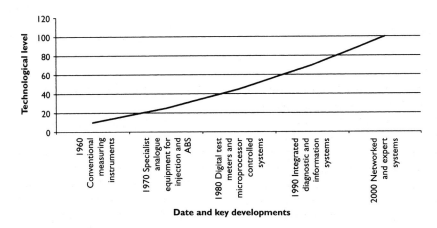

Figure 11.2 Developments of diagnostic test equipment

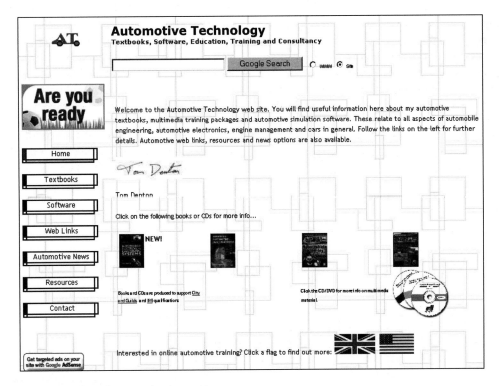

Figure 11.3 Front page from the site at the above address

www.autoelex.co.uk
Dave Rogers' web page, some useful links and general automotive information

www.picotech.com
Source of the automotive scope equipment and software used extensively in parts of this book; highly recommended

www.sae.org
Society of Automotive Engineers (as in the SAE on the oil can!). US-based but a great site with 'higher level' information, books and papers

www.howstuffworks.com/automotive
Some excellent descriptions of how automotive systems work – and lots more besides.

11.3 Future developments in diagnostic systems

11.3.1 Introduction

The latest generation of OBD is a very sophisticated and capable system for detecting emissions problems. However, it is necessary to get the driver of the vehicle to do something about the problem. With respect to this aspect, OBD2/EOBD is no improvement over OBD1 unless there is some enforcement capability.

11.3.2 OBD3

Plans for OBD3 are currently under consideration, which would take OBD2 a step further by adding the possibility of remote data transfer. This would involve using remote transmitter/transponder technology similar to that which is already being used for automatic electronic toll collection.

An OBD3 equipped vehicle would be able to report emissions problems directly back to a regulatory authority. The transmitter/transponder would communicate the Vehicle Identification Number (VIN) and any diagnostic codes that have been logged. The system could be set up to automatically report an emissions problem the instant the MIL light comes on, or alternatively, the system could respond to answer a query about its current emissions performance status.

What makes this approach so attractive is its efficiency, with remote monitoring via the on-board telemetry, the need for periodic inspections could be eliminated because only those vehicles that reported problems would have to be tested. The regulatory authorities could focus their efforts on vehicles and owners who are actually causing a violation rather than just random testing. It is clear that with a system like this, much more efficient use of available regulatory enforcement resources could be implemented, with a consequential improvement in the quality of our air.

An inevitable change that could come with OBD3 would be even closer scrutiny of vehicle emissions. The misfire detection algorithms currently required by OBD2 only watches for misfires during driving conditions that occur during the prescribed driving cycles. It does not monitor misfires during other engine operating modes like full load for example. More sophisticated methods of misfire detection (as previously discussed in Chapter 7) will become common place. These systems can feed back other information to the ECU about the combustion process; for example, the maximum cylinder pressure, detonation events or work done (via an Indicated Mean Effective Pressure (IMEP) calculation). This adds another dimension to the engine control system allowing greater efficiency and more power from any given engine design by just using more sophisticated ECU control strategy.

Future OBD system will undoubtedly incorporate new developments in sensor technology. Currently the evaluation is done via sensors monitoring emissions indirectly. Clearly an improvement would be the ability to measure exhaust gas composition directly via on-board measurement (OBM) systems. This is more in keeping with emission regulation philosophy and would overcome the inherent weakness of current OBD systems; that is, they fail to detect a number of minor faults that do not individually activate the MIL, or cause excessive emissions but whose combined effect is to cause the production of excess emissions.

The main barrier is the lack of availability of suitably durable and sensitive sensors for CO, NOx and HC. Some progress has been made with respect to this and some vehicles are now being fitted with NOx sensors. Currently there does appear to be gap between the laboratory based sensors used in research and reliable mass produced units that could form the basis of an on-board monitoring (OBM) system. The integration of combustion measurement in production vehicles produces a similar problem.

11.3.3 Diesel engines

Another development for future consideration is the further implementation of OBD for diesel

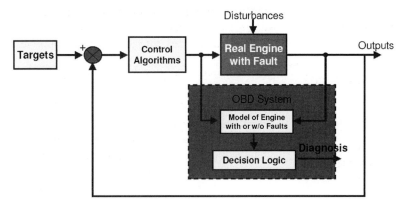

Figure 11.4 Model based calibration of an OBD system

engines. As diesel engine technology becomes more sophisticated, so does the requirement for OBD. In addition, emission legislation is driving more sophisticated requirements for after treatment of exhaust gas. All of these sub-systems are to be subjected to checking via the OBD system and present their own specific challenges. For example, the monitoring of exhaust after treatment systems (particulate filters and catalysts) in addition to more complex EGR and air management systems.

11.3.4 Rate based monitoring

Rate based monitoring will be more significant for future systems which allows in-use performance ratio information to be logged. It is a standardised method of measuring monitoring frequency and filters out the affect of short trips, infrequent journeys etc. as factors which could affect the OBD logging and reactions. It is an essential part of the evaluation where driving habits or patterns are not known and it ensures that monitors run efficiently in use and detect faults in a timely and appropriate manner. It is defined as:

$$\text{Minimum frequency} = N/D$$

Where N = number of times a monitor has run, D = number of times vehicle has been operated.

11.3.5 Model based development

A significant factor in the development of any future system will be the implementation of the latest technologies with respect to hardware and software development. Model based development and calibration of the system will dramatically reduce the testing time by reducing the number of test iterations required. This technique is quite common for developing engine specific calibrations for ECUs during the engine development phase (Figure 11.4).

Hardware-in-loop (HIL) simulation plays a part in rapid development of any hardware. New hardware can be tested and validated under a number of simulated conditions and its performance verified before it even goes near any prototype vehicle. The following tasks can be performed with this technology.

- Full automation of testing for OBD functionality.
- Testing parameter extremes.
- Testing of experimental designs.
- Regression testing of new designs of software and hardware.
- Automatic documentation of results.

11.3.6 Comment

What with the possibility of navigation systems reporting where we are, speed and traffic light cameras everywhere and monitoring systems informing the authorities about the condition of our vehicles, whatever will be next?!

11.4 Software

11.4.1 Introduction

Many web sites have information you can download. Some also include simulation programs, demonstration programs and/or electronic books. This section outlines a small selection of those available.

11.4.2 Automotive technology – Electronics

This simulation program is created by the author of this book. It is a great teaching aid and covers some complex topics in an easy to understand

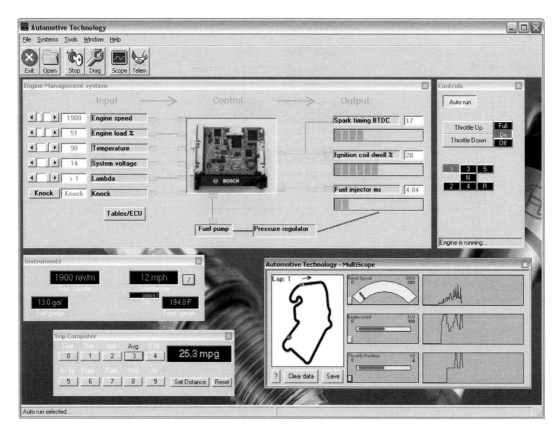

Figure 11.5 A screenshot from the AT program

way (Figure 11.5). It is even possible to set faults in the system and then, using built in test equipment, carry out diagnostic tests!

From the 'Help File':

Click on the key button in the toolbar to open the 'Controls' window (if it is not already visible). Now click on the key button in this window to start the engine. You can also right click the screen and choose to start the engine from there. Don't forget to check that the car is not in gear – or else it won't start!

Next choose from the 'Run' menu the electronic system you would like to operate or work with. I suggest 'Engine Management' is the best place to start.

In common with all of the other simulation windows you will use, you can set or control the operating inputs to the system. For engine management control, these are engine speed, engine load, temperature and so on. The system will react and control the outputs in just the same way as a real vehicle. Be warned the unregistered version runs out of fuel!

An ideal way of learning about automotive electronic systems.

Available as shareware from: www.automotive-technology.co.uk

11.4.3 Other programs

Just two programs have been selected here for the reader to consider further – they are both highly recommended as sources of information relating to diagnostics and/or ways of learning more about automotive systems.

PicoTech www.picotech.com

The software, used extensively in this book, to show automotive waveforms is a great source of information (Figure 11.6). A free demo version can be downloaded or requested on CD.

Auto Solve www.auto-solve.com

This is an e-book presentation on CD that is a very useful source of information and diagnostic

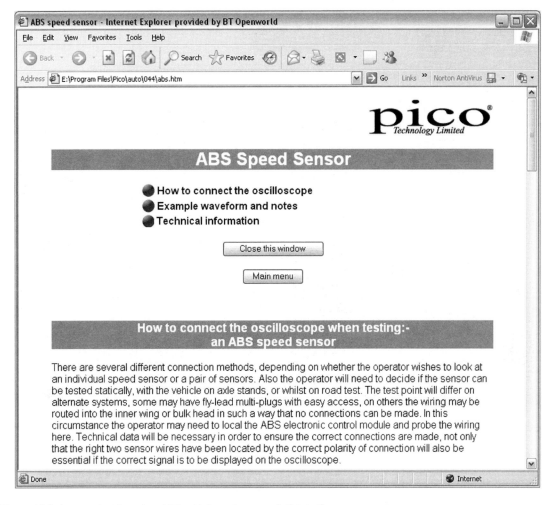

Figure 11.6 A screenshot from the additional information pages built into the scope program

methods (Figure 11.7). It even includes some useful flow charts and lists of codes.

11.5 Summary

Modern motor vehicles are highly sophisticated machines that incorporate the latest developments in electrical, electronics, software and mechanical engineering. They are a marvel of modern engineering practice and truly show how all these technologies can be integrated and harmonised for maximum benefit to the end user.

It is clear that this level of technology produces the safest, quietest and most efficient road vehicles we have ever known. The disadvantage of this level of sophistication really becomes apparent when something goes wrong! Clearly, the more sophisticated the device, the more difficult it will be to repair, or understand in order to repair. It is often the case that it may be perceived that no faults can be found or fixed without specialised manufacturer equipment which is only available at dealers.

This is not the case! The fundamental principles of diagnostics in conjunction with an applied, logical thought process are the most powerful tools that you have. Any specialist equipment will still only be as good as the person using it. Modern vehicle systems are certainly sophisticated but the fundamental principles apply. An ECU is only monitoring voltages from its sensors. These are the inputs; the outputs are voltages and currents which drive actuators (injector, idle speed control valves etc.), they are all the same and applied logic can fix most problems.

Engines and chassis are also complicated subsystems of the vehicle but in all cases the laws of physics apply, and all engines do the same thing in more or less the same way. They are just

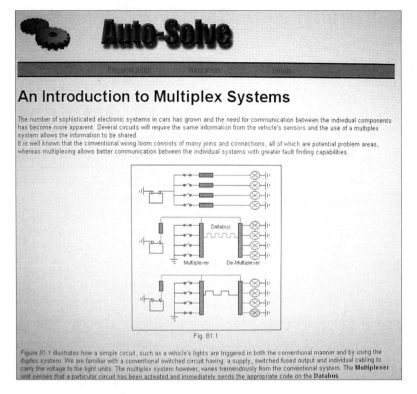

Figure 11.7 One section from the Auto-Solve CD

energy converters! The basic principles are still valid; for example, the ignition still needs to be advanced under cruise conditions when the mixture is weaker, whether this is done mechanically or electronically. Likewise, any electrical circuit not conducting a current – is broken – somewhere!

There are always a few simple rules to follow.

- Don't overlook the obvious.
- Look for simple solutions first.
- Always get as much information as possible up front.
- Never assume anything ... check it yourself.
- Be logical when diagnosing faults.
- Most of all, have confidence in your ability.

Follow these rules, never be afraid to ask for help and learn from the experience. You will build up a portfolio of useful experience and knowledge that will help develop your career as a diagnostic expert. There is nothing that quite beats the feeling of solving a problem, especially if you know that it has puzzled other people before you – to the point that they have given up!

Good Luck!

Index

ABS, 174, 175, 180
ABS speed sensor, 40, 41
AC, 164, 165
Accuracy, 25, 26
Actuator, 49, 94, 240
Advance angle, 108, 135
Aerial, 221
Air bag, 241, 242, 243, 244, 245
Air conditioning, 214, 231, 234, 235, 237, 238, 239
Air-cooled system, 148
Air cored gauge, 227
Air flow meter, 40, 42, 89, 92, 124, 141
Air flow sensor, 38, 91
Air supply, 131, 137, 144, 147, 148
Air temperature, 135, 147
Alternator, 4, 58, 159, 164, 165, 166, 167, 168, 228
Ampere hour capacity, 156
Antifreeze, 150, 151
Anti-lock brakes *see* ABS
Anti-roll bar, 192, 193
Automatic gearbox, 253, 254, 257
Automatic temperature control, 236
Auxiliaries, 209, 215
Auxiliary systems, 12, 214, 225

Back probing, 17
Battery, 58, 112, 155, 156, 157, 158, 159, 242
Battery charger, 157, 158
Battery rating, 155
Beam axles, 190
Beam setting, 209
Before top dead centre (BTDC), 101, 111, 114, 135
Belt tensioners, 241, 243, 244, 245
Bimetal, 227
Black box, 13, 14, 231, 240
Block diagram, 13, 19, 124, 175, 229, 243, 257, 258
Blower motors, 233
Bonding, 221
Boost charging, 157
(Booster) brake servo, 171
Brake, 169, 171, 172, 177
Brake adjusters, 170
Brake disk, 172
Brake fluid, 172
Brake lights, 202, 212, 213
Brake roller, 172, 173
Braking efficiency, 171, 172
Bypass filtration, 152

California Air Resources Board *see* CARB
Calliper, 170
Camber, 184, 187
Camshaft drives, 99
Camshaft sensor, 42, 114
CAN, 31, 36, 37, 59, 60, 201, 202
CAN controller, 202
CAN diagnostics, 202
CAN-High, 59, 60, 202
CAN-Low, 59, 60, 202
CAN signals, 59, 202, 203
Capacitors, 197, 221
CARB, 62, 63, 74, 75
Carbon dioxide (CO_2), 34, 62
Carbon monoxide *see* CO
Carburation, 104, 105
Carburettor, 103, 104, 106
Castor, 184, 185
Catalytic converter, 62, 121, 130, 146
Charging, 156, 157, 159, 163, 166, 167, 168, 220
Charging circuits, 166, 167
Chip, 199, 200
Choke, 104, 105, 106, 199
Circuit, 14, 166, 197, 199, 200, 204, 211, 229
Circuit testing, 66, 163, 168, 209, 215
Closed loop, 19, 46, 68, 92, 110, 111, 133
Closed loop control, 110, 133, 239
Clutch, 161, 247
CO, 34, 62, 73, 120, 130, 146, 262
Coil, 55, 56, 108, 114, 117
Cold cranking amps (CCA), 156
Colour codes, 15
Common rail (CR) diesel, 36, 51, 52, 129
Common rail, 51, 52, 129
Compression, 35, 59, 78, 98, 101, 132
Compression tester, 35, 101
Condensation, 234
Condenser, 234, 235
Constant current, 157
Constant energy, 55, 109, 111, 136
Constant velocity joints, 249
Constant voltage, 157, 165, 227
Constant voltage charging, 157, 164
Constantly variable TransaXle *see* CTX
Constantly variable transmission, 254
Control maps, 136
Controller area network *see* CAN
Coolant, 148, 149, 150
Coolant temperature, 44, 87, 111, 126, 135, 236
Cooling, 115, 148, 149, 150, 151

268 Index

Cooling fan, 149, 212, 214
CR injector, 52, 130
Crank sensor, 42, 43, 81
Crankcase ventilation, 153
Crankshaft, 42, 71, 129, 151
Crankshaft position sensor, 114
Crankshaft sensor, 70, 87, 111
Crash sensor, 93, 242, 243
Cruise control, 239, 240, 241
CTX, 254, 255
Cylinder layouts, 99
Cylinder leakage tester, 101
Cylinder pressure sensor, 83

Dampers, 194
Darlington pair, 199
Data sources, 21
DC, 164, 165
Diagnostic, 1, 2, 6, 7, 8, 20, 40, 61, 101, 106, 117, 119, 126, 130, 138, 147, 150, 153, 172, 177, 180, 186, 195, 202, 237, 240, 244, 251, 256, 260
Diagnostic-expert, 266
Diagnostic link connector *see* DLC
Diagnostic routine, 7
Diagnostic trouble code *see* DTC
Dial gauge, 172
Diesel engines, 128, 131, 263
Diesel fuel injection pump, 127
Diesel injection, 127, 129
Differential, 250, 251
Differential locks, 250
Digital instrumentation, 227
Digital multimeter, 12, 17
Diode, 197, 200, 229
Direct ignition, 55, 114
Direct ignition system *see* DIS
DIS, 57, 82, 114, 117
DIS coil, 114, 117
Disc brakes, 169, 170, 171
Display techniques, 230
Distributorless ignition, 57, 112
DLC, 80
Door locking, 223
Driveshafts, 248, 249
Drum brake disc, 169
Drum brakes, 169, 170, 171
DTC, 32, 64, 65, 67, 74
Duty cycle, 54, 95
Dwell angle, 109, 132
Dynamic position sensors, 93

ECU, 25, 40, 49, 51, 52, 53, 54, 69, 111, 123, 124, 126, 128, 135, 136, 176, 211, 222, 224, 227, 234, 236, 242, 256, 265
EGR, 71, 72, 73, 84, 120, 130, 133
EGR valve, 97, 120
Electric horns, 212
Electric seat, 221, 222
Electric window, 223, 224

Electrical and electronic symbols, 198
Electrode, 115, 116
Electrohydraulic, 258
Electronic components, 197, 200
Electronic control unit *see* ECU
Electronic diesel control (EDC), 129
Electronic ignition, 109, 111, 112
Electronic spark advance (ESA), 111
Electronically controlled automatic transmission (ECAT), 256, 257, 258
Emission, 33, 34, 61, 62, 63, 66, 120, 122, 127, 128, 129, 130, 263
Engine, 11, 14, 32, 34, 78, 83, 84, 98, 99, 101, 102, 108, 111, 124, 132, 135, 138, 143, 152, 153, 160, 233, 253, 262, 265
Engine analysers, 32
Engine control, 25, 133, 256
Engine management, 108, 124, 132, 134, 135, 143, 144
Engine noise, 11, 145
Environmental Protection Agency (EPA), 62, 65, 74
EOBD, 65, 75, 76, 78, 80, 83, 141
Epicyclic gear sets, 255
Epicyclic gearbox, 254
European on-board diagnostics *see* EOBD
EVAP, 68, 84
Evaporation, 234
Evaporative system, 68, 80
Evaporator, 234, 235
Exhaust, 9, 99, 144, 147, 148
Exhaust emissions, 20, 74, 120, 122, 129
Exhaust gas measurement, 34
Exhaust gas recirculation (EGR), 71, 72, 97, 120, 128, 129, 130
Exhaust systems, 144

Fast idle, 105, 124, 125
Fault, 1, 7, 17, 20, 30, 58, 64, 79, 84, 102, 107, 108, 117, 122, 126, 131, 138, 139, 148, 151, 154, 156, 157, 163, 167, 168, 173, 174, 178, 181, 189, 190, 195, 209, 212, 215, 219, 225, 230, 231, 237, 238, 241, 252, 253, 257
Fault code, 17, 30, 80
Feedback resistors, 222
Field effect transistor (FET), 199
Filter, 147
Final drive, 250, 253
Fixed choke carburettor, 104
Four stroke cycle, 98
Four-wheel drive (4 WD), 250, 251
Front axle, 191
Fuel filter, 125, 137
Fuel injection, 23, 122, 123, 124, 125, 126, 129
Fuel injector, 94
Fuel pressure gauge, 106
Fuel pressure regulator, 125, 137
Fuel pump, 106, 124, 136, 141
Fuel rail, 137
Full-flow filtration, 152
Future of diagnostics, 260

Gas analyser 32, 106, 107
Gas discharge headlamps (GDL), 205
Gasoline direct injection (GDI), 76, 137, 138
Gauges, 32, 89, 93, 102, 182, 226
Gearbox, 247, 250, 252, 253, 254, 257
Glow plug, 132

Hall effect, 44, 45, 49, 87, 88, 109, 110, 224
Hall effect distributor, 44, 109, 110
Hand tools, 26
Hazard lights, 212
HC, 34, 62, 73, 120, 130, 146, 262
Headlight aiming, 210
Headway sensor, 240
Healing, 79
Heating, 28, 92, 148, 231, 233, 235, 236
Homogeneous cylinder charge, 138
Hot wire, 42, 91, 92
HT, 55, 56, 134, 136
HVAC, 231, 235, 237,
Hydraulic, 169, 171, 174, 176, 183, 256
Hydraulic accumulator, 171
Hydraulic modulator, 14, 176
Hydrocarbons *see* HC

ICE, 216, 218, 219
Idle control actuator, 125
Idle speed, 105, 106, 124
Idle speed control, 53
Igniter, 241, 242
Ignition, 21, 54, 55, 59, 78, 82, 101, 108, 109, 110, 111, 112, 114, 116, 134, 141, 220
Ignition coil, 54, 55, 114, 117
Ignition secondary, 55, 57
Ignition timing, 101, 108, 109, 111, 114, 117, 133, 136
In car entertainment *see* ICE
Independent front suspension (IFS), 182, 190
Indicators, 212
Inductance, 199
Induction, 98
Inductive, 42, 45, 86, 87, 110
Inductive distributor, 45, 110
Inductive pick-up, 40, 46
Inductors, 199, 221
Inertia starters, 160
Infrared (IR), 130, 218, 223
Injector, 49, 51, 94, 125, 126, 130, 131, 135, 137, 141
Injector tester, 130
Instrumentation, 226, 231
Integrated circuits, 199
Interference suppression, 220
Ionising current, 82

Knock protection, 136
Knock sensor, 46, 47, 91

Laser alignment, 188
LCD, 229, 230
Lead-acid batteries, 155

LED, 28, 200, 205, 207, 218, 229, 230
LED lighting, 205
LH-Jetronic fuel injection, 123
Lighting, 203, 204, 205, 206, 207
Lighting circuit, 204, 205, 206, 209
Limited slip differentials, 250
Liquid crystal display *see* LCD
Loading effect, 27
Logic, 6, 8
Logic probe, 27, 28
Lubrication, 151, 153

Malfunction-indicator lamp *see* MIL
Malfunction indicator light *see* MIL
Manifold, 108, 111, 137, 145
Manifold absolute pressure (MAP), 89, 108, 114
Manual gearbox, 247, 252
Mass airflow sensor, 64
Master cylinder, 169, 170
McPherson strut, 191
Metallic substrates, 121
Meter, 12, 27, 40, 42, 43, 89, 92, 124, 130, 141, 150
MIL, 32, 37, 38, 63, 74, 78, 84
Mirrors, 222
Misfire, 12, 69, 70, 71, 80, 81, 147, 262
Misfire detection, 81, 83, 147, 262
Misfire monitor, 69, 80
Mobile communications, 219
Motorised actuators, 95
Motors, 54, 95, 96, 209, 211, 212, 222, 233
Multi-point injector, 49, 50, 51
Multimeter, 17, 27, 32
Multiplexing, 200
Multipoint, 50, 137
Multipoint injection, 51, 124

Negative temperature coefficient (NTC), 44, 86, 197
Nitrogen oxides *see* NOx
Noise, vibration and harshness (NVH), 9
NOx, 62, 71, 120, 130, 146, 262

OBD, 20, 60, 61, 63, 64, 65, 74, 82, 84, 141
OBD cycles, 79
OBD scanner, 31
OBD1, 64, 76, 262
OBD2, 65, 68, 75, 76
OBD2/EOBD, 80, 262
OBD3, 262
Off-board diagnostics, 20
Oil filters, 152
Oil pressure, 154, 228
Oil pump, 151, 153
On-board diagnostics *see* OBD
On-board monitoring (OBM), 262
Open circuit, 2
Open loop, 19
Operational strategy, 136
Optical sensor, 88
Oscilloscope, 26, 28, 29, 30, 32, 40
Overrun, 105

Oxides of nitrogen *see* NO*x*
Oxygen (O$_2$), 34
Oxygen sensor, 46, 47, 48, 92

Particulate matter (PM), 120, 130
Pattern, 29, 114, 133, 143
PCV, 62, 63, 153
Permanent magnet starters, 161, 162
Phase correction, 136
PicoScope, 29, 30
Piezo-electric, 91, 93
Position memory, 222
Power, 99
Power-assisted steering, 183
Powertrain control module (PCM), 73
Pre-engaged starter, 160, 161
Pressure conscious valve *see* PCV
Pressure gauge, 35, 106, 154, 187, 257
Pressure relief valve, 151, 153
Pressure sensing, 83
Pressure test gauge, 154
Pressure tester, 35, 150
Pressure testing, 34, 257
Primary, 54, 56, 110, 150, 152
Programmed ignition, 111, 112, 113, 136
Propshaft, 249
Pulse generator, 109, 110
Pump, 124, 149, 150, 234, 253
Pyrotechnic inflater, 240, 241

Rack and pinion steering, 183
Rack Steering box, 182
Radial tyre, 182
Radio broadcast data system (RBDS), 217
Radio data system (RDS), 217
Radio interference, 221
Rain sensor, 93, 94
Rate based monitoring, 263
Receiver drier, 238, 239
Rectification, 84, 165
Refrigerant, 234, 235
Relative compression, 59
Report writing, 3
Reserve capacity, 156
Resistors, 91, 197, 221, 222, 233
Resolution, 26, 63
Revcounter, 256
Ripple, 58
Road speed (Hall effect), 49
Road test, 10, 64, 150, 187
ROM, 111, 123, 126, 133, 135, 228
Rotary idle actuator, 95

SAE, 21, 36, 64, 66, 156
Scanner/fault code-reader, 30
Scope, 14, 29, 33, 60, 265
Screening, 221
Scrub radius, 186
Seat belt tensioners, 243
Seat heating, 236

Security, 176, 182, 216, 218, 219
Security and communications, 216
Semi-trailing arms, 192
Sensors, 13, 19, 31, 40, 44, 46, 47, 68, 74, 86, 111, 121, 176, 218, 224, 228, 236, 257, 265
Sensors and actuators, 86, 124, 141, 257
Serial communication, 30
Shock absorbers *see* Dampers
Short circuit, 2, 8, 13
Sight glass, 238, 239
Silencers, 145
Single-point injection, 49, 123, 143
Single point injector, 49, 50
Six-stage process, 6
Smog, 61, 65
Smoke, 131
Smoke meter, 130, 131
Society of Automotive Engineers *see* SAE
Software, 30, 31, 32, 36, 263
Solenoid, 94, 96, 160
Solenoid actuators, 94
Spark ignition, 59, 78
Spark plug, 56, 115, 116, 117, 119
Spark plug heat range, 115
Speakers, 216, 217
Speed sensor, 40, 49, 141, 176, 240
Springs, 193
Starter, 59, 158, 159, 160
Starter circuit, 159, 160, 161, 163
Starting, 159, 163
Stationary current cut off, 110
Stator, 253
Steering rack, 182, 183
Stepper motor, 54, 55, 96, 106
Stethoscope, 252
Strain gauge pressure sensor, 90
Stratified charge, 138
Sunroof, 222, 223
Supplementary restraint system (SRS), 243, 244
Suspension, 11, 190, 193, 195
Swivel axis, 185
Symbols, 63, 197, 198
Symptom(s), 1, 2, 8, 33
Synchromesh, 248
System, 2, 14, 18, 19, 54, 68, 72, 78, 80, 98, 169, 197, 247, 262

Temperature sensor, 44, 45, 87, 125, 236
Terminal designation, 16
Terminal numbers, 15
Test lights, 12
Thermal actuator, 96, 97
Thermal gauge, 226, 227, 230
Thermistor, 86, 125, 126
Thermo-valve, 147
Thermostat, 148, 149
Thermostatic expansion valve, 235
Throttle butterfly position, 135
Throttle position, 47, 88, 126, 180
Throttle position sensor, 48, 88

Throttle position switch, 125
Timing belt, 100
Tires, 182
Titania, 46, 47, 48
Torque converter, 253, 254, 256
Trace, 29
Tracking, 185, 186, 188
Tracking gauges, 187
Traction control, 178, 180
Traction control modulator, 180
Trailing arms, 191, 192, 193
Transfer gearbox, 251
Transistors, 197, 199, 200, 201
Transmission, 88, 217, 247
Trip computer, 228, 229
Turbine, 136, 253

Vacuum gauge, 154
Valve and ignition timing, 101
Valve mechanisms, 100, 112
Variable capacitance, 90
Variable resistance, 88
Variable venturi carburettor, 104, 106
Vehicle condition monitoring, 228
Vehicle systems, 18

Ventilation, 153, 231, 233, 235
Volt, amp tester (VAT), 159
Volt drop, 13, 159, 167
Volt drop testing, 13, 163
Voltage regulator, 165, 166
VR Pump, 128

Water cooled system, 148
Water pump, 148, 149
Waveform, 29, 33, 40, 42, 49, 58, 82, 86
Web, 260
Wheel alignment, 187
Wheel balancer, 187
Wheel speed sensors, 40, 176
Wide band lambda sensor, 93
Window lift motor, 96
Wiper circuits, 211
Wiper motor, 209, 210, 211, 215, 216

Xenon lighting, 207

Zener diode, 51, 166, 197
Zirconia, 47, 48, 73, 74
Zirconium dioxide, 92